VOLUME ONE HUNDRED AND TWENTY FIVE

ADVANCES IN PROTEIN CHEMISTRY AND STRUCTURAL BIOLOGY

Apoptosis in Health and Disease - Part A

VOLUME ONE HUNDRED AND TWENTY FIVE

ADVANCES IN PROTEIN CHEMISTRY AND STRUCTURAL BIOLOGY

Apoptosis in Health and Disease - Part A

Edited by

ROSSEN DONEV

MicroPharm Ltd,
United Kingdom

ELSEVIER

ACADEMIC PRESS

An imprint of Elsevier

Academic Press is an imprint of Elsevier
50 Hampshire Street, 5th Floor, Cambridge, MA 02139, United States
525 B Street, Suite 1650, San Diego, CA 92101, United States
The Boulevard, Langford Lane, Kidlington, Oxford OX5 1GB, United Kingdom
125 London Wall, London, EC2Y 5AS, United Kingdom

First edition 2021

ISBN: 978-0-323-85315-6
ISSN: 1876-1623

For information on all Academic Press publications
visit our website at https://www.elsevier.com/books-and-journals

Publisher: Zoe Kruze
Acquisitions Editor: Ashlie Jackman
Developmental Editor: Leticia Lima
Production Project Manager: Abdulla Sait
Cover Designer: Matthew Limbert

Typeset by SPi Global, India

Contents

Contributors

Mohammad Abid
Medicinal Chemistry Laboratory, Department of Biosciences, Jamia Millia Islamia, New Delhi, India

Kwang Seok Ahn
Department of Science in Korean Medicine, College of Korean Medicine, Kyung Hee University, Seoul, Republic of Korea

Mohamed F. Alajmi
Department of Pharmacognosy, College of Pharmacy, King Saud University, Riyadh, Saudi Arabia

B. Anu
Cancer Research Program, Rajiv Gandhi Centre for Biotechnology (RGCB), Thiruvananthapuram, KL, India

Tanu Atriwal
Medicinal Chemistry Laboratory, Department of Biosciences, Jamia Millia Islamia, New Delhi, India

Anupam Bishayee
Lake Erie College of Osteopathic Medicine, Bradenton, FL, United States

Cintia Cevallos
Instituto de Investigaciones Biomédicas en Retrovirus y Sida (INBIRS), Universidad de Buenos Aires, CONICET, Buenos Aires, Argentina

Meenal Chawla
Medicinal Chemistry Laboratory, Department of Biosciences, Jamia Millia Islamia, New Delhi, India

Biswajit Das
Tumor Microenvironment and Animal Models Lab, Department of Cancer Biology, Institute of Life Sciences, Bhubaneswar, Odisha; Manipal Academy of Higher Education, Manipal, Karnataka, India

María Victoria Delpino
Instituto de Inmunología, Genética y Metabolismo (INIGEM), Universidad de Buenos Aires, CONICET, Buenos Aires, Argentina

Arunasalam Dharmarajan
Department of Biomedical Sciences, Faculty of Biomedical Sciences Technology and Research, Sri Ramachandra Institute of Higher Education & Research, Chennai, India

Ran Ding
Department of Physiology and Pathophysiology and Sino-UK Joint Laboratory of Brain Function and Injury of Henan Province, Xinxiang Medical University, Xinxiang, China

Vivek Kumar Garg
Punjab Biotechnology Incubator, Mohali, Punjab, India

Neelam Goel
Department of Information Technology, UIET, Panjab University, Chandigarh, India

Ashutosh Gupta
Department of Biochemistry, University of Allahabad, Prayagraj, Uttar Pradesh, India

Satyajit Halder
Division of Molecular Medicine, Bose Institute, Kolkata, India

K.B. Harikumar
Cancer Research Program, Rajiv Gandhi Centre for Biotechnology (RGCB), Thiruvananthapuram, KL, India

Afzal Hussain
Department of Pharmacognosy, College of Pharmacy, King Saud University, Riyadh, Saudi Arabia

Kuladip Jana
Division of Molecular Medicine, Bose Institute, Kolkata, India

Lin-Hua Jiang
Faculty of Biological Sciences, School of Biomedical Sciences, University of Leeds, Leeds, United Kingdom; Department of Physiology and Pathophysiology and Sino-UK Joint Laboratory of Brain Function and Injury of Henan Province, Xinxiang Medical University, Xinxiang, China

Nikhilesh Joardar
Parasitology Laboratory, Department of Zoology, Siksha-Bhavana, Visva-Bharati, Santiniketan, West Bengal, India

Soundarapandian Kannan
Division of Cancer Nanomedicine, Department of Zoology, School of Life Science, Periyar University, Salem, Tamil Nadu, India

Dharambir Kashyap
Department of Histopathology, Postgraduation Institute of Medical Education and Research (PGIMER), Chandigarh, India

Alan Prem Kumar
Department of Pharmacology, Yong Loo Lin School of Medicine; Cancer Science Institute of Singapore, National University of Singapore, Singapore, Singapore

Mariam Loka
Lake Erie College of Osteopathic Medicine, Bradenton, FL, United States

Philippa Malko
Faculty of Biological Sciences, School of Biomedical Sciences, University of Leeds, Leeds, United Kingdom

Niladri Mukherjee
Parasitology Laboratory, Department of Zoology, Siksha-Bhavana, Visva-Bharati, Santiniketan, West Bengal, India

N.N. Namitha
Cancer Research Program, Rajiv Gandhi Centre for Biotechnology (RGCB), Thiruvananthapuram, KL, India

Abhay Kumar Pandey
Department of Biochemistry, University of Allahabad, Prayagraj, Uttar Pradesh, India

Jorge Quarleri
Instituto de Investigaciones Biomédicas en Retrovirus y Sida (INBIRS), Universidad de Buenos Aires, CONICET, Buenos Aires, Argentina

Shantibhusan Senapati
Tumor Microenvironment and Animal Models Lab, Department of Cancer Biology, Institute of Life Sciences, Bhubaneswar, Odisha, India

Gautam Sethi
Department of Pharmacology, Yong Loo Lin School of Medicine, National University of Singapore, Singapore, Singapore

Muthu K. Shanmugam
Department of Pharmacology, Yong Loo Lin School of Medicine, National University of Singapore, Singapore, Singapore

Amit Kumar Singh
Department of Biochemistry, University of Allahabad, Prayagraj, Uttar Pradesh, India

Santi P. Sinha Babu
Parasitology Laboratory, Department of Zoology, Siksha-Bhavana, Visva-Bharati, Santiniketan, West Bengal, India

Karuppaiya Vimala
Division of Cancer Nanomedicine, Department of Zoology, School of Life Science, Periyar University, Salem, Tamil Nadu, India

Sudha Warrier
Division of Cancer Stem Cells and Cardiovascular Regeneration, Manipal Institute of Regenerative Medicine, Manipal University, Bangalore, India

CHAPTER ONE

Apoptosis in infectious diseases as a mechanism of immune evasion and survival

Jorge Quarleri[a], Cintia Cevallos[a], and María Victoria Delpino[b,]*

[a]Instituto de Investigaciones Biomédicas en Retrovirus y Sida (INBIRS), Universidad de Buenos Aires, CONICET, Buenos Aires, Argentina

[b]Instituto de Inmunología, Genética y Metabolismo (INIGEM), Universidad de Buenos Aires, CONICET, Buenos Aires, Argentina

*Corresponding author: e-mail address: mdelpino@ffyb.uba.ar

Contents

Abstract

In pluricellular organisms, apoptosis is indispensable for the development and homeostasis. During infection, apoptosis plays the main role in the elimination of infected cells. Infectious diseases control apoptosis, and this contributes to disease pathogenesis. Increased apoptosis may participate in two different ways. It can assist the dissemination of intracellular pathogens or induce immunosuppression to favor pathogen dissemination. In other conditions, apoptosis can benefit eradicate infectious agents from the host. Accordingly, bacteria, viruses, fungi, and parasites have developed strategies to inhibit host cell death by apoptosis to allow intracellular survival and persistence of the pathogen. The clarification of the intracellular signaling pathways, the receptors involved and the pathogen factors that interfere with apoptosis could disclose new therapeutic targets for blocking microbial actions on apoptotic pathways. In this review, we summarize the current knowledge on pathogen anti-apoptotic and apoptotic approaches and the mechanisms involving in disease.

Advances in Protein Chemistry and Structural Biology, Volume 125
ISSN 1876-1623
https://doi.org/10.1016/bs.apcsb.2021.01.001

1. Pathways involved in regulated cell death with a focus on apoptosis

1.1 Introduction-mechanisms of apoptosis

The first observations of programmed cell death date from 1842, when Karl Vogt observed that during the metamorphosis of the toad, the notochord is eliminated by cell death. However, research on programmed cell death started in 1972 when the term apoptosis began to be used by John Keer et al. Apoptosis was differentiated from necrosis as a defined form of programmed cell death (PCD) with particular morphological changes (Kerr, Wyllie, & Currie, 1972). Apoptosis occurs in several physiological and pathological conditions. As a physiological process, apoptosis is critical for tissue homeostasis including tissue remodeling and the regulation of immune responses (Singh, Letai, & Sarosiek, 2019). The era of molecular apoptosis started in the 1990s with the identification, during *Caenorhabditis elegans* development of Cell death abnormality gene 9 (CED9) and CED4, known in mammalian cells, as B-cell lymphoma 2 (BCL2) and apoptotic peptidase-activating factor 1 [APAF1], respectively (Hengartner, Ellis, & Horvitz, 1992; Hengartner & Horvitz, 1994; Yuan & Horvitz, 1992). Over the last 30 years, the molecular mechanisms regulating apoptosis have been extensively investigated in several organisms. It is currently well-known that apoptosis consists of two main subtypes, named extrinsic and intrinsic apoptosis. In both pathways, the main regulators are caspases, BCL2 family, P53, tumor necrosis factor (TNF) family, and inhibitors of apoptosis proteins (IAPs).

Extrinsic apoptosis is facilitated mainly by Fas cell surface death ligand (FasL) and its interaction with Fas receptor (also known as CD95) or tumor necrosis factor receptor superfamily member 6 (TNFRSF6). Both processes involved the TNF receptors (TNFR) family together with extrinsic signals. The process is driven by initiator caspases CASP8 (caspase 8) and CASP 10 (Schulze-Osthoff, Ferrari, Los, Wesselborg, & Peter, 1998). Extrinsic apoptosis is also triggered by the activation of the initiator CASP9 or the dephosphorylation of death-associated protein kinase 1 (DAPK1) by dependence receptors—a family of more than a dozen membrane receptors that have common functional trails but are not related by their structure—e.g., unc-5 netrin receptor B and DCC netrin 1 receptor (Bredesen, Mehlen, & Rabizadeh, 2004).

The intrinsic apoptosis, also known as mitochondrial apoptosis, is started by the release of pro-apoptotic factors such as procaspase-9, cytochrome *c*, diablo IAP-binding mitochondrial protein (DIABLO/Smac), as well as endonuclease G from mitochondrial intermembrane space ignited by mitochondrial outer membrane permeabilization (Tait & Green, 2010) following by activation of the initiator CASP9. BCL2 family closely regulated the mitochondrial outer membrane permeabilization. The process involved pro-apoptotic members (e.g., BCL2 associated X, BAX, BAK/BAK1) and also anti-apoptotic members (e.g., BCL2, BCLXL, NOXA, BCLB) (Czabotar, Lessene, Strasser, & Adams, 2014; Singh et al., 2019).

Caspase activation does not assurance cell death; but due to their function in substrate cleavage and the destruction of subcellular structures CASP3, CASP6, and CASP7 are considered as important executioners lead to apoptosis (Galluzzi, Lopez-Soto, Kumar, & Kroemer, 2016).

1.1.1 Caspases

Caspases are cysteine-dependent aspartate-specific proteases. Like many proteases, caspases initially exist as inactive zymogens, namely, procaspases (McIlwain, Berger, & Mak, 2015). Initiator and effector caspases regulate apoptosis, whereas inflammatory caspases control pyroptosis. CASP3, CASP6, and CASP7 are essential executioner caspases in various types of apoptosis. They are usually activated by CASP8 and CASP9 in the extrinsic and intrinsic pathways, respectively. CASP8 manages the response to TNF in the initiation of inflammation, apoptosis, and necroptosis. TNF is one of the most effective physiological inducers of the NF-κB pathway to trigger genes coding for cytokines and pro-survival factors. This effect is realized through the TNFRSF1A complex including FADD. Active CASP8 deactivates the TNFRSF1A complex activity by cleaving RIPK1, thus favoring the activation of CASP3 or CASP7 and subsequent apoptosis. CASP2 and CASP10 are another originator caspases contributing to regulated cell death under certain conditions, but the underlying mechanism remains unclear. CASP1, CASP4, CASP5, and CASP11 ignite pyroptosis by cleaving members of the gasdermin (GSDM) family, especially GSDMD, to induce pore formation and plasma membrane rupture (Liu et al., 2016). Moreover, another member of the gasdermin family—GSDME—after its cleavage by CASP3 also permeabilizes the mitochondrial membrane, releasing cytochrome *c* and activating the apoptosome, thus linking inflammasome activation to augment the mitochondrial apoptotic pathway (Rogers et al., 2019). CASP12 is involved in endoplasmic reticulum stress-associated RCD (Yoneda et al., 2001) (although this finding did not result in follow-up papers and has been disputed (Kalai et al., 2003)) and functions as an anti-inflammatory regulator partly owing to the inhibition of CASP1 inflammasome and the NF-κB pathway (McIlwain et al., 2015).

1.2 Classification of cell death and immunological consequences of cell death

Once the first classifications of cell death patterns depended on the morphological and structural details observed in individual cells and tissues. Using as a model prenatal tissues treated with toxic agents, Schewichel and Merker established in 1973, the classification of cell death into types I, II, and III (Condensation and fragmentation of single cells, the primary formation of lysosomes in dying cells, and disintegration of cells into fragments without implicating the lysosomes, respectively) (Schweichel & Merker, 1973). Apoptosis corresponds to the type I cell death characterized by irreversible condensation of chromatin, DNA fragmentation, membrane blebbing, and apoptotic body formation. Apoptosis was considered a non-pathological cell death, also designed "shrinkage necrosis" (Kerr, 1971).

Autophagy-dependent cell death is related to type II cell death and involved the formation of large-scale autophagic vacuolization-containing cytosolic constituents and organelles. Even though there is no doubt that in most cases autophagy promotes cell survival (Kroemer, Marino, & Levine, 2010), autophagy can also cause cell death in particular circumstances (Liu et al., 2013; Nassour et al., 2019).

Finally, type III cell death, well known as necrosis, implicates the loss of membrane integrity and enlargement of subcellular organelles, a process named oncosis. Conventionally, necrosis—also named inflammatory cell death—is referred to as unprogrammed cell death involving cellular damage. Nowadays a form of necrosis, necroptosis arises in a controlled manner mimicking features of apoptosis and necrosis (Linkermann & Green, 2014; Linkermann, Stockwell, Krautwald, & Anders, 2014; Vanden Berghe, Linkermann, Jouan-Lanhouet, Walczak, & Vandenabeele, 2014; Weinlich, Oberst, Beere, & Green, 2017).

In 2005, the Nomenclature Committee on Cell Death (NCCD) proposed a set of recommendations for the definition and interpretation of cell death (Kroemer et al., 2005). Henceforward, five reports from NCCD has been published to defined cell death types. This included the molecular descriptions of cell death procedures (Galluzzi et al., 2012), the discrimination between essential and the accessory aspects of regulated cell death that are those who etiologically mediate its occurrence and the change in morphologic, and biochemical manifestation (Galluzzi et al., 2015). Focus on the signal transduction molecules implicated in the initiation, execution, and propagation step including the pathophysiological relevance of each type of regulated cell death were approached in 2018 (Galluzzi et al., 2018). It has been established that cell death can be mainly shared into accidental cell death and regulated cell death. We refer to accidental cell death when cells exposed to extreme chemical, physical or mechanical stimuli succumb in an uncontrolled manner with concomitant loss of integrity. On the contrary, regulated cell death can be initiated by a genetically encoded mechanism to preserve physiologic tissue homeostasis (Galluzzi et al., 2015).

Regulated cell death also named programmed cell death, when occurs in physiological conditions, can be categorized into multiple subclasses dependent on the steps involved. Given the morphological characteristics and molecular mechanisms, programmed cell death can be further characterized as apoptosis and non-apoptotic cell death. Among the last, necroptosis and pyroptosis, are frequently observed during viral infection and in the context of physiological development. Whereas others including ferroptosis, netotic, entotic, parthanatos, autophagy, alkaliptosis, oxeiptosis, and lysosome-dependent cell death are less studied until now and current knowledge is limited to cellular responses to different toxins that do not reproduce normal physiology (Galluzzi et al., 2015).

The immune system can be activated by some stimulus-induced cell death mechanisms. The term immunogenic cell death was coined in 2005 (Casares et al., 2005). It is opposite to the process of silent efferocytosis in which dying or dead cells are empty by phagocytosis without immune reaction or as occurs in tolerogenic cell death are activated deploy mechanisms to inhibit the immune response (Green, Ferguson, Zitvogel, & Kroemer, 2009).

Classically, apoptosis was considered as tolerogenic cell death, but accumulating evidence strongly suggests that apoptosis could be classified into immunogenic cell death in some cases (Obeid et al., 2007). Dying cells not only induce tissue regeneration and limits infection when it occurs, but also elicited an acute or a chronic inflammatory response (Galluzzi, Buque, Kepp, Zitvogel, & Kroemer, 2017).

Inflammation has a main role in a variety of human diseases. Dead or damaged cells release damage-associated molecular patterns (DAMPs) modulating the balance between immunogenic, and tolerogenic cell death.

The immune system can also recognize other danger signals, the pathogen-associated molecular patterns (PAMPs). DAMPs and PAMPs are recognized by the same pattern recognition receptors (PPRs) and can be proteins (histones, mitochondrial transcription factor A—TFAM, high mobility group box 1 protein—HMGB1, among others) and non-proteinaceous factors (RNA, DNA, extracellular ATP, LPS) (Tang, Kang, Coyne, Zeh, & Lotze, 2012).

DAMPs are released in various types of cell death, with different expression profiles accordingly to the stimuli (Hou et al., 2013). DAMPs stimulate different PPRs such as membrane-bound PPRs (e.g., the mannose receptors, asialoglycoprotein receptor family, toll-like receptors (TLR), C-type lectin receptors) and cytoplasmic PPRs (e.g., NOD-like receptors (NLR), NODs, NLRPs, RIG-I like receptors (RLR)) expressed extensively in immune cells among other cell types (Tang et al., 2012). Immune cell death is associated with the activation of several inflammation-related pathways including RIPK1-NF-κB, IL-17A-IL-17R, and DNA-TMEM pathway (Ahn, Xia, Rabasa Capote, Betancourt, & Barber, 2018; Ma et al., 2017; Yatim et al., 2015). The release of DAMPs supplies adequate signals to stimulate local inflammatory and systemic immune responses.

1.3 Mechanisms of apoptosis

1.3.1 Introduction

Before discussing the specific roles of apoptosis in infections, we will first introduce the major apoptotic factors that control the fate of host cells. Apoptosis involved the synchronized disintegration of damaged cells into fragments named apoptotic bodies, small membrane-surrounded intracellular content, containing DAMPs and PAMPs (is the process involved an infectious agent). TAM receptors present in macrophages recognize apoptotic cells and remove these damaged cells to prevent an inflammatory response.

During apoptosis, cellular destruction is mediated by caspases produced as inactive precursors, zymogens, or pro-enzymes (Van Opdenbosch & Lamkanfi, 2019). Diversity of cell death signaling is responsible for the activation of initiator and then effector caspases that encourage the various routes involved in cell destruction. These processes include DNA fragmentation, chromatin condensation, and membrane blebbing to finally lead to the formation of apoptotic bodies (Kerr et al., 1972; Segawa et al., 2014). Two pathways conduce to apoptosis depending on signals from inside and outside the cells, intrinsic and extrinsic pathway, respectively. Intrinsic pathway (also called mitochondrial or BCL2 regulated) is characterized by the release of cytochrome *c* from mitochondria. In contrast, the extrinsic pathway requires the activation of death receptors present on the cell surface.

1.3.2 Intrinsic pathway

During intrinsic apoptosis, the mitochondria have received sufficient damage to secrete proteins that exist in the intermembrane space of mitochondria. This apoptotic pathway is mediated by BCL2 family members (Wei et al., 2001). In healthy cells, BCL2 associated X protein (BAX) is located in the cytosol and BCL2 antagonist-killer named BAK is located in the outer mitochondrial membrane. When BAX and BAK are activated, they oligomerize inside the outer membrane of mitochondria with concomitant membrane permeabilization in a mechanism that was not completely elucidated until now (Czabotar et al., 2013). Pro-survival BCL2 family members including BCL2, BCLW, BCLXL, A1, and MCL1 control the activities of BAX and BAK. Specifically, BCL2 members avoid their mitochondrial targeting, oligomerization, and activation by bind BAX and BAK (Uren et al., 2017). Among other factors, infection activates the upregulation of the BCL2 homology factors (BH3)-apoptotic factors, such as BIM, BAD, PUMA, BID (BCL2 domain 2 interacting-domain death agonist), BIK, NOXA, HRK, and BMF (O'Neill, Huang, Zhang, Chen, & Luo, 2016). BAX and BAK could be directly activated by BH3—only proteins (proteins that share one motif, the BH3 domain, with BCL2 family proteins) to inhibit pro-survival BCL2 members (Llambi et al., 2011). Additionally, BH3-only proteins could promote different interactions between the pro-survival BCL2 members and BAX and BAK to induce or prevent apoptosis, depending on surrounding conditions (Youle & Strasser, 2008).

Activation of BAX and BAK is known as a step of no return, which inevitably conduces to cell death. This is in part due to the release of cytochrome *c* into the cytosol mediated by BAX and BAK. The presence of cytochrome *c* is essential to maintain the mitochondrial respiratory chain and its membrane potential to scavenge reactive oxygen species (Li et al., 2000; Liu, Kim, Yang, Jemmerson, & Wang, 1996). In the cytosol, cytochrome *c* activates a signaling pathway that activates caspases (Hakem et al., 1998; Zou, Henzel, Liu, Lutschg, & Wang, 1997). The second mitochondria-derived activator of caspases (SMAC), also called DIABLO is another released mitochondrial protein and antagonizes the cytosolic caspase X-linked inhibitor of apoptosis protein (XIAP) with concomitant activation of CASP-9 (Verhagen et al., 2000). Then, active CASP-9 cleaves CASP-3, -6, and-7 among others, inducing DNAses and inactivating proteins involved in DNA repair (Kuida et al., 1998; Yu et al., 2002). A total of 19 members of the human BCL2 family were detected by 2008. Studies performed in mice revealed that loos of individual BH3-only proteins outcome in mild abnormal phenotypes, indicating redundancy among death factors (Youle & Strasser, 2008).

A large number of BCL-2 proteins and variances in their mechanisms of action and binding affinities make a prediction of the apoptotic response difficult. Studies to determine the role of BCL2 family members in infections have been limited to cell lines due to the absence of knockout mice. Experiments performed in mice that overexpress BCL2 suggest that apoptosis promotes pathogen clearance and/or inhibits immune response during infections (Gamradt et al., 2016).

1.3.3 Extrinsic pathway

Apoptosis can also be induced by the activation of cell death receptors by extracellular signals. The receptors involved are members of proteins that belong to the TNF superfamily (Youle & Strasser, 2008). In addition to its function as a protein involved in proliferation, cell survival, differentiation, and inflammation; TNF-α can also induce apoptosis (Aggarwal, Gupta, & Kim, 2012). Many cell types produce TNF-α in response to invading pathogens (Baud & Karin, 2001). This TNF-α can binding to the surface receptor TNFR1 to elicit the recruitment of TNFR-associated death domain (TRADD) protein via its death domain (DD), TNF-receptor associated factor 2 (TRAF2), the receptor-interacting protein (RIPK1), and the cellular inhibitor of apoptosis protein (cIAP) (Hsu, Shu, Pan, & Goeddel, 1996).

Procaspase-8 is recruited by TRADD to constitute the death-inducing signaling complex or the complex of TRADD, TRAF2, RIPK1, FADD, and CASP-8 can be formed. Both ways induce caspase-8 autoactivation with the concomitant cleave of CASP-3 and apoptosis cell death (Vince & Silke, 2016). CASP-8 can also cleave BID (a BH3-only protein) in some cell types. The cleaved-BID activates BAX and BAK to lead mitochondria-mediated apoptosis (Li, Zhu, Xu, & Yuan, 1998). While CASP-8 has long been implicated in apoptosis death, it has recently been shown to be involved in other types of cell death, necroptosis, and pyroptosis (Fritsch et al., 2019). This constitutes one example of multiple pieces of evidence of crosstalk between cell death pathways.

Most pathogens can disrupt apoptosis, the activation of alternative programmed cell death could act as a backup mechanism that allows altered cells to alert immune cells to the presence of a potential pathogen (Jorgensen, Rayamajhi, & Miao, 2017).

In the next section, we will review the current knowledge of immunological consequences of regulated cell death and the relationship of regulated cell death in the survival of pathogens during infection with a focus on apoptosis.

2. Apoptosis in survival and immune evasion of pathogens

2.1 Introduction

Mechanisms of cell death are used by multicellular organisms to sacrifice infected cells for the profit of the surrounding cells. For example, in reaction to bacterial infection, programmed cell death, such as apoptosis, is produced as an innate immune response of the host (Zitvogel, Kepp, & Kroemer, 2010). This allows pathogens to be eliminated at an early stage of infection. Then, apoptotic bodies can be phagocytized by dendritic cells and thereby contribute to the generation of a protective immune response (Elliott & Ravichandran, 2010). On the other hand, cell death could also advantage pathogens. Some pathogens are capable of inhibiting cell death to survive and multiply within the host cell; and others induce cell death as a mechanism for escaping a cell and infecting an adjacent cell (Behar, Divangahi, & Remold, 2010; Lamkanfi & Dixit, 2010). In the following sections, the specific roles of apoptosis in infections caused by different pathogens will be discussed.

2.2 Pathogen-mediated inhibition of apoptosis

Extrinsic apoptosis can be activated by PAMPs and death ligands secreted during infections. Moreover, infections also activate intrinsic apoptosis due to DNA damage, calcium signaling, and metabolic stress. NF-κB transcription factor family is involved in the modulation of the expression of death factor and survival, and successful pathogens display mechanisms to disrupt apoptosis.

During infection, the bacteria *Legionella pneumophila* replicates in lung macrophages by blocking apoptosis. This process is due to the transcriptional upregulation of pro-survival BCL2 family members, including A1 and BCL2 in a mechanism dependent on NF-κB (Abu-Zant et al., 2007; Losick & Isberg, 2006). In infected macrophages was no observed increase in the expression levels of pro-survival BCL2 members. However, *Legionella* spp. blocks host protein translation (Speir et al., 2016). The BH3-mimetic d that inhibits BCL2 fails to activate apoptosis, but the loss of the related protein BCLXL makes *Legionella*-infected macrophages death and prevents bacterial growth (Speir et al., 2016). BCLXL prevents apoptosis in a mechanism dependent on MCL1 (Speir et al., 2016). At a difference of other pro-survival factors, MCL1 has a half-life is around 30 min (Adams & Cooper, 2007). In this condition, inhibition of host protein translation by the effector's proteins of *Legionella*, Lgt1, and Lgt3 is enough to induce the quick loss of MCL1 (Speir et al., 2016). Together, that evidence indicates that BCLXL is the unique pro-survival BCL2 family member capable to effectively inhibit apoptosis.

The obligate intracellular bacteria *Chlamydia trachomatis* infects macrophages and translocate a serine protease to degrade BH3-only factors (Fischer et al., 2004). The infection also induces the phosphorylation of BAD that is internalized by the bacteria-containing vacuole to prevent mitochondria interaction and consequent apoptosis (Verbeke et al., 2006).

Besides, *Chlamydia* spp. leads the expression of pro-survival MCL1 (Rajalingam et al., 2008). Since MCL1 is quickly degraded via the ubiquitin-proteasome pathway, the bacteria express a deubiquitination enzyme (Cdu1), which sequester ubiquitinated MCL1 to prevent its proteasomal degradation promoting the survival of infected cells to favor intracellular replication of the bacteria (Fischer et al., 2017). Other bacterial pathogens induce the increase of the expression of prosurvival BCL2 family members via activating the NK-κB signaling ways. *Mycobacterium tuberculosis* is a bacteria that replicates at intracellular and prevents apoptosis via the secretion of MPT64 protein that modulates the expression of host miRNA-21 with a concomitant increase of prosurvival BCL2 expression (Wang, Liu, Tang, Liu, & Yao, 2014). Also, *M. tuberculosis* induces the upregulation of MCL1 in macrophages. This molecule has a key role in *M. tuberculosis* replication since MCL1 knockdown activates apoptosis and inhibits bacterial survival and replication (Sly, Hingley-Wilson, Reiner, & McMaster, 2003).

In the case of viruses, adenovirus, murine γ-herpesvirus, Epstein-Barr virus, and Kaposi's sarcoma-associated herpesvirus (human herpesvirus-8, HHV-8) produce proteins that partake similarities with mammalian BCL2 inhibiting BAX and BAK mediated apoptosis (Henderson et al., 1993; Rao et al., 1992; Sarid, Sato, Bohenzky, Russo, & Chang, 1997). The orthologous of BCL2 from γ-herpesvirus induces the survival of immature B lymphocytes allowing the permanence of viral infection for long times (Coleman et al., 2014). These orthologous of BCL2 appear to have other roles. Since the replacement with pro-survival BCL2 members is not sufficient to reverse the viral effect (Gallo, Lampe, Gunther, & Brune, 2017).

Other viruses (e.g., Deerpox, Sheeppox) inhibit BAX and BAK activation and consequent BAX/BAK-mediated apoptosis (Banadyga et al., 2011; Kvansakul et al., 2007; Okamoto et al., 2012).

Other groups of viruses inhibit apoptotic caspases by molecular imitation. For example, baculoviruses express IAP homologs to inhibit host caspases (Huang et al., 2000). Human herpesvirus-8 (HHV-8) associated with sarcoma of Kaposi inhibits not only BAX and BAK, but also CASP8 activation (Mesri, Cesarman, & Boshoff, 2010; Thome et al., 1997). Also, viruses can express pan-caspase inhibitors that do not appear to have similarities to human genes involved in inhibiting caspases (Best, 2008). Studies indicate that activating BAX and BAK in the absence of caspases causes antiviral immune responses because of the DNA release from mitochondria. Cyclic GMP-AMP synthase senses the cytosolic DNA inducing inflammatory responses (Rongvaux et al., 2014; White et al., 2014). Thus, we can suspect that inhibition of BAX and BAK in conjunction with inhibition of caspases may be an evolutionary pressure to prevent host cell death and also inflammatory responses that can eliminate viral pathogens. As a hepatocellular carcinoma (HCC)-associated virus, hepatitis B virus (HBV) has drawn a lot of attention. Although the HCC development is usually featured by the inhibition of apoptosis, the outcome of cell apoptosis is a result of a complex biological process, as there are different regulations of multiple cell signaling pathways by the various HBV proteins. Even the same apoptotic signaling pathway can be affected by the opposite consequences in the cells, such as the mitochondrial apoptotic pathway. This pathway can be activated by the BAX insertion into the mitochondrial membrane, which is driven by HBV. In contrast, this pathway can be inhibited during HBV infection involving the HBx protein. Activation of NF-κB signaling is a double-edged sword: promoting the expression of pro-survival genes to facilitate cell proliferation and upregulating the expression of the death-associated receptor to sensitize cells to apoptotic stimuli. Therefore, the HBV effect on cell apoptosis varies depending on the cellular context, different signaling pathways, HBV genotypes, protein mutations, and possibly different clinical stages. Although apoptosis induction is widely considered as a positive strategy against cancer, the apoptotic cells can promote the proliferation of the surrounding tumor cells by affecting the microenvironment. Since HBV is frequently detected in clinical HCC cancers, it is more relevant that HBV/HBx accelerates cell transformation and facilitates apoptosis inhibition as a long-term effect. In most studies, HBV/HBx is shown to promote the anti-apoptotic proteins or inhibit the function of pro-apoptotic proteins. However, in some experiments, the higher level of ROS or the upregulated death associated receptors in the hepatocytes might be factors for apoptosis induction (Lin & Zhang, 2017).

In the gut, infected epithelial cells are removed by apoptosis, and block caspases activity is also a strategy employed by enteric pathogens to prevent cell death. *E. coli enteropathogenic* and *E. coli enterohemorrhagic* (EPEC/EHEC, respectively) prevent the activation of apoptotic caspases via the secretion of the effector NleF (Pollock et al., 2017). Additionally EPEC produce the effector non-Lee-encoded effector B1 (NleB1) that interacts with the Fas-associated death domain (FADD), TRADD, TNFR1, and RIPK1 to inhibit extrinsic apoptosis (Li et al., 2013; Pearson et al., 2013).

EPEC secretes another effector protein, EPEC secreted/signaling protein L (EspL). This effector cleaves RIPK1 and RIPK3 to inhibit necroptosis (Pearson et al., 2017).

The strategies described above have allowed the identification of new host cell death pathways and their roles in preventing infectious disease. These studies about the different strategies employed by pathogens to inhibit host cell apoptosis shed light on possible targets for therapeutic intervention (Fig. 1).

2.3 Activation of apoptosis by pathogens

In some circumstances, pathogens including those that inhibit apoptosis can target mitochondria to induce apoptotic cell death. Apoptosis can be a mechanism to eliminate immune or epithelial cells that appear as an initial barrier to prevent infections. Since apoptosis generally does not elicit inflammatory responses, is understood to be the preferred pathway rather than other forms of programmed cell death. Accordingly, hemolysin secreted by EHEC induces the release of cytochrome *c* from mitochondria and activation of caspases (Bielaszewska et al., 2013, 2017).

The human pathogen *Neisseria gonorrhoeae* encourages host cell apoptosis by releasing the outer membrane protein PorB into the host mitochondria membrane compromising its integrity due to pore formation (Jiang, Davies, Lithgow, Strugnell, & Gabriel, 2011; Muller et al., 2002). In contrast, PorB from *N. meningitidis* interacts with voltage-dependent anion channel (VDAC) mimicking the interaction of antiapoptotic BCL2 with VDAC with the consequent inhibition of apoptosis (Massari, Gunawardana, Liu, & Wetzler, 2010).

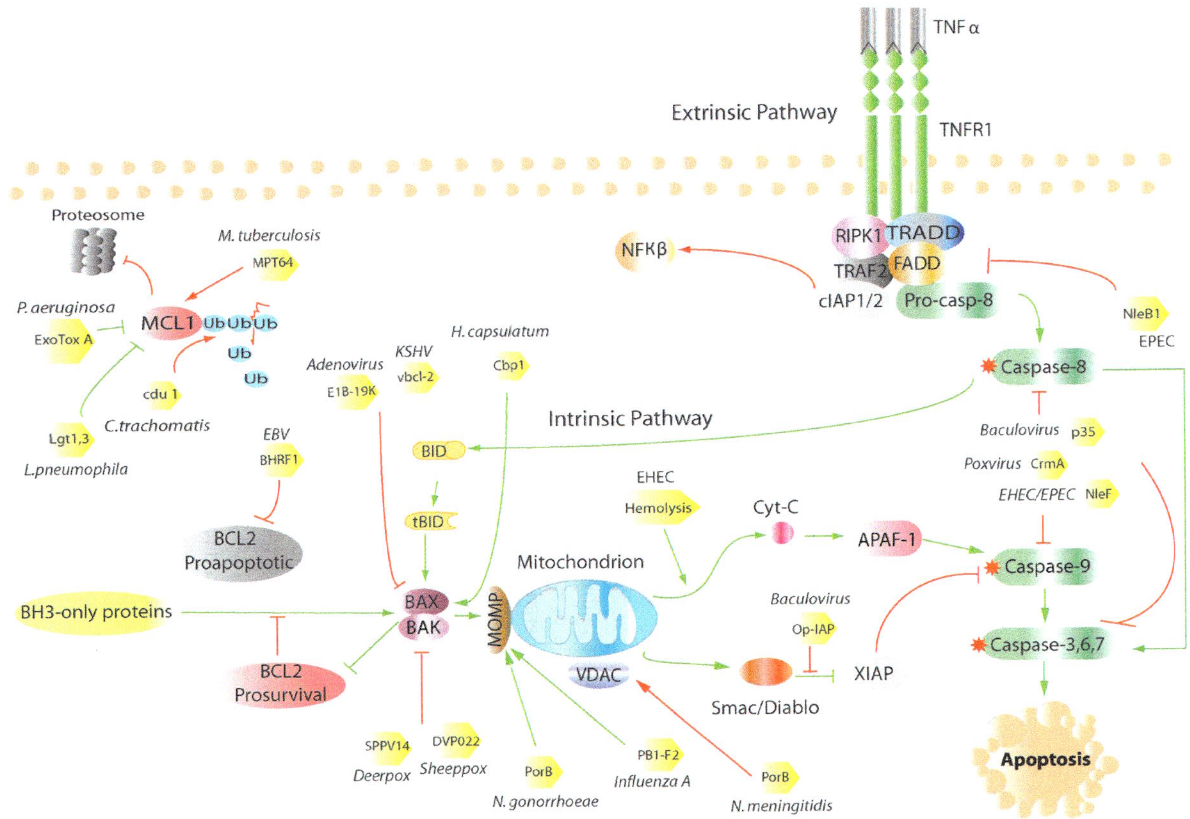

Fig. 1 See figure legend on next page.

Activation of apoptosis can also be unfavorable to the host by stimulating viral dissemination due to immune cell death. Apoptosis can be activated by human immunodeficiency virus (HIV) in infected activated-T lymphocytes, while it induces pyroptosis in nonproductive infected resting T lymphocytes (Monroe et al., 2014). Experiments performed in the murine model revealed that reovirus triggers Fas-mediated apoptosis whit consequent apoptotic neuronal cell death and encephalitis (Clarke, Beckham, Leser, Hoyt, & Tyler, 2009). Loss of T-cells depending on BIM- and Fas-mediated apoptosis has been also observed during chronic murine γ-herpesvirus infection (Hughes et al., 2008).

During viral infections, the increase in the expression of Fas and FasL regulates the life-span of immune cells including macrophages and neutrophils (O'Donnell et al., 2015). Virus influenza infection induces pro-survival factors in neutrophils that prevent Fas-mediated apoptosis with the consequent inflammatory complications observed during the disease. This increase in neutrophils survival is also observed during respiratory syncytial virus, human cytomegalovirus, and herpes simplex virus infection. On the contrary, an increase of neutrophils apoptosis origins neutropenia, increasing the susceptibility to

Fig. 1 The apoptosis pathways and their regulation by pathogens. Apoptosis can be triggered via intrinsic or extrinsic pathways. In the intrinsic pathway caspase 8 can cleave to BID and generate tBID, which activates BAX and/or BAK to trigger mitochondrial outer-membrane permeabilization (MOMP), resulting in the release of cytochrome *c* (CytC), and SMAC/DIABLO into the cytosol. CytC binds to apoptotic protease activating factor 1 (APAF1). APAF1 then oligomerizes into the apoptosome, producing caspase 9 activation. SMAC/DIABLO blocks the caspase inhibitor XIAP promoting caspase 9 activation and regulating the caspase cascade. Either caspases 8 or 9 will cleave the effector caspases 3 and 7 into their active forms, leading to apoptosis. Moreover, the upregulation of BH3-only proteins (BH3) prevents the BAX, and BAK inhibition produced by pro-survival BCL-2 family members. The extrinsic pathway is initiated by the oligomerization of death receptors, such as TNFR1, after ligand binding. This contact leads to the interaction with TRADD, TRAF2 and RIPK1, and cellular inhibitor of apoptosis protein (cIAP1/2). cIAP1/2 are recruited to an activated TNF receptor and participate in the regulation of NF-κB. Additionally, the cIAPs can target caspases and Smac for degradation by mediating their ubiquitination. RIPK1 then recruits FADD who recruits caspase 8 who is activated and being capable to cleave the effector caspase to induce apoptosis. Pathogens are capable of inhibiting or inducing apoptosis through their intervention in some point of the pathway, regulating the balance of pro-survival BCL-2 family members, and pro-apoptotic factors, such as BAD, BAX and BAK, targeting mitochondria, caspases, or death receptor signaling. Different pathogens are capable to inhibit apoptosis by activating the NK-κB signaling pathways. Yellow hexagons indicate the virulence factor. Green lines indicate the mechanisms that induce apoptosis and red lines those that inhibit apoptosis.

several infections as occurs during respiratory syncytial virus, human cytomegalovirus, and herpes simplex virus infection (Croker et al., 2011).

Immune cell apoptosis can be directly induced by the impact of mitochondria by viral proteins. Accordingly, the influenza A virus protein polymerase basic protein 1-frame 2 (PB1-F2) impacts mitochondria and induces the loss of mitochondrial membrane potential with consequent apoptosis cell death (Chen et al., 2001). The preferential target cell for PB1-F2 are alveolar macrophages and then the infection is favored since macrophages are critical to start an appropriate immune response in early infections (Jaworska et al., 2014).

Apoptosis death of host cells also occurs after intracellular pathogens successful replication to help in the escape and infect new cells. Apoptosis of macrophages infected whit *Leishmania* activates the phagocytosis of apoptotic bodies loaded with parasites by uninfected macrophages to propagate the infection (Real et al., 2014).

Histoplasma capsulatum is a fungal pathogen that secretes calcium-binding protein 1 (Cbp1) to induce activation of caspases mediated by BAX and BAK. This mechanism promotes the lytic escape from macrophages (Isaac et al., 2015).

Legionella inhibits cell death during macrophage infection but activates apoptosis to escape from spent cells. In contrast to macrophages that allow Legionella survival, dendritic cells suffer apoptosis immediately after infection (Dolezal et al., 2012). These findings propose that some pathogens can sense the infected cell to modulate survival or apoptosis; however, the mechanisms involved have not been completely elucidated. The inhibition of protein translation can induce a cellular stress response, which conduces to the posttranslational activation of BH3-only factors; also, it causes the loss of pro-survival factors with concomitant cell death by apoptosis (Goodall et al., 2016).

Accordingly, translation inhibitors origin delayed cell death allowing the gradual loss of cell viability in mechanisms that dependent on caspases (Speir et al., 2017). Intracellular pathogens can replicate under these conditions. But the blocking of protein translation by chemicals or pathogen infection share phenotypes that may lead to the activation of caspase 3/7 activity, as occurred in Legionella infections (Speir et al., 2017).

Apoptosis is also induced by extracellular pathogens that arrest host protein translation by target ribosomes (Lemaitre & Girardin, 2013). *Pseudomonas aeruginosa* and *Corynebacterium diphtheriae* secrete exotoxin A and diphtheria toxin, respectively (Lemaitre & Girardin, 2013). These toxins catalyze the transfer of adenosine diphosphate (ADP) ribose to elongation factor 2 inhibiting protein synthesis. Additionally, exotoxin A also induces rapid depletion of MCL1 in macrophages and overdue cell death associated with caspase3 activity (Lemaitre & Girardin, 2013). While these toxins are known to be important virulence factors, their specific functions have not yet been fully clarified. Such toxin-induced apoptosis death in macrophages allows the evasion of the immune response by extracellular pathogens.

Parasites also evade the immune response inducing apoptotic cell death. Helminths can extend their infections inducing the inhibition of Th1 or Th2 responses promoting the increase of immunoregulatory cells, targeting PPRs, and diminishing the number of immune cells via induction of apoptosis. Helminth-triggered apoptosis in immune cells depletes host immunity, which flattens on the way to creating a tolerant milieu and chronic infection (Zakeri, 2017).

The *Plasmodium falciparum* sporozoite induces apoptosis of Kupffer cells in the liver, contributing to the induction of T cell tolerance (Gomes, Bhardwaj, Rivera-Correa, Freire-De-Lima, & Morrot, 2016).

This suggests coevolution between infectious agents and the immune system, in which pathogens destroy defense mechanisms inducing the death of immune cells by apoptosis, among others, while the immune system has developed mechanisms to recognize damaged cells and alert other cells to a potential pathogen (Fig. 1).

3. Concluding remarks

Apoptosis is indispensable for the development and homeostasis of pluricellular organisms. During infection disease, modulation of apoptosis enables pathogens to replicate and survive in the host. The current understanding of the different mechanisms employed by pathogens to subvert host apoptotic pathways reveals the presence of new potential therapeutic targets. However, regulated cell death occurs in a variety of pathways that produce different morphological variations and immunological concerns; and apoptosis is only one of them. Despite the diversity of regulated cell death, the evolutionary relationship between distinct regulated cell death pathways remains largely unknown. Additionally, regulated cell death can occur in mixed variants, and distinct lethal subroutines arise in action in parallel and also in a hierarchized manner. More research on apoptosis and its interlay with other regulated cell death is needed to determine the interplay between the different signaling pathways. Additionally, more research is required to define the main point of no return of regulated cell death to investigate the role of excessive or deficient regulated cell death in human diseases. Clarification of the mechanisms, receptors, and microbial factors involved in apoptosis and another regulated cell death could reveal new insights about the host-pathogen relationship and new potential therapeutic targets.

References

Abu-Zant, A., Jones, S., Asare, R., Suttles, J., Price, C., Graham, J., et al. (2007). Anti-apoptotic signalling by the Dot/Icm secretion system of L. pneumophila. *Cellular Microbiology*, *9*(1), 246–264. https://doi.org/10.1111/j.1462-5822.2006.00785.x.

Adams, K. W., & Cooper, G. M. (2007). Rapid turnover of mcl-1 couples translation to cell survival and apoptosis. *The Journal of Biological Chemistry*, *282*(9), 6192–6200. https://doi.org/10.1074/jbc.M610643200.

Aggarwal, B. B., Gupta, S. C., & Kim, J. H. (2012). Historical perspectives on tumor necrosis factor and its superfamily: 25 years later, a golden journey. *Blood*, *119*(3), 651–665. https://doi.org/10.1182/blood-2011-04-325225.

Ahn, J., Xia, T., Rabasa Capote, A., Betancourt, D., & Barber, G. N. (2018). Extrinsic phagocyte-dependent STING signaling dictates the immunogenicity of dying cells. *Cancer Cell*, *33*(5), 862–873.e865. https://doi.org/10.1016/j.ccell.2018.03.027.

Banadyga, L., Lam, S. C., Okamoto, T., Kvansakul, M., Huang, D. C., & Barry, M. (2011). Deerpox virus encodes an inhibitor of apoptosis that regulates Bak and Bax. *Journal of Virology*, *85*(5), 1922–1934. https://doi.org/10.1128/JVI.01959-10.

Baud, V., & Karin, M. (2001). Signal transduction by tumor necrosis factor and its relatives. *Trends in Cell Biology*, *11*(9), 372–377. https://doi.org/10.1016/s0962-8924(01)02064-5.

Behar, S. M., Divangahi, M., & Remold, H. G. (2010). Evasion of innate immunity by Mycobacterium tuberculosis: Is death an exit strategy? *Nature Reviews. Microbiology*, *8*(9), 668–674. https://doi.org/10.1038/nrmicro2387.

Best, S. M. (2008). Viral subversion of apoptotic enzymes: Escape from death row. *Annual Review of Microbiology*, *62*, 171–192. https://doi.org/10.1146/annurev.micro.62.081307.163009.

Bielaszewska, M., Mellmann, A., Bletz, S., Zhang, W., Kock, R., Kossow, A., et al. (2013). Enterohemorrhagic Escherichia coli O26:H11/H-: A new virulent clone emerges in Europe. *Clinical Infectious Diseases*, *56*(10), 1373–1381. https://doi.org/10.1093/cid/cit055.

Bielaszewska, M., Ruter, C., Bauwens, A., Greune, L., Jarosch, K. A., Steil, D., et al. (2017). Host cell interactions of outer membrane vesicle-associated virulence factors of enterohemorrhagic Escherichia coli O157: Intracellular delivery, trafficking and mechanisms of cell injury. *PLoS Pathogens*, *13*(2), e1006159. https://doi.org/10.1371/journal.ppat.1006159.

Bredesen, D. E., Mehlen, P., & Rabizadeh, S. (2004). Apoptosis and dependence receptors: A molecular basis for cellular addiction. *Physiological Reviews*, *84*(2), 411–430. https://doi.org/10.1152/physrev.00027.2003.

Casares, N., Pequignot, M. O., Tesniere, A., Ghiringhelli, F., Roux, S., Chaput, N., et al. (2005). Caspase-dependent immunogenicity of doxorubicin-induced tumor cell death. *The Journal of Experimental Medicine*, *202*(12), 1691–1701. https://doi.org/10.1084/jem.20050915.

Chen, W., Calvo, P. A., Malide, D., Gibbs, J., Schubert, U., Bacik, I., et al. (2001). A novel influenza A virus mitochondrial protein that induces cell death. *Nature Medicine*, 7(12), 1306–1312. https://doi.org/10.1038/nm1201-1306.

Clarke, P., Beckham, J. D., Leser, J. S., Hoyt, C. C., & Tyler, K. L. (2009). Fas-mediated apoptotic signaling in the mouse brain following reovirus infection. *Journal of Virology*, *83*(12), 6161–6170. https://doi.org/10.1128/JVI.02488-08.

Coleman, C. B., McGraw, J. E., Feldman, E. R., Roth, A. N., Keyes, L. R., Grau, K. R., et al. (2014). A gammaherpesvirus Bcl-2 ortholog blocks B cell receptor-mediated apoptosis and promotes the survival of developing B cells in vivo. *PLoS Pathogens*, *10*(2), e1003916. https://doi.org/10.1371/journal.ppat.1003916.

Croker, B. A., O'Donnell, J. A., Nowell, C. J., Metcalf, D., Dewson, G., Campbell, K. J., et al. (2011). Fas-mediated neutrophil apoptosis is accelerated by Bid, Bak, and Bax and inhibited by Bcl-2 and Mcl-1. *Proceedings of the National Academy of Sciences of the United States of America*, *108*(32), 13135–13140. https://doi.org/10.1073/pnas.1110358108.

Czabotar, P. E., Lessene, G., Strasser, A., & Adams, J. M. (2014). Control of apoptosis by the BCL-2 protein family: Implications for physiology and therapy. *Nature Reviews. Molecular Cell Biology*, *15*(1), 49–63. https://doi.org/10.1038/nrm3722.

Czabotar, P. E., Westphal, D., Dewson, G., Ma, S., Hockings, C., Fairlie, W. D., et al. (2013). Bax crystal structures reveal how BH3 domains activate Bax and nucleate its oligomerization to induce apoptosis. *Cell*, *152*(3), 519–531. https://doi.org/10.1016/j.cell.2012.12.031.

Dolezal, P., Aili, M., Tong, J., Jiang, J. H., Marobbio, C. M., Lee, S. F., et al. (2012). Legionella pneumophila secretes a mitochondrial carrier protein during infection. *PLoS Pathogens*, *8*(1), e1002459. https://doi.org/10.1371/journal.ppat.1002459.

Elliott, M. R., & Ravichandran, K. S. (2010). Clearance of apoptotic cells: Implications in health and disease. *The Journal of Cell Biology*, *189*(7), 1059–1070. https://doi.org/10.1083/jcb.201004096.

Fischer, A., Harrison, K. S., Ramirez, Y., Auer, D., Chowdhury, S. R., Prusty, B. K., et al. (2017). Chlamydia trachomatis-containing vacuole serves as deubiquitination platform to stabilize Mcl-1 and to interfere with host defense. *eLife*, *6*, e21465. https://doi.org/10.7554/eLife.21465.

Fischer, S. F., Vier, J., Kirschnek, S., Klos, A., Hess, S., Ying, S., et al. (2004). Chlamydia inhibit host cell apoptosis by degradation of proapoptotic BH3-only proteins. *The Journal of Experimental Medicine*, *200*(7), 905–916. https://doi.org/10.1084/jem.20040402.

Fritsch, M., Gunther, S. D., Schwarzer, R., Albert, M. C., Schorn, F., Werthenbach, J. P., et al. (2019). Caspase-8 is the molecular switch for apoptosis, necroptosis and pyroptosis. *Nature*, *575*(7784), 683–687. https://doi.org/10.1038/s41586-019-1770-6.

Gallo, A., Lampe, M., Gunther, T., & Brune, W. (2017). The viral Bcl-2 homologs of Kaposi's sarcoma-associated herpesvirus and rhesus rhadinovirus share an essential role for viral replication. *Journal of Virology*, *91*(6), 1–16. https://doi.org/10.1128/JVI.01875-16.

Galluzzi, L., Bravo-San Pedro, J. M., Vitale, I., Aaronson, S. A., Abrams, J. M., Adam, D., et al. (2015). Essential versus accessory aspects of cell death: Recommendations of the NCCD 2015. *Cell Death and Differentiation*, *22*(1), 58–73. https://doi.org/10.1038/cdd.2014.137.

Galluzzi, L., Buque, A., Kepp, O., Zitvogel, L., & Kroemer, G. (2017). Immunogenic cell death in cancer and infectious disease. *Nature Reviews. Immunology*, *17*(2), 97–111. https://doi.org/10.1038/nri.2016.107.

Galluzzi, L., Lopez-Soto, A., Kumar, S., & Kroemer, G. (2016). Caspases connect cell-death signaling to organismal homeostasis. *Immunity*, *44*(2), 221–231. https://doi.org/10.1016/j.immuni.2016.01.020.

Galluzzi, L., Vitale, I., Aaronson, S. A., Abrams, J. M., Adam, D., Agostinis, P., et al. (2018). Molecular mechanisms of cell death: Recommendations of the Nomenclature Committee on Cell Death 2018. *Cell Death and Differentiation*, *25*(3), 486–541. https://doi.org/10.1038/s41418-017-0012-4.

Galluzzi, L., Vitale, I., Abrams, J. M., Alnemri, E. S., Baehrecke, E. H., Blagosklonny, M. V., et al. (2012). Molecular definitions of cell death subroutines: Recommendations of the Nomenclature Committee on Cell Death 2012. *Cell Death and Differentiation*, *19*(1), 107–120. https://doi.org/10.1038/cdd.2011.96.

Gamradt, P., Xu, Y., Gratz, N., Duncan, K., Kobzik, L., Hogler, S., et al. (2016). The influence of programmed cell death in myeloid cells on host resilience to infection with Legionella pneumophila or Streptococcus pyogenes. *PLoS Pathogens*, *12*(12), e1006032. https://doi.org/10.1371/journal.ppat.1006032.

Gomes, P. S., Bhardwaj, J., Rivera-Correa, J., Freire-De-Lima, C. G., & Morrot, A. (2016). Immune escape strategies of malaria parasites. *Frontiers in Microbiology*, 7, 1617. https://doi.org/10.3389/fmicb.2016.01617.

Goodall, K. J., Finch-Edmondson, M. L., van Vuuren, J., Yeoh, G. C., Gentle, I. E., Vince, J. E., et al. (2016). Cycloheximide can induce Bax/Bak dependent myeloid cell death independently of multiple BH3-only proteins. *PLoS One*, *11*(11), e0164003. https://doi.org/10.1371/journal.pone.0164003.

Green, D. R., Ferguson, T., Zitvogel, L., & Kroemer, G. (2009). Immunogenic and tolerogenic cell death. *Nature Reviews. Immunology*, *9*(5), 353–363. https://doi.org/10.1038/nri2545.

Hakem, R., Hakem, A., Duncan, G. S., Henderson, J. T., Woo, M., Soengas, M. S., et al. (1998). Differential requirement for caspase 9 in apoptotic pathways in vivo. *Cell*, *94*(3), 339–352. https://doi.org/10.1016/s0092-8674(00)81477-4.

Henderson, S., Huen, D., Rowe, M., Dawson, C., Johnson, G., & Rickinson, A. (1993). Epstein-Barr virus-coded BHRF1 protein, a viral homologue of Bcl-2, protects human B cells from programmed cell death. *Proceedings of the National Academy of Sciences of the United States of America*, *90*(18), 8479–8483. https://doi.org/10.1073/pnas.90.18.8479.

Hengartner, M. O., Ellis, R. E., & Horvitz, H. R. (1992). Caenorhabditis elegans gene ced-9 protects cells from programmed cell death. *Nature*, *356*(6369), 494–499. https://doi.org/10.1038/356494a0.

Hengartner, M. O., & Horvitz, H. R. (1994). C. elegans cell survival gene ced-9 encodes a functional homolog of the mammalian proto-oncogene bcl-2. *Cell*, *76*(4), 665–676. https://doi.org/10.1016/0092-8674(94)90506-1.

Hou, W., Zhang, Q., Yan, Z., Chen, R., Zeh Iii, H. J., Kang, R., et al. (2013). Strange attractors: DAMPs and autophagy link tumor cell death and immunity. *Cell Death & Disease*, *4*, e966. https://doi.org/10.1038/cddis.2013.493.

Hsu, H., Shu, H. B., Pan, M. G., & Goeddel, D. V. (1996). TRADD-TRAF2 and TRADD-FADD interactions define two distinct TNF receptor 1 signal transduction pathways. *Cell*, *84*(2), 299–308. https://doi.org/10.1016/s0092-8674(00)80984-8.

Huang, Q., Deveraux, Q. L., Maeda, S., Salvesen, G. S., Stennicke, H. R., Hammock, B. D., et al. (2000). Evolutionary conservation of apoptosis mechanisms: Lepidopteran and baculoviral inhibitor of apoptosis proteins are inhibitors of mammalian caspase-9. *Proceedings of the National Academy of Sciences of the United States of America*, *97*(4), 1427–1432. https://doi.org/10.1073/pnas.97.4.1427.

Hughes, P. D., Belz, G. T., Fortner, K. A., Budd, R. C., Strasser, A., & Bouillet, P. (2008). Apoptosis regulators Fas and Bim cooperate in shutdown of chronic immune responses and prevention of autoimmunity. *Immunity*, *28*(2), 197–205. https://doi.org/10.1016/j.immuni.2007.12.017.

Isaac, D. T., Berkes, C. A., English, B. C., Murray, D. H., Lee, Y. N., Coady, A., et al. (2015). Macrophage cell death and transcriptional response are actively triggered by the fungal virulence factor Cbp1 during H. capsulatum infection. *Molecular Microbiology*, *98*(5), 910–929. https://doi.org/10.1111/mmi.13168.

Jaworska, J., Coulombe, F., Downey, J., Tzelepis, F., Shalaby, K., Tattoli, I., et al. (2014). NLRX1 prevents mitochondrial induced apoptosis and enhances macrophage antiviral immunity by interacting with influenza virus PB1-F2 protein. *Proceedings of the National Academy of Sciences of the United States of America, 111*(20), E2110–E2119. https://doi.org/10.1073/pnas.1322118111.

Jiang, J. H., Davies, J. K., Lithgow, T., Strugnell, R. A., & Gabriel, K. (2011). Targeting of Neisserial PorB to the mitochondrial outer membrane: An insight on the evolution of beta-barrel protein assembly machines. *Molecular Microbiology, 82*(4), 976–987. https://doi.org/10.1111/j.1365-2958.2011.07880.x.

Jorgensen, I., Rayamajhi, M., & Miao, E. A. (2017). Programmed cell death as a defence against infection. *Nature Reviews. Immunology, 17*(3), 151–164. https://doi.org/10.1038/nri.2016.147.

Kalai, M., Lamkanfi, M., Denecker, G., Boogmans, M., Lippens, S., Meeus, A., et al. (2003). Regulation of the expression and processing of caspase-12. *The Journal of Cell Biology, 162*(3), 457–467. https://doi.org/10.1083/jcb.200303157.

Kerr, J. F. (1971). Shrinkage necrosis: A distinct mode of cellular death. *The Journal of Pathology, 105*(1), 13–20. https://doi.org/10.1002/path.1711050103.

Kerr, J. F., Wyllie, A. H., & Currie, A. R. (1972). Apoptosis: A basic biological phenomenon with wide-ranging implications in tissue kinetics. *British Journal of Cancer, 26*(4), 239–257. https://doi.org/10.1038/bjc.1972.33.

Kroemer, G., El-Deiry, W. S., Golstein, P., Peter, M. E., Vaux, D., Vandenabeele, P., et al. (2005). Classification of cell death: Recommendations of the Nomenclature Committee on Cell Death. *Cell Death and Differentiation, 12*(Suppl. 2), 1463–1467. https://doi.org/10.1038/sj.cdd.4401724.

Kroemer, G., Marino, G., & Levine, B. (2010). Autophagy and the integrated stress response. *Molecular Cell, 40*(2), 280–293. https://doi.org/10.1016/j.molcel.2010.09.023.

Kuida, K., Haydar, T. F., Kuan, C. Y., Gu, Y., Taya, C., Karasuyama, H., et al. (1998). Reduced apoptosis and cytochrome c-mediated caspase activation in mice lacking caspase 9. *Cell, 94*(3), 325–337. https://doi.org/10.1016/s0092-8674(00)81476-2.

Kvansakul, M., van Delft, M. F., Lee, E. F., Gulbis, J. M., Fairlie, W. D., Huang, D. C., et al. (2007). A structural viral mimic of prosurvival Bcl-2: A pivotal role for sequestering proapoptotic Bax and Bak. *Molecular Cell, 25*(6), 933–942. https://doi.org/10.1016/j.molcel.2007.02.004.

Lamkanfi, M., & Dixit, V. M. (2010). Manipulation of host cell death pathways during microbial infections. *Cell Host & Microbe, 8*(1), 44–54. https://doi.org/10.1016/j.chom.2010.06.007.

Lemaitre, B., & Girardin, S. E. (2013). Translation inhibition and metabolic stress pathways in the host response to bacterial pathogens. *Nature Reviews. Microbiology, 11*(6), 365–369. https://doi.org/10.1038/nrmicro3029.

Li, K., Li, Y., Shelton, J. M., Richardson, J. A., Spencer, E., Chen, Z. J., et al. (2000). Cytochrome c deficiency causes embryonic lethality and attenuates stress-induced apoptosis. *Cell, 101*(4), 389–399. https://doi.org/10.1016/s0092-8674(00)80849-1.

Li, S., Zhang, L., Yao, Q., Li, L., Dong, N., Rong, J., et al. (2013). Pathogen blocks host death receptor signalling by arginine GlcNAcylation of death domains. *Nature, 501*(7466), 242–246. https://doi.org/10.1038/nature12436.

Li, H., Zhu, H., Xu, C. J., & Yuan, J. (1998). Cleavage of BID by caspase 8 mediates the mitochondrial damage in the Fas pathway of apoptosis. *Cell, 94*(4), 491–501. https://doi.org/10.1016/s0092-8674(00)81590-1.

Lin, S., & Zhang, Y. J. (2017). Interference of apoptosis by hepatitis B Virus. *Viruses, 9*(8), 230. https://doi.org/10.3390/v9080230.

Linkermann, A., & Green, D. R. (2014). Necroptosis. *The New England Journal of Medicine, 370*(5), 455–465. https://doi.org/10.1056/NEJMra1310050.

Linkermann, A., Stockwell, B. R., Krautwald, S., & Anders, H. J. (2014). Regulated cell death and inflammation: An auto-amplification loop causes organ failure. *Nature Reviews. Immunology, 14*(11), 759–767. https://doi.org/10.1038/nri3743.

Liu, X., Kim, C. N., Yang, J., Jemmerson, R., & Wang, X. (1996). Induction of apoptotic program in cell-free extracts: Requirement for dATP and cytochrome c. *Cell*, *86*(1), 147–157. https://doi.org/10.1016/s0092-8674(00)80085-9.

Liu, Y., Shoji-Kawata, S., Sumpter, R. M., Jr., Wei, Y., Ginet, V., Zhang, L., et al. (2013). Autosis is a Na+,K+-ATPase-regulated form of cell death triggered by autophagy-inducing peptides, starvation, and hypoxia-ischemia. *Proceedings of the National Academy of Sciences of the United States of America*, *110*(51), 20364–20371. https://doi.org/10.1073/pnas.1319661110.

Liu, X., Zhang, Z., Ruan, J., Pan, Y., Magupalli, V. G., Wu, H., et al. (2016). Inflammasome-activated gasdermin D causes pyroptosis by forming membrane pores. *Nature*, *535*(7610), 153–158. https://doi.org/10.1038/nature18629.

Llambi, F., Moldoveanu, T., Tait, S. W., Bouchier-Hayes, L., Temirov, J., McCormick, L. L., et al. (2011). A unified model of mammalian BCL-2 protein family interactions at the mitochondria. *Molecular Cell*, *44*(4), 517–531. https://doi.org/10.1016/j.molcel.2011.10.001.

Losick, V. P., & Isberg, R. R. (2006). NF-kappaB translocation prevents host cell death after low-dose challenge by Legionella pneumophila. *The Journal of Experimental Medicine*, *203*(9), 2177–2189. https://doi.org/10.1084/jem.20060766.

Ma, Y. F., Chen, C., Li, D., Liu, M., Lv, Z. W., Ji, Y., et al. (2017). Targeting of interleukin (IL)-17A inhibits PDL1 expression in tumor cells and induces anticancer immunity in an estrogen receptor-negative murine model of breast cancer. *Oncotarget*, *8*(5), 7614–7624. https://doi.org/10.18632/oncotarget.13819.

Massari, P., Gunawardana, J., Liu, X., & Wetzler, L. M. (2010). Meningococcal porin PorB prevents cellular apoptosis in a toll-like receptor 2- and NF-kappaB-independent manner. *Infection and Immunity*, *78*(3), 994–1003. https://doi.org/10.1128/IAI.00156-09.

McIlwain, D. R., Berger, T., & Mak, T. W. (2015). Caspase functions in cell death and disease. *Cold Spring Harbor Perspectives in Biology*, 7(4), 1–26. https://doi.org/10.1101/cshperspect.a026716.

Mesri, E. A., Cesarman, E., & Boshoff, C. (2010). Kaposi's sarcoma and its associated herpesvirus. *Nature Reviews. Cancer*, *10*(10), 707–719. https://doi.org/10.1038/nrc2888.

Monroe, K. M., Yang, Z., Johnson, J. R., Geng, X., Doitsh, G., Krogan, N. J., et al. (2014). IFI16 DNA sensor is required for death of lymphoid CD4 T cells abortively infected with HIV. *Science*, *343*(6169), 428–432. https://doi.org/10.1126/science.1243640.

Muller, A., Rassow, J., Grimm, J., Machuy, N., Meyer, T. F., & Rudel, T. (2002). VDAC and the bacterial porin PorB of Neisseria gonorrhoeae share mitochondrial import pathways. *The EMBO Journal*, *21*(8), 1916–1929. https://doi.org/10.1093/emboj/21.8.1916.

Nassour, J., Radford, R., Correia, A., Fuste, J. M., Schoell, B., Jauch, A., et al. (2019). Autophagic cell death restricts chromosomal instability during replicative crisis. *Nature*, *565*(7741), 659–663. https://doi.org/10.1038/s41586-019-0885-0.

Obeid, M., Tesniere, A., Ghiringhelli, F., Fimia, G. M., Apetoh, L., Perfettini, J. L., et al. (2007). Calreticulin exposure dictates the immunogenicity of cancer cell death. *Nature Medicine*, *13*(1), 54–61. https://doi.org/10.1038/nm1523.

O'Donnell, J. A., Kennedy, C. L., Pellegrini, M., Nowell, C. J., Zhang, J. G., O'Reilly, L. A., et al. (2015). Fas regulates neutrophil lifespan during viral and bacterial infection. *Journal of Leukocyte Biology*, 97(2), 321–326. https://doi.org/10.1189/jlb.3AB1113-594RR.

Okamoto, T., Campbell, S., Mehta, N., Thibault, J., Colman, P. M., Barry, M., et al. (2012). Sheeppox virus SPPV14 encodes a Bcl-2-like cell death inhibitor that counters a distinct set of mammalian proapoptotic proteins. *Journal of Virology*, *86*(21), 11501–11511. https://doi.org/10.1128/JVI.01115-12.

O'Neill, K. L., Huang, K., Zhang, J., Chen, Y., & Luo, X. (2016). Inactivation of prosurvival Bcl-2 proteins activates Bax/Bak through the outer mitochondrial membrane. *Genes & Development*, *30*(8), 973–988. https://doi.org/10.1101/gad.276725.115.

Pearson, J. S., Giogha, C., Muhlen, S., Nachbur, U., Pham, C. L., Zhang, Y., et al. (2017). EspL is a bacterial cysteine protease effector that cleaves RHIM proteins to block necroptosis and inflammation. *Nature Microbiology*, *2*, 16258. https://doi.org/10.1038/nmicrobiol.2016.258.

Pearson, J. S., Giogha, C., Ong, S. Y., Kennedy, C. L., Kelly, M., Robinson, K. S., et al. (2013). A type III effector antagonizes death receptor signalling during bacterial gut infection. *Nature*, *501*(7466), 247–251. https://doi.org/10.1038/nature12524.

Pollock, G. L., Oates, C. V. L., Giogha, C., Wong Fok Lung, T., Ong, S. Y., Pearson, J. S., et al. (2017). Distinct roles of the antiapoptotic effectors NleB and NleF from enteropathogenic Escherichia coli. *Infection and Immunity*, *85*(4), e01071-16. https://doi.org/10.1128/IAI.01071-16.

Rajalingam, K., Sharma, M., Lohmann, C., Oswald, M., Thieck, O., Froelich, C. J., et al. (2008). Mcl-1 is a key regulator of apoptosis resistance in Chlamydia trachomatis-infected cells. *PLoS One*, *3*(9), e3102. https://doi.org/10.1371/journal.pone.0003102.

Rao, L., Debbas, M., Sabbatini, P., Hockenbery, D., Korsmeyer, S., & White, E. (1992). The adenovirus E1A proteins induce apoptosis, which is inhibited by the E1B 19-kDa and Bcl-2 proteins. *Proceedings of the National Academy of Sciences of the United States of America*, *89*(16), 7742–7746. https://doi.org/10.1073/pnas.89.16.7742.

Real, F., Florentino, P. T., Reis, L. C., Ramos-Sanchez, E. M., Veras, P. S., Goto, H., et al. (2014). Cell-to-cell transfer of Leishmania amazonensis amastigotes is mediated by immunomodulatory LAMP-rich parasitophorous extrusions. *Cellular Microbiology*, *16*(10), 1549–1564. https://doi.org/10.1111/cmi.12311.

Rogers, C., Erkes, D. A., Nardone, A., Aplin, A. E., Fernandes-Alnemri, T., & Alnemri, E. S. (2019). Gasdermin pores permeabilize mitochondria to augment caspase-3 activation during apoptosis and inflammasome activation. *Nature Communications*, *10*(1), 1689. https://doi.org/10.1038/s41467-019-09397-2.

Rongvaux, A., Jackson, R., Harman, C. C., Li, T., West, A. P., de Zoete, M. R., et al. (2014). Apoptotic caspases prevent the induction of type I interferons by mitochondrial DNA. *Cell*, *159*(7), 1563–1577. https://doi.org/10.1016/j.cell.2014.11.037.

Sarid, R., Sato, T., Bohenzky, R. A., Russo, J. J., & Chang, Y. (1997). Kaposi's sarcoma-associated herpesvirus encodes a functional bcl-2 homologue. *Nature Medicine*, *3*(3), 293–298. https://doi.org/10.1038/nm0397-293.

Schulze-Osthoff, K., Ferrari, D., Los, M., Wesselborg, S., & Peter, M. E. (1998). Apoptosis signaling by death receptors. *European Journal of Biochemistry*, *254*(3), 439–459. https://doi.org/10.1046/j.1432-1327.1998.2540439.x.

Schweichel, J. U., & Merker, H. J. (1973). The morphology of various types of cell death in prenatal tissues. *Teratology*, 7(3), 253–266. https://doi.org/10.1002/tera.1420070306.

Segawa, K., Kurata, S., Yanagihashi, Y., Brummelkamp, T. R., Matsuda, F., & Nagata, S. (2014). Caspase-mediated cleavage of phospholipid flippase for apoptotic phosphatidylserine exposure. *Science*, *344*(6188), 1164–1168. https://doi.org/10.1126/science.1252809.

Singh, R., Letai, A., & Sarosiek, K. (2019). Regulation of apoptosis in health and disease: The balancing act of BCL-2 family proteins. *Nature Reviews. Molecular Cell Biology*, *20*(3), 175–193. https://doi.org/10.1038/s41580-018-0089-8.

Sly, L. M., Hingley-Wilson, S. M., Reiner, N. E., & McMaster, W. R. (2003). Survival of Mycobacterium tuberculosis in host macrophages involves resistance to apoptosis dependent upon induction of antiapoptotic Bcl-2 family member Mcl-1. *Journal of Immunology*, *170*(1), 430–437. https://doi.org/10.4049/jimmunol.170.1.430.

Speir, M., Lawlor, K. E., Glaser, S. P., Abraham, G., Chow, S., Vogrin, A., et al. (2016). Eliminating Legionella by inhibiting BCL-XL to induce macrophage apoptosis. *Nature Microbiology*, *1*, 15034. https://doi.org/10.1038/nmicrobiol.2015.34.

Speir, M., Vogrin, A., Seidi, A., Abraham, G., Hunot, S., Han, Q., et al. (2017). Legionella pneumophila strain 130b evades macrophage cell death independent of the effector SidF in the absence of flagellin. *Frontiers in Cellular and Infection Microbiology*, *7*, 35. https://doi.org/10.3389/fcimb.2017.00035.

Tait, S. W., & Green, D. R. (2010). Mitochondria and cell death: Outer membrane permeabilization and beyond. *Nature Reviews. Molecular Cell Biology*, *11*(9), 621–632. https://doi.org/10.1038/nrm2952.

Tang, D., Kang, R., Coyne, C. B., Zeh, H. J., & Lotze, M. T. (2012). PAMPs and DAMPs: Signal 0s that spur autophagy and immunity. *Immunological Reviews*, *249*(1), 158–175. https://doi.org/10.1111/j.1600-065X.2012.01146.x.

Thome, M., Schneider, P., Hofmann, K., Fickenscher, H., Meinl, E., Neipel, F., et al. (1997). Viral FLICE-inhibitory proteins (FLIPs) prevent apoptosis induced by death receptors. *Nature*, *386*(6624), 517–521. https://doi.org/10.1038/386517a0.

Uren, R. T., O'Hely, M., Iyer, S., Bartolo, R., Shi, M. X., Brouwer, J. M., et al. (2017). Disordered clusters of Bak dimers rupture mitochondria during apoptosis. *eLife*, *6*, 1–23. https://doi.org/10.7554/eLife.19944.

Van Opdenbosch, N., & Lamkanfi, M. (2019). Caspases in cell death, inflammation, and disease. *Immunity*, *50*(6), 1352–1364. https://doi.org/10.1016/j.immuni.2019.05.020.

Vanden Berghe, T., Linkermann, A., Jouan-Lanhouet, S., Walczak, H., & Vandenabeele, P. (2014). Regulated necrosis: The expanding network of non-apoptotic cell death pathways. *Nature Reviews. Molecular Cell Biology*, *15*(2), 135–147. https://doi.org/10.1038/nrm3737.

Verbeke, P., Welter-Stahl, L., Ying, S., Hansen, J., Hacker, G., Darville, T., et al. (2006). Recruitment of BAD by the Chlamydia trachomatis vacuole correlates with host-cell survival. *PLoS Pathogens*, *2*(5), e45. https://doi.org/10.1371/journal.ppat.0020045.

Verhagen, A. M., Ekert, P. G., Pakusch, M., Silke, J., Connolly, L. M., Reid, G. E., et al. (2000). Identification of DIABLO, a mammalian protein that promotes apoptosis by binding to and antagonizing IAP proteins. *Cell*, *102*(1), 43–53. https://doi.org/10.1016/s0092-8674(00)00009-x.

Vince, J. E., & Silke, J. (2016). The intersection of cell death and inflammasome activation. *Cellular and Molecular Life Sciences*, *73*(11-12), 2349–2367. https://doi.org/10.1007/s00018-016-2205-2.

Wang, Q., Liu, S., Tang, Y., Liu, Q., & Yao, Y. (2014). MPT64 protein from Mycobacterium tuberculosis inhibits apoptosis of macrophages through NF-kB-miRNA21-Bcl-2 pathway. *PLoS One*, *9*(7), e100949. https://doi.org/10.1371/journal.pone.0100949.

Wei, M. C., Zong, W. X., Cheng, E. H., Lindsten, T., Panoutsakopoulou, V., Ross, A. J., et al. (2001). Proapoptotic BAX and BAK: A requisite gateway to mitochondrial dysfunction and death. *Science*, *292*(5517), 727–730. https://doi.org/10.1126/science.1059108.

Weinlich, R., Oberst, A., Beere, H. M., & Green, D. R. (2017). Necroptosis in development, inflammation and disease. *Nature Reviews. Molecular Cell Biology*, *18*(2), 127–136. https://doi.org/10.1038/nrm.2016.149.

White, M. J., McArthur, K., Metcalf, D., Lane, R. M., Cambier, J. C., Herold, M. J., et al. (2014). Apoptotic caspases suppress mtDNA-induced STING-mediated type I IFN production. *Cell*, *159*(7), 1549–1562. https://doi.org/10.1016/j.cell.2014.11.036.

Yatim, N., Jusforgues-Saklani, H., Orozco, S., Schulz, O., Barreira da Silva, R., Reis E Sousa, C., et al. (2015). RIPK1 and NF-kappaB signaling in dying cells determines cross-priming of CD8(+) T cells. *Science*, *350*(6258), 328–334. https://doi.org/10.1126/science.aad0395.

Yoneda, T., Imaizumi, K., Oono, K., Yui, D., Gomi, F., Katayama, T., et al. (2001). Activation of caspase-12, an endoplastic reticulum (ER) resident caspase, through tumor necrosis factor receptor-associated factor 2-dependent mechanism in response to the ER stress. *The Journal of Biological Chemistry, 276*(17), 13935–13940. https://doi.org/10.1074/jbc.M010677200.

Youle, R. J., & Strasser, A. (2008). The BCL-2 protein family: Opposing activities that mediate cell death. *Nature Reviews. Molecular Cell Biology, 9*(1), 47–59. https://doi.org/10.1038/nrm2308.

Yu, S. W., Wang, H., Poitras, M. F., Coombs, C., Bowers, W. J., Federoff, H. J., et al. (2002). Mediation of poly(ADP-ribose) polymerase-1-dependent cell death by apoptosis-inducing factor. *Science, 297*(5579), 259–263. https://doi.org/10.1126/science.1072221.

Yuan, J., & Horvitz, H. R. (1992). The Caenorhabditis elegans cell death gene ced-4 encodes a novel protein and is expressed during the period of extensive programmed cell death. *Development, 116*(2), 309–320. Retrieved from https://www.ncbi.nlm.nih.gov/pubmed/1286611.

Zakeri, A. (2017). Helminth-induced apoptosis: A silent strategy for immunosuppression. *Parasitology, 144*(13), 1663–1676. https://doi.org/10.1017/S0031182017000841.

Zitvogel, L., Kepp, O., & Kroemer, G. (2010). Decoding cell death signals in inflammation and immunity. *Cell, 140*(6), 798–804. https://doi.org/10.1016/j.cell.2010.02.015.

Zou, H., Henzel, W. J., Liu, X., Lutschg, A., & Wang, X. (1997). Apaf-1, a human protein homologous to C. elegans CED-4, participates in cytochrome c-dependent activation of caspase-3. *Cell, 90*(3), 405–413. https://doi.org/10.1016/s0092-8674(00)80501-2.

CHAPTER TWO

Reactive oxygen mediated apoptosis as a therapeutic approach against opportunistic *Candida albicans*

Tanu Atriwal[a], Meenal Chawla[a], Afzal Hussain[b], Mohamed F. Alajmi[b], and Mohammad Abid[a,*]

[a]Medicinal Chemistry Laboratory, Department of Biosciences, Jamia Millia Islamia, New Delhi, India
[b]Department of Pharmacognosy, College of Pharmacy, King Saud University, Riyadh, Saudi Arabia
*Corresponding author: e-mail address: mabid@jmi.ac.in

Contents

Abstract

Candida albicans are polymorphic fungal species commonly occurs in a symbiotic association with the host's usual microflora. Certain specific changes in its usual microenvironment can lead to diseases ranging from external mucosal to severally lethal systemic infections like invasive candidiasis hospital-acquired fatal infection caused by different species of *Candida*. The patient acquired with this infection has a high mortality and morbidity rate, ranging from 40% to 60%. This is an ill-posed problem by its very nature. Hence, early diagnosis and management is a crucial part. Antifungal drug resistance against the first and second generation of antifungal drugs has made it difficult to treat such fatal diseases. After a few dormant years, recently, there has been a rapid turnover of identifying novel drugs with low toxicity to limit the problem of drug resistance. After an initial overview of related work, we examine specific prior work on how a change in oxidative stress can facilitate apoptosis in *C. albicans*. Subsequently, it was investigated

Advances in Protein Chemistry and Structural Biology, Volume 125
ISSN 1876-1623
https://doi.org/10.1016/bs.apcsb.2020.12.004

that *Candida* spp. suppresses the production of ROS mediated host defense system. Here, we have reviewed possibly all the small molecule inhibitors, natural products, antimicrobial peptide, and some naturally derived semi-synthetic compounds which are known to influence oxidative stress, to generate a proper apoptotic response in *C. albicans* and thus might be a novel therapeutic approach to augment the current treatment options.

1. Introduction

C. albicans is a part of the normal flora of the human body, usually present on the skin, oral cavity, urogenital and gastrointestinal tracts of healthy beings. Naturally, *C. albicans* is found in a planktonic form. Still, due to changes in its environmental condition or immune dysfunction, it becomes virulent and capable enough to cause a variety of superficial to deep-seated mycotic infections with a more than 40% of mortality rate. Observational studies have demonstrated a high prevalence of approximately 4 lac life-threatening infections globally each year in severely immune-deficient patients. Several studies have examined the resistance toward the four major classes of known antifungal drugs like Azoles, Echinocandins, Polyenes, and Flucytosine due variety of reasons like immune system evasion, change in sterol composition, asymptomatic biofilm formation, and drug overuse. Recent data by Centres for Disease Control and Prevention (CDC) has reported an increase in the number of antifungal drug resistance cases with more than 34,000 cases with 1700 deaths. This alarming global crisis in fungal drug resistance needs attention, which should guarantee judicious antifungal prescribing by medical practitioners, utilization of combinatorial strategies, and continual antifungal drug discovery and development (Perfect, 2017; Pfaller & Diekema, 2007; Silva, Rodrigues, Araújo, Rodrigues, & Henriques, 2017; Spampinato & Leonardi, 2013).

The fungicidal mechanism employed by the first line of defense of our body, the innate immune system, involves the production of reactive oxygen species (ROS) like superoxide, hydrogen peroxide, etc., which are toxic to *C. albicans*. Some previous studies also suggested that amphotericin B and fluconazole activity involves the production of endogenous ROS induced by oxidative damage. ROS causes apoptosis, a programmed cell death (PCD), essential for normal cell functioning, via different pathways (Simon, Haj-Yehia, & Levi-Schaffer, 2000). Therefore, if, by specific means, we can manipulate the apoptotic response of *C. albicans by targeting and increasing* ROS production, then we might be able to control this opportunistic pathogen.

Several investigators have probed inhibitory molecules that can manipulate the oxidative stress by increasing the yeast cell's ROS production. As ROS is one of the critical regulators of apoptosis; thus, in this review article, we will briefly discuss *C. albicans*, apoptosis, and some oxidative stress modulating inhibitors.

2. *C. albicans*

C. albicans is the most prevalent fungal pathogen, which causes superficial mucosal to life-threatening systemic infection in humans. They are prime fungal species isolated from infected medical devices, which account for ~15% of hospital-acquired cases of sepsis (Nobile & Johnson, 2015). It is a polymorphic fungus that takes the most common morphologies like predominant yeast form and pseudohyphae form and true hyphae form-the invasive form. This opportunistic fungal pathogen is usually present in a symbiotic association with the normal microflora of humans (Douglas, 2003; Lohse, Gulati, Johnson, & Nobile, 2018).

The ability of this fungal pathogen to transit from planktonic form to highly virulent form depends on different factors like variations in the homeostatic conditions in the host's body such as the change in the normal body temperature, or pH causes colonization of this opportunistic pathogen leading to invasive infection in severely immune-compromised individuals, adherence to the biomedical devices, host immune invasion (Aminahlkner, 2004; Tsui, Kong, & Jabra-Rizk, 2016).

2.1 Oxidative stress in *C. albicans*

The healthy individuals fight infection against *C. albicans* by phagocytic cells, neutrophils, and macrophages through the production of reactive oxygen species (ROS) by the process known as Respiratory burst. These phagocytic cells trigger the NADPH oxidase complex's association by stimulating cytokines, which affects superoxide production ($O_2^{-}\bullet$). The activation of NADPH oxidase, the multi-subunit enzyme, is highly regulated (Babior, 2004). The superoxide produced is then dismutated by enzyme superoxide dismutase to hydroxyl anions (OH^-) or hydrogen peroxide (H_2O_2) and hydroxyl radicals (OH) through Haber-Weiss reaction. ROS can further from various toxic chemicals in phagosome such as hypochlorous acid (HOCl), when H_2O_2 reacts with chloride ions (Cl^-) in a reaction catalyzed by myeloperoxidase and toxic peroxynitrite (ONOO), when the superoxide interacts with nitric oxide radical generated by inducible nitric oxide synthase (iNOS). It was also discovered that the combination of

ROS and cationic stress induced during maturation of phagocyte together results in *C. albicans* killing. The ROS formed within the phagosome generates the toxic environment, which causes oxidative stress in *C. albicans,* and the ROS generated by phagocytes resulted in cell death.

However, sometimes *C. albicans* can colonize the host's body as it sense and respond to ROS effectively. It does so by inducing the antioxidant genes following exposure to H_2O_2, which include catalase (*CAT1*), glutathione peroxidase (*GPX*), and superoxide dismutase (*SOD*) antioxidant-encoding genes, and also the genes encoding components of the glutathione/glutaredoxin (*GSH1, TTR1*) and thioredoxin (*TSA1, TRX1, TRR1*) systems, that helps in restoring oxidatively-damaged proteins, protein folding and sulfur metabolism. Transcriptional activation of these genes leads to resistance of *C. albicans* to oxidative stress. Despite this, extensively induced production of ROS can leads to apoptosis in *C. albicans* (Chakraborti, Chakraborti, Chattopadhyay, & Shaha, 2019; Dantas et al., 2015; Ikner & Shiozaki, 2005).

3. Apoptosis

Cell death usually can be classified into necrosis and apoptosis. Apoptosis is the form of programmed cell death that takes place in multicellular organisms. Several regulatory pathways and genes are there to ensure normal apoptosis in the cells, i.e., highly regulated. But sometimes its deregulations occur because of a variety of internal or external stimulus which leads to diverse diseases such as uncontrolled propagation of cells and failure of cells to undergo apoptosis in the response of cellular damage, results in the progression of deadly diseases like cancer, but an unnecessarily higher level of apoptosis can result into diseases like Parkinson's or Alzheimer's, rheumatoid arthritis (D'Arcy, 2019). Induction of apoptosis in diseases where it is usually inhibited or where its induction leads to removing the harmful pathogen is extensively discussed therapeutic approach among scientists. We will study in brief about the biochemical features and the apoptotic pathways.

3.1 Biochemical features

The biochemical changes like protein cleavage, protein cross-linking, DNA fragmentation, cell shrinkage, the formation of apoptotic bodies, and phagocytic recognition by neighboring cells are readily observed (Poon, Lucas, Rossi, & Ravichandran, 2014). The Specific cell surface markers that aid in the phagocytic recognition are phosphatidylserine, usually facing inward in the cell's plasma membrane but at apoptosis. It is present in the plasma membrane's outer side, where it acts as a ligand for phagocytes. Proteins such as annexin I and calreticulin also serve as recognition markers (Brown, Yang, & Ray, 2014).

3.2 Pathways

The apoptosis pathways were distinguished according to whether caspases, a family of cysteine proteases, are involved or not. Caspases are expressed as inactive enzymes, which, once activated, leads to a protease cascade in which one caspase activates another. They are mainly grouped into initiators (caspases-2, -8, -9, -10), effectors or executioners (caspases-3, -6, -7), and inflammatory caspases (caspases-1, -4, -5). They can cleave substrate at aspartic acid residues, and their specificities are mainly due to amino acids recognition (Elmore, 2007). Caspase activation is highly regulated and includes key regulator proteins, Bcl-2 family of proteins. These include anti-apoptotic proteins Bcl-2, Bcl-x_L, Bcl-w, Mcl-1 and A1, which inhibits apoptosis during cellular stress and promotes cell growth. They constitute three conserved BH (Bcl-2 homology) domains, i.e., BH1, BH2 and BH3, which fold together leading to the formation of hydrophobic groove on pro-survival molecules. This groove acts as a target for binding pro-apoptotic proteins Bik/Nbk/Blk, Bid, Bad, Hrk/DP5, Bim/Bod, Noxa, Puma/Bbc3, Bmf and Bcl-G_s, which promotes cell death. These pro-apoptotic proteins contain only BH3 domain, so-called BH-3-only proteins. The other class of pro-apoptotic Bcl-2 proteins constitutes multi-domain Bax-like proteins. These multi-domain proteins act at step, distinct from the BH3-only proteins (Willis, 2003). Various stimulus acts to trigger the apoptotic pathway via multi-signal pathways and which in turn synchronized by multi-complicated intrinsic and extrinsic ligands. The apoptotic process can be initiated by the cell via a mechanism known as the intrinsic pathway that detects cellular damage through various intracellular sensors. It can also be initiated from the contact of an immune system cell and the damaged cell, which is the extrinsic pathway of apoptosis (D'Arcy, 2019). We will discuss the caspase-dependent and caspase-independent pathways and the stimulus that leads to their activation.

3.2.1 Caspase-dependent apoptosis pathway

This classic PCD pathway involves the participation of caspase-3, caspase-8, caspase-9, caspase-12, caspase-7 cascade, which is activated by various receptors such as the TNF-alpha receptor, FasL receptor, TLR, Death receptor, etc. Iron channel like calcium channel is also known to be involved in this pathway as calcium concentration plays a significant role in determining the cell fate via opening and closing its channel and participating in cell death and proliferation.

The various stimuli are involved in the activation of this pathway. NO and ROS production stimulated by ER (Endoplasmic Reticulum) stress or ROS production by exposure of some radiations like UV and X-rays causes mitochondrial membrane depolarization, and permeabilization which results in the release of one of pro-apoptosis signal molecule, cytochrome *c* which then triggers caspase-9, caspase-3 activation and in the end leads to apoptosis. External pathogens can induce and activate caspase-8 dependent apoptotic pathway. DNA damage also enhances ROS production. Besides this, various peptides or molecules induce apoptosis by disturbing the calcium concentration in cytosol, leading to calpain activation, calpain activation, and activated caspase-12; calpain activation, in turn, activating caspase-12 activates caspase-12 inactivates the Bcl-Xl. The increased calcium concentration can also result in enhanced ROS and simultaneously also trigger the ER stress, and then activation of the JNK pathway, thus finally Bax activation.

3.2.2 Caspase-independent apoptosis pathway

AIF (apoptosis-inducing factor), a significant caspase-independent pro-apoptosis factor and ROS, both have been found to trigger this pathway. AIF is known to release from mitochondria and translocate to the nucleus leading to cleavage of the DNA and apoptosis. ROS is known to play a role in the release of AIF from mitochondria by activating poly (ADP-ribose) polymerase-1 (PARP-1), and this activation is necessary for AIF release. As ROS also participated in the caspase-dependent apoptosis pathway it might act as a connecting link between two apoptosis types. ROS mainly comes from mitochondria, so this organelle plays an imperative role in the crosstalk of apoptosis pathways. Besides AIF and ROS, other factors, for instance, lysosomal membrane permeabilization; virus's protein; drugs; p53 suppression tumor factors or many other compounds also act as apoptotic factors using this pathway (Hongmei, 2012). The short-lived and highly reactive molecules ROS are usually detoxified by cells GSH; if detoxification does not occur, it might lead to apoptosis. So, GSH was thought to act early in the apoptosis pathway. This was also proved by studies where depletion of mitochondrial GSH by treatment with L-buthionine sulfoximine in the human B lymphoma cell line leads to apoptosis by inducing caspase-3 activation (Armstrong et al., 2002). This allowed us to conclude that ROS production at a higher amount and its failure to undergo detoxification affects the apoptotic process directly and indirectly, either through

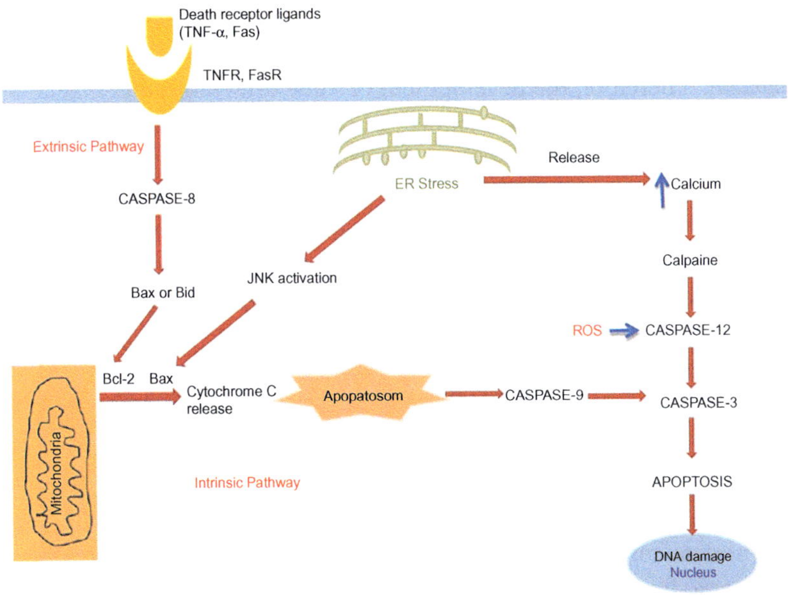

Fig. 1 Summary of apoptotic pathway.

caspase-dependent or caspase-independent pathways. Thus, increasing ROS in harmful pathogens via inhibitor molecules is a popular approach among several scientists. The diagrammatic representation of apoptotic pathways is shown in Fig. 1.

4. ROS inducing inhibitor/oxidative stress modulating inhibitors in *C. albicans*

Globally, the number of people affected by fungal diseases is 1.5 billion per annum, which is a massive burden on society. Additionally, the mortality rate estimated was 1.5 million deaths a year (Bongomin, Gago, Oladele, & Denning, 2017). Since prevalence and mortality rates due to invasive fungal infection are proliferating, the resistance against conventional drugs is also increasing. There is an emergence to look for some alternative therapies, drug molecules that can target specific sites and molecules that can robust the host immune response against the pathogenic organism *C. albicans*. One such strategy could be targeting an increase in ROS production. Since ROS production plays a significant role in phagocytosis-mediated host responses. Phagocytes like macrophages, dendritic cells and neutrophils play a crucial role in the host response to *C. albicans*. Since the cell wall is the first point of contact of *Candida* with phagocytes, the cell wall's structural components play a major role in increasing the immune host response towards *C. albicans*. Therefore, the development of some drug molecules that could increase the host's ROS production can be a promising new strategy for developing novel antifungal agents. Since there have been large arrays of improvements achieved in the last few years for the role of antimicrobial peptides and naturally derived small molecule inhibitors (SMI) as a curative approach. Therefore, some small molecular inhibitors and naturally occurring compounds have been studied by the researchers elaborated in this article's review (Delattin, Cammue, & Thevissen, 2014; Wellington, Dolan, & Krysan, 2009).

4.1 Antimicrobial peptides

Antimicrobial peptides (AMP) are small, cationic and evolutionarily conserved host defense oligo-peptide molecules with a broad spectrum of targets like bacteria, fungus and viruses. They were first in the 1940s when Dubos first isolated antimicrobial compounds from soil *Bacillus* strain, which prevented mice from pneumococci infection (Bahar & Ren, 2013). They are the key modulators of innate immune response (Brown & Hancock, 2006). Over the past three decades, there has been a sustained research activity in identifying different antimicrobial peptides as they are recognized as a possible source of pharmaceuticals for the treatment of antibiotic-microbial resistance (Mahlapuu, Håkansson, Ringstad, & Björn, 2016; Zhang & Gallo, 2016).

Melittin is naturally occurring antimicrobial peptide derived from European honey bee *Apis mellifera* venom, which has high cytolytic potency. It is a hexacosa peptide with a peptide sequence (GIGAVLKVLTTGLPA LISWIKRKRQQ-CONH2), constituting a large hydrophobic region with a predominant presence of hydrophilic amino acids at the -COOH terminal. This peptide usually disrupts the membrane integrity and leads to forming a voltage-dependent ion channel across the lipid bilayer. In order to understand its mechanism on *C. albicans,* the cells were treated with melittin. The study displayed the fungicidal activity against the yeast cell with apoptosis as the mechanism of action. Extensive data shows the increase in the production of ROS, which is a major stimulus of apoptosis (Park & Lee, 2010). Psacotheasin is a knottin-type antimicrobial peptide isolated from the larvae of *Psacothea hilaris,* a long-horned beetle with a peptide sequence (CIAKGNGCQPSGVQGNCCSGHCHKEPGWVAGYC K-NH2).

Previous research typically investigated the antifungal activity of Psacotheasin against *C. albicans* strain by perturbing and disrupting the plasma membrane of *C. albicans*. Following this research, Psacotheasin was investigated for its action mechanism, which elucidated anti-candida activity via an apoptotic pathway. Additionally, Psacotheasin-exposed cells displayed a significant increase of intracellular ROS level as compared to untreated cells (Hwang, Hwang, Lee, & Lee, 2011).

Hepcidin 25, also known as Hepc25, is a human antimicrobial peptide. It is a cysteine-rich amino acid sequence (DTHFPICIFCCGCCHRSKCG MCCKT), forming eight disulfide bonds and a stable β hairpin structure with amino-terminal ATCUN motif (H2N-X-X-H, where X is variable). This ATCUN motif can bind to Cu(II) and Ni(II) metals. It regulates the transportation of ferroportin, a Fe^{2+} transporter. In addition to this, it plays a vital role in ROS production in *C. albicans* (Chen & Lan, 2020).

Scolopendin is an antimicrobial peptide isolated from *Scolopendra subspinipes mutilans* (SPSEKAGLQPVGRIGRMLKK). Recently, scolopendin was investigated for its antifungal activity against *C. albicans* cells elucidated strong potential against the fungal cells with series of characteristic changes related to apoptosis-like phosphotidylserine externalization, chromatin condensation accompanied with ROS accumulation in the yeast cells (Lee, Hwang, & Lee, 2017).

Furthermore, Periplanetasin-2, an antimicrobial peptide derived from *Periplaneta Americana,* an American cockroach put forth a potent antifungal effect against pathogenic fungi *C. albicans* without causing hemolytic toxicity. Periplanetasin-2 prompted oxidative stress by increasing ROS concentration and lipid peroxidation in yeast cells, thus, causing apoptosis by externalizing phosphatidylserine and fragmentation of chromatin material (Yun, Hwang, & Lee, 2017).

Concerning research conducted, Lactoferrin B, like peptide (LBLP), a 23 mer antimicrobial peptide derived from *Scolopendra subspinipes mutilans* with a sequence (RMKKLGNHKVSCERNTKRCRKAI), displayed an excellent potentiality against *C. albicans* with hemolytic toxicity. This study suggests that LBLP mediates antifungal property by increasing Nitric Oxide (NO) in the yeast cell. Increase in NO caused nitrosative stress, which disrupted Ca^{2+} homeostasis, mitochondrial ROS accumulation, and destabilization of the NAD^{+}/NADH ratio, thus leading to apoptosis yeast cells (Kim, Hwang, & Lee, 2020).

In another study, Arenicin-1, (RWCVYAYVRVRGVLVRYRRCW) a 21 mer AMP derived from *Arenicola marina* (Lugworm) displayed effective antifungal property against C. albicans. Arenicin-1 treatment to yeast cells showed ROS accumulation. ROS accumulation leads to metacaspase activation, which in turn caused metacaspase dependent apoptosis in a yeast cell (Cho & Lee, 2011b).

Moreover, Cecropin A, a 37 amino acid residue AMP derived from *Hyalophora cecropia*, the giant silk moth, represented a potent antimicrobial activity. When tested on *C. albicans* cells. On further investigation, it was established that cecropin-A induced apoptosis, linked with redox state and ion balance of *C. albicans* (Yun & Lee, 2016b).

A recent study reported the antifungal *property of* Papiliocin, a 37-mer peptide derived from the swallowtail butterfly *Papilio xuthus against C. albicans.* Closer examination of the results revealed that Papiliocin induced the accretion of reactive oxygen species (ROS) and hydroxyl radicals, which lead to the induction of apoptosis in yeast cells (Hwang, Hwang, Lee, Kim, et al., 2011).

After rigorous examination, it was discovered that Pleurocidin, an antimicrobial peptide with 25 amino acid residues (GWGSFFKKAAHVGK HVGKAALTHYL-NH_2), found in mucosal secretions of the skin of *Pleuronectes americanus*, a winter flounder displays broad-spectrum antimicrobial activity without cytotoxicity. Additionally, Pleurocidin treatment on *C. albicans* caused morphological and physiological changes in the yeast cell. On further investigation, it was revealed that there was an accumulation of intracellular ROS (Cho & Lee, 2011a).

In another study, a previously reported natural peptide with broad-spectrum antifungal activity MCh-AMP1, derived from *Matricaria chamomilla* L. flowers, was investigated for its mode of action against *C. albicans*. The outcome suggests that MCh-AMP1 showed its fungicidal activity at MIC 32–64 μg/mL. Notably, MCh-AMP1 increases the ROS concentration to a maximum of 64.8% and disrupted the yeast cell's potassium ion channels; thus, results from this study suggest that MCh-AMP1 caused programmed cell death due to ROS accumulation in the yeast cell (Seyedjavadi et al., 2020).

Defensins are small cysteine-rich cationic, broad-spectrum antimicrobial peptides obtained from different sources. One such defensin is *Raphanus sativus* antifungal peptide 2 (RsAFP2), an antifungal plant defensin derived from *Raphanus sativus* seeds. It acts on glucosylceramides (GlcCer), a sphingolipid with great biological importance in fungal cells as it contributes toward the fungal cell division, spore formation, and hyphal transition (Del Poeta, Nimrichter, Rodrigues, & Luberto, 2014). RsAFP2 induces membrane permeabilization and fungal cell death by increasing ROS level in yeast cells in a dose-dependent manner (Aerts et al., 2007). Through recent reports, 43 mer defensin-Coprisin obtained dung beetle,

Copris tripartitus was studied for its antifungal property and the mode of action it pertains to the yeast cells of *C. albicans*. Coprisin showed a robust antifungal property with low hemolytic cytotoxicity. Furthermore, it elucidated that it disrupted the redox metabolism in the yeast cells which leads to metacaspase activation. Hence, was confirmed as a molecule with strong apoptotic activity against *C. albicans* (Lee et al., 2012). In a recent study, another plant defensin HsAFP1 obtained from *Heuchera sanguinea* (coral bells) displayed antifungal property against different pathogenic fungus. Likewise, HsAF1 treatment to *C. albicans* yeast cells showed ROS accumulation, thus induced apoptosis in *C. albicans* (Aerts et al., 2011).

Furthermore, Cathelicidins, cationic peptides composed of a conserved N-terminal cathelin-like domain and a variable C-terminal antimicrobial domain was investigated for its anti-candida properties. In the recent study, Human Cathelicidins LL-37(LLGDFFRKSKEKIGKEFKRIVQRIKDFL RNLVPRTES) and two of its fragments like LL13–37 (GKEFKRIVQRI KDFLRNLVPRTES) and LL17–32 (FKRIVQRIKDFLRNLV) were studied for their antifungal property against *C. albicans*. Human cathelicidin LL37 has emerged from the pro-protein human cationic antimicrobial protein (hCAP-18) stored in neutrophil granules and epithelial cells. Active LL37 is generated after the cleavage of the pro-protein hCAP-18 by the enzyme serine protease (Reinholz, Ruzicka, & Schauber, 2012). Closer examination of the results revealed that LL-37 was the best antifungal compound that induces accumulation of ROS in the fungal cell and binds to the negatively charged biomolecules of the fungal membrane thus, disrupting the membrane.

4.2 Naturally derived inhibitors

It is only in the last two decades that the discovery of naturally derived small molecule inhibitors have garnered worldwide attention in the past decade in the field of drug discovery due to an advantage over traditional cytotoxic drugs. Here, we have discussed some naturally derived molecules, crude extracts and small molecule inhibitors, which could induce some significant amount of ROS in the pathogenic yeast cell, *C. albicans*.

Lycopene **1**, a naturally occurring red carotenoid pigment majorly found in tomatoes, have drawn scientific interest because of its potential as a significant biological molecule. The current study reports that lycopene induces apoptosis by increasing intracellular ROS, majorly hydroxyl ions in the fungal cells. Additionally, lycopene treatment leads to the intracellular calcium burden, which was accompanied by mitochondrial dysfunction (Choi & Lee, 2015).

In addition, Nerol **2**, a naturally derived monoterpenoid alcohol obtained from essential oils of lemongrass shows strong antimicrobial activity against different bacterial species and fungal species. Furthermore, a study depicts a potent antifungal activity against *C. albicans* strain at MIC of 0.77 μL/mL (Wang et al., 2020). Notably, it induces apoptosis, which could be easily observed by certain morphological changes like an increase in ROS, phosphotidylserine externalization, mitochondrial dysfunction, and metacaspase activation (Tian, Lu, et al., 2017).

Moreover, from the recent study Baicalein (BE) **3** -5,6,7-trihydroxyflavone naturally derived flavonoid, formerly isolated from the roots *Scutellaria baicalensis* is a Chinese medicinal plant. A previous study suggests that Baicalein is a potentially active biofilm inhibitor w.r.t *C. albicans*. It can decrease by about 70% biofilm inhibition at 4–32 μg/mL concentration. It also exhibits a synergistic effect with already known antifungal drugs like fluconazole and amphotericin B (Serpa et al., 2012). To further add on to these studies, a group of researchers studied the effect of BE in the role of apoptosis. They found that 2 μg/mL BE treatment on *C. albicans* cells for 4 h leads to 11% mortality rate which includes 7% as apoptosis and 4% as necrosis while 0.5 μg/mL Amphotericin B treatment resulted in 24% of death rate out of which 15% was due to apoptosis and 9% of it was because of necrosis. On the other hand when BE and AmB when used in a combination, the mortality rate increased up to 89%–28% because of necrosis while 61% as apoptosis, thus, representing a synergistic action of BE and AmB on *C. albicans* apoptosis accompanied by an increase in the ROS production (Fu et al., 2011).

Another study on natural polyphenol, Curcumin **4** had elucidated a potent antifungal activity against *C. albicans* by raising the ROS concentration inside fungal cells. Not only had this increased ROS production, but it has also decreased the hyphal formation by targeting thymidine uptake 1 (TUP1) global inhibitor (Neelofar et al., 2011; Sharma, Manoharlal, Puri, & Prasad, 2010).

Another study revealed that a lignan compound, Nortrachelogenin **5** isolated from *Partrinia scabiosaefolia* displayed strong antifungal properties toward *C. albicans*. It leads to the ROS accumulation in the fungal cells; thus, inducing programmed cell death in the candida cells accompanied by mitochondrial dysfunction. Furthermore, this study revealed that the cells treated with Nortrachelogenin were unable to redevelop their cell wall due to an incapability to regenerate their cell wall on the protoplast (Lee, Woo, & Lee, 2016).

Another study performed on the natural compound *Apigenin* **6** isolated from Aster yomena, a perennial herb cultivated in South Korea and traditionally known for its antimicrobial property, showed a potent antifungal property against *C. albicans*. Interestingly, biochemical analysis of apigenin on *C. albicans* showed redox imbalance in fungal cells directed by glutathione depletion and ROS accumulation inside the cells. Furthermore, apigenin-induced apoptosis is marked with the externalization of phosphotidylserine marker (Lee, Woo, & Lee, 2019).

Moreover, Floricolin **7** a *p*-terphenyl molecule isolated from an endolichenic fungus, showed some promising results against both planktonic state and biofilm-forming cells of *C. albicans* by disrupting the redox homeostasis, leading to ROS accumulation accompanied by membrane disintegration, mitochondrial disruption and ultimately cell death (Zhang et al., 2018).

Another well-known naturally occurring monoterpenoid, Perillaldehyde **8** derived from *Perilla frutescens* depicted antifungal mode of action at MIC of 4 μL/mL via increasing in the level of ROS supplemented by programmed cell death (Qu et al., 2019; Tian, Qu, et al., 2017).

An antifungal active macrocyclic bis(bibenzyl) compound Plagiochin E **9** isolated from *Marchantia polymorpha*, a liverwort showed strong antifungal capacity against *C. albicans* via increasing the ROS concentration, thus, leading to apoptosis through metacaspase dependent pathway (Wu, Chang, Cheng, Sun, & Lou, 2010).

Farnesol **10**, a well-known quorum sensing inhibitory molecule which decreases the pathogenicity of *C. albicans* strain by different means like negatively regulating yeast to hyphal morphogenesis, biofilm inhibition also elucidated the property to disrupt redox metabolism as one of its mechanism (Dižová & Bujdáková, 2017).

A recent study reported, Hibicuslide C **11**, a compound extracted from *Abutilon theophrasti* through column chromatography, elucidated a strong capacity to inhibit *C. albicans* treated cells by ROS dependent apoptosis. The experimental work suggests that Hibicuslide C cells showed 33.64% increase in intracellular ROS levels than untreated cells. To further examine the role of ROS in antifungal activity, NAC(*N*-acetyl cysteine) was added to the Hibicuslide C treated cells, which displayed an increase in the survival rate of the fungal cells, thus confirming the fungicidal effect of hibicuslide C on *C. albicans* (Hwang et al., 2014). The structures of inhibitors **1–11** are shown in Fig. 2 and MIC values are given in Table 1.

In another study, Ebselen **12**, an organo-selenium compound exhibited a strong antifungal potential against *C. albicans* at MIC of 0.5–2 μg/mL. On further examination, it was revealed that ebselen exerts its fungicidal activity by accumulating the the fungal cells ROS (Thangamani et al., 2017).

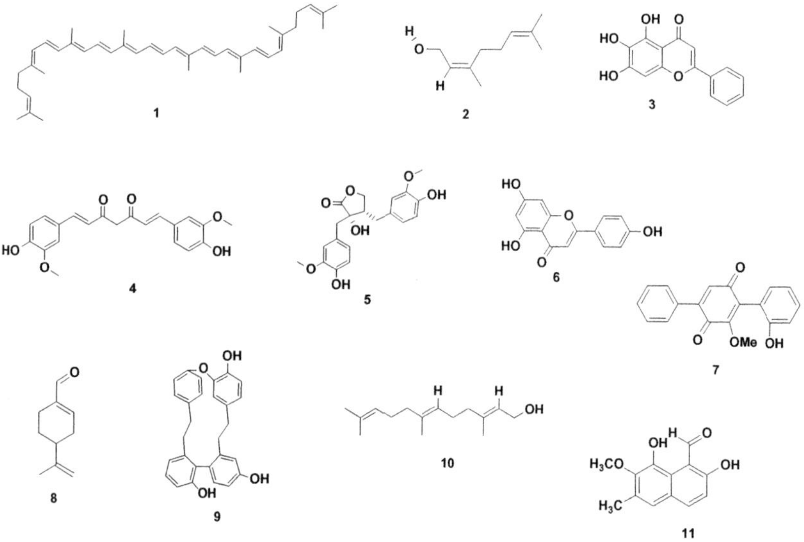

Fig. 2 Chemical structures of naturally derived small molecule inhibitors of apoptosis (1–11): 1—Lycopene; 2—Nerol; 3—Baicalein; 4—Curcumin; 5—Nortrachelogenin; 6—Apigenin; 7—Floricolin; 8—Perillaldehyde; 9—Plagiochin E; 10—Farnesol; 11—Hibicuslide C.

Table 1 MIC values of compounds (1 – 11).

Compound	MIC (μg/mL)
1	5
2	0.77
3	1–8
4	25–2000
5	25
6	100–300
7	8
8	4
9	16
10	0.2–12
11	5

In another study, a diphenyl ether derivative, Dorcinol D **13** isolated from endolichenic fungus, displayed a strong fungicidal effect against *C. albicans*. Furthermore, when a possible mechanism of action was investigated, it was elucidated that this molecule disrupts the cell membrane permeability accompanied by an intracellular increase in ROS level (Li et al., 2015).

In addition to all these studies, another study performed on a phytochemical compound, Medioresinol **14** already known to exhibit leishmanicidal activity, isolated from a bark of a traditionally known medicinal plant *Sambucus williamsii*, presented antifungal property against *C. albicans*. In this study, when bio-physiological aspects were studied, they depicted cell cycle arrest, increase in ROS species, mitochondrial dysfunction and chromatin fragmentation which are the key morphological markers of apoptosis. Hence, indicating that medioresinol exhibit its fungicidal effect by programmed cell death (Hwang, Hwang, Liu, Woo, & Lee, 2012). Furthermore, another naturally known molecule Cinnamaldehyde **15** showed antifungal property against *C. albicans* via ROS mediated apoptosis as one of the mechanisms of fungal inhibition (Chen et al., 2019).

Accordingly, a naturally derived compound Silibinin **16** isolated from *Silybum marianum* (milk thistle) showed a decisive inhibitory action against pathogenic fungus *C. albicans*. On the treatment of silibinin at 100 μM concentration against *Candida* cells showed a 24.17% increase in apoptosis than the untreated cells. Successively, it was found that silibinin not only resulted in ROS accumulation, but it also resulted in a change in ion homeostasis by releasing intracellular K^+ ions and accumulation of calcium ions inside the cytoplasm and mitochondria. Among these phenomena, ROS accumulation and Ca^{2+} overload are the markers of apoptosis. Notably, another natural compound obtained from *Silybum marianum*, Silymarin **17,** is known to show a strong anti-fungicidal effect against *C. albicans* by apoptosis as its mechanism of action (Lee & Lee, 2018; Yun & Lee, 2016a).

Additionally, another study performed on naturally-derived alcohol Honokiol **18** isolated from *Magnolia Officinalis*, disclosed an antifungal property against *C. albicans* mitochondrial dysfunction and accumulation of ROS in the fungal cells (Sun, Liao, & Wang, 2017).

Previously studied antifungal agent Amentoflavone **19**, isolated from Selaginella tamariscina, a traditional herb earlier used to treat chronic tracheitis, was further investigated to elucidate the mode of action by which amentoflavone depicts its antifungal effects. The study revealed that amentoflavone treated cells exhibited a series of cellular changes associated with apoptosis, such as DNA and nuclear fragmentation externalization of phosphatidylserine, accumulation of intracellular ROS and hydroxyl radicals, and activation of metacaspase related pathway (Hwang, Lee, Jin, Woo, & Lee, 2012).

Further, Isoquercitrin **20**, a phenolic compound, belongs to the class organic compound flavonoid 3-O-glycosides, also known as isotrifolinn or hirsutrin; when studied for the antifungal activity, it showed strong potentiality against *C. albicans*. The further study elucidated that Isoquercitrin triggers ROS mediated apoptosis in *C. albicans* as a mode of action (Yun, Woo, & Lee, 2016). The structures of inhibitors from **12 to 20** are shown in Fig. 3 and biological data is shown in Table 2.

In another study, Resveratrol **21** (3,5,4′-trihydroxy-*trans*-stilbene), a polyphenolic compound is widely known for its antimicrobial activity, was further investigated to uncover the details of its mechanism of action against *C. albicans*. Hence, the study revealed that resveratrol exerts its effect from the early to late apoptosis stage. Furthermore, it increases the ROS concentration in fungal cells and regulates the activation of metacaspase release in response to mitochondrial dysfunction (Lee & Lee, 2015).

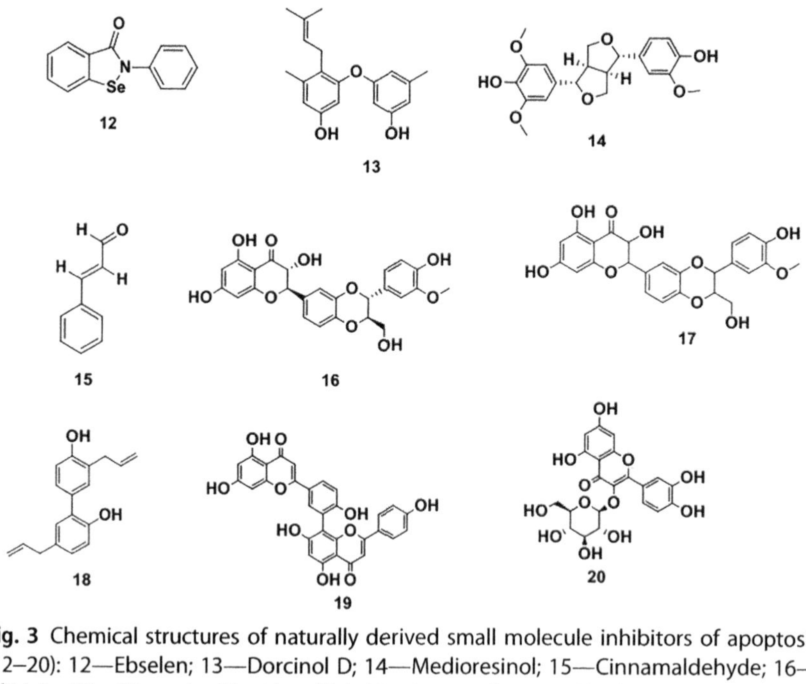

Fig. 3 Chemical structures of naturally derived small molecule inhibitors of apoptosis (12–20): 12—Ebselen; 13—Dorcinol D; 14—Medioresinol; 15—Cinnamaldehyde; 16—Silibinin; 17—Silymarin; 18—Honokiol; 19—Amentoflavone; 20—Isoquercitrin.

Table 2 MIC values of compounds (12 – 20).

Compound	MIC (μg/mL)
12	0.5–2
13	8
14	3.125
15	130
16	20–40 μM/mL
17	10
18	16–32
19	5
20	80 μM/mL

Further, Retigeric acid B **22**, a triterpene acid procured from *Lobaria* kurokawae, exerts an antifungal effect against pathogenic fungus, *C. albicans* by apoptosis as an underlying mechanism of action (Chang et al., 2011). Previous studies outlined Coumarin's antifungal potential (1,2 benzopyrone) **23** and its derivatives against *C. albicans*. However, the exact mechanism by which it acts was still not known. However, after rigorous examination, it was discovered that coumarin works by inducing ROS mediated apoptosis.

In addition, recently conducted studies investigated the anti-candida activity of silver-coumarin complex molecules. They also displayed, substantial antifungal property against *C. albicans* cell by disrupting the membrane integrity, ergosterol biosynthesis, and cytochrome biosynthesis. Additionally, this alteration in cytochrome biosynthesis further leads to apoptosis of the fungal cells (Jia et al., 2019; Thati et al., 2007).

Recently, a series of small molecule inhibitors derived from 1,2-benzisothiazolinone (BZT) scaffolds: DFD-VI-15 **24**, BD-I-186, DFD-V-49, and DFD-V-66 were identified. They showed substantial antifungal properties against *C. albicans*. Furthermore, when the mode of action study was performed on the compound (DFD-VI-15), it was found that this compound targets the respiratory pathway and raises ROS in the yeast cells (Alex et al., 2012).

From previous reports, Limonene **25**, a terpene, the essential compound of citrus oil widely used in the food and beverage industry for its aromatic fragrance. However, the Limonene on *C. albicans* exhibited a antifungal property compared to fluconazole, a well-established antifungal drug.

Following this, molecular analysis was performed by Sequential Windowed Acquisition of All Theoretical Fragment Ion Mass Spectra (SWATH-MS). This study revealed that Limonene regulates various types of transcription factors which were related to cell membrane damage, nucleolar stress, cytoskeleton organization, and apoptosis, subsequently it was suggested that Limonene inhibits *C. albicans* by inducing oxidative stress, leading to DNA damage and membrane disintegration (Thakre et al., 2018).

Through recent investigation, Carvacrol **26**, a monoterpenoid phenol, well known for its antibacterial activity, when was scrutinized for its antifungal activity against *C. albicans* exhibited an intense antifungal activity. On further examination, carvacrol elucidated fungal inhibition in association with ROS mediated apoptosis. Carvacrol treatments led to membrane permeability, DNA fragmentation and increased Ca^{2+} level in mitochondria (Lone, Wani, Fru, & Ahmad, 2020).

A recent study reported that Eugenol **27**, a naturally derived alcohol with an intense potent antifungal activity with MIC of 500–100 μg/mL (Table 3). However, to improve eugenol efficacy, it was tosylate at OH- group, which resulted in seven eugenol-tosylate congeries. Out of these seven derivatives, three compounds ET-5 **27a**, ET-6 **27b**, ET-7 **27c** showed excellent

Table 3 IC_{50}/MIC values of compounds (21 – 30).

Compound	IC_{50} (μg/mL)	MIC (μg/mL)
21	400	–
22	–	8–64
23	–	200
24	1.6	–
25	–	20 mM
26	–	16–32
27	–	500–1000
27a	–	0.125
27b	–	0.25
27c	–	1
28	–	5–15
29	–	65–250
30	–	6

antifungal potential by inducing ROS mediated apoptosis as a mode of action. Mitochondrial depolarization, phosphorylation externalization DNA fragmentation, and apoptosis's major markers supported the induction of apoptosis in *C. albicans*. More recently, it was reported that Piperine **28** is an excellent antifungal agent. It inhibits not only planktonic form of *C. albicans* but also showed anti-biofilm activity in a stage-dependent manner. Further studies performed on *C. albicans* confirm that piperine disrupts membrane integrity, induces oxidative stress and apoptosis (Thakre et al., 2020). Citronellal **29**, a monoterpenoid derived from the *Cymbopogon* plants, showed antifungal activity with an increase in the ROS levels to elicit cell necrosis, DNA damage and mitochondrial dysfunction in *C. albicans* (de Almeida et al., 2016). Additionally, HSAF **30** (heat-stable antifungal factor), a polycyclic tetramate macrolactam isolated from *Lyaobacter enzymogenes C3*, also persuaded the apoptosis of *C. albicans* by prompting the production of ROS (Ding et al., 2016; Saibabu, Singh, Ansari, Fatima, & Hameed, 2017). The structural representation of inhibitors **21–30** is shown in Fig. 4.

Furthermore, a crude extract of *Allium sativum*, commonly known as garlic also presented antifungal property against *C. albicans* complemented with ROS accumulation and glutathione depletion (Lemar et al., 2005).

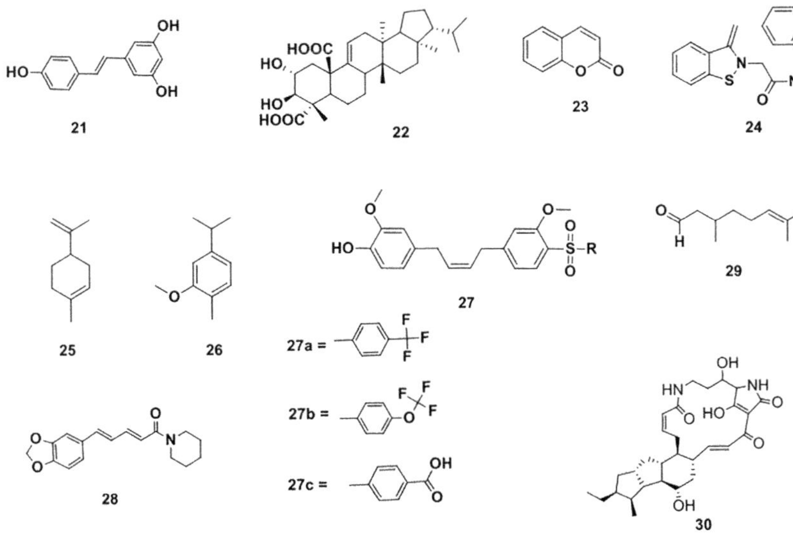

Fig. 4 Chemical structures of naturally derived small molecule inhibitors of apoptosis (21–30): 21—Resveratrol; 22—Retigeric acid B; 23—Coumarin; 24—DFD-VI-15; 25—Limonene; 26—Carvacrol; 27—Eugenol (27a—ET-5; 27b—ET-6; 27c—ET-7); 28—Piperine; 29—Citronellal; 30—HSAF.

Previously, Dill seed essential oils (DESO), obtained from *Anethum graveolens* L had been known for anti-candida property. To further add on to this study, the contribution of DESO in the mechanism of apoptosis was elucidated (Table 3). The study revealed that DESO induced DNA fragmentation, phosphotidylserine exposure, ROS accumulation and metacaspases activation in fungal cells (Chen et al., 2014).

5. Conclusion and future perspective

Since the frequency of invasive fungal infections by an opportunistic pathogen, *C. albicans* is expected to increase in the future the repertoire and antifungal activity is limited; there is an emergence of novel antifungal drugs with broad-spectrum activity and low toxicity. Clear, strong and well-documented evidence shown by the extensive studies performed in the past few decades regarding the identification of new antimicrobial drugs from natural resources have stated that they are non-toxic and readily available. However, it is still highly challenging to translate them into clinical applications, as several vital issues remain unresolved and demands some attention. Here, several studies have been discussed that identified active, non-toxic, and induce ROS accumulation and subsequent apoptosis in *C. albicans,* thus, can be regarded as a promising alternative therapeutic approach to combat the problem of antifungal drug resistance. Conversely, this antifungal strategy is still restricted by a narrow range of action. Therefore, future research should focus on the further optimization of the known lead compounds to improve their pharmacokinetic activity. In vivo studies of these inhibitors should be performed on suitable animal model and synergistic studies of these identified drug molecules with the conventional antifungal drugs should also be validated. Future research should also focus on the identification of specific targets of these inhibitors. Through a better understanding of fungal pathophysiology and our continuous efforts, we can succeed in developing a novel drug.

References

Aerts, A. M., Bammens, L., Govaert, G., Carmona-Gutierrez, D., Madeo, F., Cammue, B. P. A., et al. (2011). The antifungal plant Defensin HsAFP1 from Heuchera Sanguinea induces apoptosis in Candida Albicans. *Frontiers in Microbiology, 2*. https://doi.org/10.3389/fmicb.2011.00047.

Aerts, A. M., François, I. E. J. A., Meert, E. M. K., Li, Q.-T., Cammue, B. P. A., & Thevissen, K. (2007). The antifungal activity of RsAFP2, a plant Defensin from Raphanus sativus, involves the induction of reactive oxygen species in Candida albicans. *Journal of Molecular Microbiology and Biotechnology, 13*(4), 243–247. https://doi.org/10.1159/000104753.

Alex, D., Gay-Andrieu, F., May, J., Thampi, L., Dou, D., Mooney, A., et al. (2012). Amino acid-derived 1,2-benzisothiazolinone derivatives as novel small-molecule antifungal inhibitors: Identification of potential genetic targets. *Antimicrobial Agents and Chemotherapy*, *56*(9), 4630–4639. https://doi.org/10.1128/AAC.00477-12.
Aminahlkner, K. (2004). *Yeast signaling pathways in the oxidative stress response*. https://doi.org/10.1016/j.mrfmmm.2004.09.006.
Armstrong, J. S., Steinauer, K. K., Hornung, B., Irish, J. M., Lecane, P., Birrell, G. W., et al. (2002). Role of glutathione depletion and reactive oxygen species generation in apoptotic signaling in a human B lymphoma cell line. *Cell Death & Differentiation*, *9*(3), 252–263. https://doi.org/10.1038/sj.cdd.4400959.
Babior, B. M. (2004). NADPH oxidase. *Current Opinion in Immunology*, *16*(1), 42–47. https://doi.org/10.1016/j.coi.2003.12.001.
Bahar, A., & Ren, D. (2013). Antimicrobial Peptides. *Pharmaceuticals*, *6*(12), 1543–1575. https://doi.org/10.3390/ph6121543.
Bongomin, F., Gago, S., Oladele, R. O., & Denning, D. W. (2017). Global and multi-national prevalence of fungal diseases—Estimate precision. *Journal of Fungi*, *3*(4), 57. https://doi.org/10.3390/jof3040057.
Brown, K. L., & Hancock, R. E. (2006). Cationic host defense (antimicrobial) peptides. *Current Opinion in Immunology*, *18*(1), 24–30. https://doi.org/10.1016/j.coi.2005.11.004.
Brown, D. A., Yang, N., & Ray, S. D. (2014). Apoptosis. In *Encyclopedia of Toxicology* (pp. 287–294). Elsevier. https://doi.org/10.1016/B978-0-12-386454-3.00242-6.
Chakraborti, S., Chakraborti, T., Chattopadhyay, D., & Shaha, C. (2019). Oxidative stress in microbial diseases. In *Oxidative Stress in Microbial Diseases*. Springer Nature. https://doi.org/10.1007/978-981-13-8763-0.
Chang, W.-Q., Wu, X.-Z., Cheng, A.-X., Zhang, L., Ji, M., & Lou, H.-X. (2011). Retigeric acid B exerts antifungal effect through enhanced reactive oxygen species and decreased cAMP. *Biochimica et Biophysica Acta (BBA) - General Subjects*, *1810*(5), 569–576. https://doi.org/10.1016/j.bbagen.2011.02.001.
Chen, R.-C., & Lan, C.-Y. (2020). Human antimicrobial peptide Hepcidin 25-induced apoptosis in Candida albicans. *Microorganisms*, *8*(4), 585. https://doi.org/10.3390/microorganisms8040585.
Chen, L., Wang, Z., Liu, L., Qu, S., Mao, Y., Peng, X., et al. (2019). Cinnamaldehyde inhibits Candida albicans growth by causing apoptosis and its treatment on vulvovaginal candidiasis and oropharyngeal candidiasis. *Applied Microbiology and Biotechnology*, *103*(21 – 22), 9037–9055. https://doi.org/10.1007/s00253-019-10119-3.
Chen, Y., Zeng, H., Tian, J., Ban, X., Ma, B., & Wang, Y. (2014). Dill (Anethum graveolens L.) seed essential oil induces Candida albicans apoptosis in a metacaspase-dependent manner. *Fungal Biology*, *118*(4), 394–401. https://doi.org/10.1016/j.funbio.2014.02.004.
Cho, J., & Lee, D. G. (2011a). Oxidative stress by antimicrobial peptide pleurocidin triggers apoptosis in Candida albicans. *Biochimie*, *93*(10), 1873–1879. https://doi.org/10.1016/j.biochi.2011.07.011.
Cho, J., & Lee, D. G. (2011b). The antimicrobial peptide arenicin-1 promotes generation of reactive oxygen species and induction of apoptosis. *Biochimica et Biophysica Acta (BBA) - General Subjects*, *1810*(12), 1246–1251. https://doi.org/10.1016/j.bbagen.2011.08.011.
Choi, H., & Lee, D. G. (2015). Lycopene induces apoptosis in Candida albicans through reactive oxygen species production and mitochondrial dysfunction. *Biochimie*, *115*, 108–115. https://doi.org/10.1016/j.biochi.2015.05.009.
Dantas, A., Day, A., Ikeh, M., Kos, I., Achan, B., & Quinn, J. (2015). Oxidative stress responses in the human fungal pathogen, Candida albicans. *Biomolecules*, *5*(1), 142–165. https://doi.org/10.3390/biom5010142.
D'Arcy, M. S. (2019). Cell death: A review of the major forms of apoptosis, necrosis and autophagy. *Cell Biology International*, *43*(6), 582–592. https://doi.org/10.1002/cbin.11137.

de Almeida, L. F. D., de Paula, J. F., de Almeida, R. V. D., de Williams, D. W., Hebling, J., & Cavalcanti, Y. W. (2016). Efficacy of citronella and cinnamon essential oils on Candida albicans biofilms. *Acta Odontologica Scandinavica*, *74*(5), 393–398. https://doi.org/10.3109/00016357.2016.1166261.

Del Poeta, M., Nimrichter, L., Rodrigues, M. L., & Luberto, C. (2014). Synthesis and biological properties of fungal glucosylceramide. *PLoS Pathogens*, *10*(1), e1003832. https://doi.org/10.1371/journal.ppat.1003832.

Delattin, N., Cammue, B. P., & Thevissen, K. (2014). Reactive oxygen species-inducing antifungal agents and their activity against fungal biofilms. *Future Medicinal Chemistry*, *6*(1), 77–90. https://doi.org/10.4155/fmc.13.189.

Ding, Y., Li, Z., Li, Y., Lu, C., Wang, H., Shen, Y., et al. (2016). HSAF-induced antifungal effects in Candida albicans through ROS-mediated apoptosis. *RSC Advances*, *6*(37), 30895–30904. https://doi.org/10.1039/C5RA26092B.

Dižová, S., & Bujdáková, H. (2017). *Properties and role of the quorum sensing molecule farnesol in relation to the yeast Candida albicans*. https://doi.org/10.1691/ph.2017.6174.

Douglas, L. J. (2003). Candida biofilms and their role in infection. *Trends in Microbiology*, *11*(1), 30–36. https://doi.org/10.1016/S0966-842X(02)00002-1.

Elmore, S. (2007). Apoptosis: A review of programmed cell death. *Toxicologic Pathology*, *35*(4), 495–516. https://doi.org/10.1080/01926230701320337.

Fu, Z., Lu, H., Zhu, Z., Yan, L., Jiang, Y., & Cao, Y. (2011). Combination of Baicalein and amphotericin B accelerates Candida albicans apoptosis. *Biological & Pharmaceutical Bulletin*, *34*(2), 214–218. https://doi.org/10.1248/bpb.34.214.

Hongmei, Z. (2012). Extrinsic and intrinsic apoptosis signal pathway review. In *Apoptosis and medicine*. InTech. https://doi.org/10.5772/50129.

Hwang, J. H., Choi, H., Kim, A. R., Yun, J. W., Yu, R., Woo, E.-R., et al. (2014). Hibicuslide C-induced cell death in Candida albicans involves apoptosis mechanism. *Journal of Applied Microbiology*, *117*(5), 1400–1411. https://doi.org/10.1111/jam.12633.

Hwang, B., Hwang, J.-S., Lee, J., Kim, J.-K., Kim, S. R., Kim, Y., et al. (2011). Induction of yeast apoptosis by an antimicrobial peptide, Papiliocin. *Biochemical and Biophysical Research Communications*, *408*(1), 89–93. https://doi.org/10.1016/j.bbrc.2011.03.125.

Hwang, B., Hwang, J.-S., Lee, J., & Lee, D. G. (2011). The antimicrobial peptide, psacotheasin induces reactive oxygen species and triggers apoptosis in Candida albicans. *Biochemical and Biophysical Research Communications*, *405*(2), 267–271. https://doi.org/10.1016/j.bbrc.2011.01.026.

Hwang, J. H., Hwang, I., Liu, Q.-H., Woo, E.-R., & Lee, D. G. (2012). (+)-Medioresinol leads to intracellular ROS accumulation and mitochondria-mediated apoptotic cell death in Candida albicans. *Biochimie*, *94*(8), 1784–1793. https://doi.org/10.1016/j.biochi.2012.04.010.

Hwang, I. S., Lee, J., Jin, H. G., Woo, E. R., & Lee, D. G. (2012). Amentoflavone stimulates mitochondrial dysfunction and induces apoptotic cell death in Candida albicans. *Mycopathologia*, *173*(4), 207–218. https://doi.org/10.1007/s11046-011-9503-x.

Ikner, A., & Shiozaki, K. (2005). Yeast signaling pathways in the oxidative stress response. *Mutation Research - Fundamental and Molecular Mechanisms of Mutagenesis*, *569*(1–2), 13–27. https://doi.org/10.1016/j.mrfmmm.2004.09.006.

Jia, C., Zhang, J., Yu, L., Wang, C., Yang, Y., Rong, X., et al. (2019). Antifungal activity of coumarin against Candida albicans is related to apoptosis. *Frontiers in Cellular and Infection Microbiology*, *8*, 445. https://doi.org/10.3389/fcimb.2018.00445.

Kim, S., Hwang, J. S., & Lee, D. G. (2020). Lactoferricin B like peptide triggers mitochondrial disruption-mediated apoptosis by inhibiting respiration under nitric oxide accumulation in <scp> *Candida albicans* </scp>. *IUBMB Life*, *72*(7), 1515–1527. https://doi.org/10.1002/iub.2284.

Lee, J., Hwang, J.-S., Hwang, I., Cho, J., Lee, E., Kim, Y., et al. (2012). Coprisin-induced antifungal effects in Candida albicans correlate with apoptotic mechanisms. *Free Radical Biology and Medicine*, *52*(11 – 12), 2302–2311. https://doi.org/10.1016/j.freeradbiomed.2012.03.012.

Lee, H., Hwang, J.-S., & Lee, D. G. (2017). Scolopendin, an antimicrobial peptide from centipede, attenuates mitochondrial functions and triggers apoptosis in Candida albicans. *Biochemical Journal*, *474*(5), 635–645. https://doi.org/10.1042/BCJ20161039.

Lee, J., & Lee, D. G. (2015). Novel antifungal mechanism of resveratrol: Apoptosis inducer in Candida albicans. *Current Microbiology*, *70*(3), 383–389. https://doi.org/10.1007/s00284-014-0734-1.

Lee, W., & Lee, D. G. (2018). Potential role of potassium and chloride channels in regulation of silymarin-induced apoptosis in Candida albicans. *IUBMB Life*, *70*(3), 197–206. https://doi.org/10.1002/iub.1716.

Lee, H., Woo, E.-R., & Lee, D. G. (2016). (−)-Nortrachelogenin from Partrinia scabiosaefolia elicits apoptotic response in Candida albicans. *FEMS Yeast Research*, *16*(3), 1–10. fow013 https://doi.org/10.1093/femsyr/fow013.

Lee, W., Woo, E.-R., & Lee, D. G. (2019). Effect of apigenin isolated from Aster yomena against Candida albicans: Apigenin-triggered apoptotic pathway regulated by mitochondrial calcium signaling. *Journal of Ethnopharmacology*, *231*, 19–28. https://doi.org/10.1016/j.jep.2018.11.005.

Lemar, K. M., Passa, O., Aon, M. A., Cortassa, S., Müller, C. T., Plummer, S., et al. (2005). Allyl alcohol and garlic (Allium sativum) extract produce oxidative stress in Candida albicans. *Microbiology*, *151*(10), 3257–3265. https://doi.org/10.1099/mic.0.28095-0.

Li, Y., Chang, W., Zhang, M., Li, X., Jiao, Y., & Lou, H. (2015). Diorcinol D exerts fungicidal action against Candida albicans through cytoplasm membrane destruction and ROS accumulation. *PLoS One*, *10*(6), e0128693. https://doi.org/10.1371/journal.pone.0128693.

Lohse, M. B., Gulati, M., Johnson, A. D., & Nobile, C. J. (2018). Development and regulation of single-and multi-species Candida albicans biofilms. *Nature Reviews Microbiology*, *16*(1), 19–31. https://doi.org/10.1038/nrmicro.2017.107.

Lone, S. A., Wani, M. Y., Fru, P., & Ahmad, A. (2020). Cellular apoptosis and necrosis as therapeutic targets for novel eugenol Tosylate congeners against Candida albicans. *Scientific Reports*, *10*(1), 1191. https://doi.org/10.1038/s41598-020-58256-4.

Mahlapuu, M., Håkansson, J., Ringstad, L., & Björn, C. (2016). Antimicrobial peptides: An emerging category of therapeutic agents. *Frontiers in Cellular and Infection Microbiology*, *6*, 194. https://doi.org/10.3389/fcimb.2016.00194.

Neelofar, K., Shreaz, S., Rimple, B., Muralidhar, S., Nikhat, M., & Khan, L. A. (2011). Curcumin as a promising anticandidal of clinical interes. *Canadian Journal of Microbiology*, *57*, 204–210. https://doi.org/10.1139/W10-117.

Nobile, C. J., & Johnson, A. D. (2015). Candida albicans biofilms and human disease. *Annual Review of Microbiology*, *69*(1), 71–92. https://doi.org/10.1146/annurev-micro-091014-104330.

Park, C., & Lee, D. G. (2010). Melittin induces apoptotic features in Candida albicans. *Biochemical and Biophysical Research Communications*, *394*(1), 170–172. https://doi.org/10.1016/j.bbrc.2010.02.138.

Perfect, J. R. (2017). The antifungal pipeline: A reality check. *Nature Reviews Drug Discovery*, *16*(9), 603–616. https://doi.org/10.1038/nrd.2017.46.

Pfaller, M. A., & Diekema, D. J. (2007). Epidemiology of invasive candidiasis: A persistent public health problem. *Clinical Microbiology Reviews*, *20*(1), 133–163. https://doi.org/10.1128/CMR.00029-06.

Poon, I. K. H., Lucas, C. D., Rossi, A. G., & Ravichandran, K. S. (2014). Apoptotic cell clearance: Basic biology and therapeutic potential. *Nature Reviews Immunology*, *14*(3), 166–180. https://doi.org/10.1038/nri3607.

Qu, S., Chen, L., Tian, H., Wang, Z., Wang, F., Wang, L., et al. (2019). Effect of Perillaldehyde on prophylaxis and treatment of vaginal candidiasis in a murine model. *Frontiers in Microbiology, 10,* 1466. https://doi.org/10.3389/fmicb.2019.01466.

Reinholz, M., Ruzicka, T., & Schauber, J. (2012). Cathelicidin LL-37: An antimicrobial peptide with a role in inflammatory skin disease. *Annals of Dermatology, 24*(2), 126. https://doi.org/10.5021/ad.2012.24.2.126.

Saibabu, V., Singh, S., Ansari, M. A., Fatima, Z., & Hameed, S. (2017). Insights into the intracellular mechanisms of citronellal in Candida albicans: Implications for reactive oxygen species-mediated necrosis, mitochondrial dysfunction, and DNA damage. *Revista da Sociedade Brasileira de Medicina Tropical, 50*(4), 524–529. https://doi.org/10.1590/0037-8682-0114-2017.

Serpa, R., França, E. J. G., Furlaneto-Maia, L., Andrade, C. G. T. J., Diniz, A., & Furlaneto, M. C. (2012). In vitro antifungal activity of the flavonoid baicalein against Candida species. *Journal of Medical Microbiology, 61*(12), 1704–1708. https://doi.org/10.1099/jmm.0.047852-0.

Seyedjavadi, S. S., Khani, S., Eslamifar, A., Ajdary, S., Goudarzi, M., Halabian, R., et al. (2020). The antifungal peptide MCh-AMP1 derived from Matricaria chamomilla inhibits Candida albicans growth via inducing ROS generation and altering fungal cell membrane permeability. *Frontiers in Microbiology, 10,* 3150. https://doi.org/10.3389/fmicb.2019.03150.

Sharma, M., Manoharlal, R., Puri, N., & Prasad, R. (2010). Antifungal curcumin induces reactive oxygen species and triggers an early apoptosis but prevents hyphae development by targeting the global repressor TUP1 in Candida albicans. *Bioscience Reports, 30*(6), 391–404. https://doi.org/10.1042/BSR20090151.

Silva, S., Rodrigues, C. F., Araújo, D., Rodrigues, M. E., & Henriques, M. (2017). Candida species biofilms' antifungal resistance. *Journal of Fungi, 3*(1). https://doi.org/10.3390/jof3010008.

Simon, H.-U., Haj-Yehia, A., & Levi-Schaffer, F. (2000). Role of reactive oxygen species (ROS) in apoptosis induction. *Apoptosis, 5*(5), 415–418. https://doi.org/10.1023/A:1009616228304.

Spampinato, C., & Leonardi, D. (2013). Candida infections, causes, targets, and resistance mechanisms: Traditional and alternative antifungal agents. *BioMed Research International, 2013,* 204237. https://doi.org/10.1155/2013/204237.

Sun, L., Liao, K., & Wang, D. (2017). Honokiol induces superoxide production by targeting mitochondrial respiratory chain complex I in Candida albicans. *PLoS One, 12*(8), e0184003. https://doi.org/10.1371/journal.pone.0184003.

Thakre, A., Jadhav, V., Kazi, R., Shelar, A., Patil, R., Kharat, K., et al. (2020). Oxidative stress induced by piperine leads to apoptosis in Candida albicans. *Medical Mycology.* Epub ahead of print https://doi.org/10.1093/mmy/myaa058. 32658959.

Thakre, A., Zore, G., Kodgire, S., Kazi, R., Mulange, S., Patil, R., et al. (2018). Limonene inhibits Candida albicans growth by inducing apoptosis. *Medical Mycology, 56*(5), 565–578. https://doi.org/10.1093/mmy/myx074.

Thangamani, S., Eldesouky, H. E., Mohammad, H., Pascuzzi, P. E., Avramova, L., Hazbun, T. R., et al. (2017). Ebselen exerts antifungal activity by regulating glutathione (GSH) and reactive oxygen species (ROS) production in fungal cells. *Biochimica et Biophysica Acta (BBA) - General Subjects, 1861*(1), 3002–3010. https://doi.org/10.1016/j.bbagen.2016.09.029.

Thati, B., Noble, A., Rowan, R., Creaven, B. S., Walsh, M., McCann, M., et al. (2007). Mechanism of action of coumarin and silver(I)–coumarin complexes against the pathogenic yeast Candida albicans. *Toxicology In Vitro, 21*(5), 801–808. https://doi.org/10.1016/j.tiv.2007.01.022.

Tian, J., Lu, Z., Wang, Y., Zhang, M., Wang, X., Tang, X., et al. (2017). Nerol triggers mitochondrial dysfunction and disruption via elevation of Ca2+ and ROS in Candida albicans. *The International Journal of Biochemistry & Cell Biology*, *85*, 114–122. https://doi.org/10.1016/j.biocel.2017.02.006.

Tian, H., Qu, S., Wang, Y., Lu, Z., Zhang, M., Gan, Y., et al. (2017). Calcium and oxidative stress mediate perillaldehyde-induced apoptosis in Candida albicans. *Applied Microbiology and Biotechnology*, *101*(8), 3335–3345. https://doi.org/10.1007/s00253-017-8146-3.

Tsui, C., Kong, E. F., & Jabra-Rizk, M. A. (2016). Pathogenesis of Candida albicans biofilm. *Pathogens and Disease*, *74*(4), ftw018. https://doi.org/10.1093/femspd/ftw018.

Wang, Z., Yang, K., Chen, L., Yan, R., Qu, S., Li, Y., et al. (2020). Activities of Nerol, a natural plant active ingredient, against Candida albicans in vitro and in vivo. *Applied Microbiology and Biotechnology*, *104*(11), 5039–5052. https://doi.org/10.1007/s00253-020-10559-2.

Wellington, M., Dolan, K., & Krysan, D. J. (2009). Live Candida albicans suppresses production of reactive oxygen species in phagocytes. *Infection and Immunity*, 77(1), 405–413. https://doi.org/10.1128/IAI.00860-08.

Willis, S. (2003). The Bcl-2-regulated apoptotic pathway. *Journal of Cell Science*, *116*(20), 4053–4056. https://doi.org/10.1242/jcs.00754.

Wu, X.-Z., Chang, W.-Q., Cheng, A.-X., Sun, L.-M., & Lou, H.-X. (2010). Plagiochin E, an antifungal active macrocyclic bis(bibenzyl), induced apoptosis in Candida albicans through a metacaspase-dependent apoptotic pathway. *Biochimica et Biophysica Acta (BBA) - General Subjects*, *1800*(4), 439–447. https://doi.org/10.1016/j.bbagen.2010.01.001.

Yun, J., Hwang, J.-S., & Lee, D. G. (2017). The antifungal activity of the peptide, periplanetasin-2, derived from American cockroach Periplaneta americana. *Biochemical Journal*, *474*(17), 3027–3043. https://doi.org/10.1042/BCJ20170461.

Yun, D. G., & Lee, D. G. (2016a). Silibinin triggers yeast apoptosis related to mitochondrial Ca2+ influx in Candida albicans. *The International Journal of Biochemistry & Cell Biology*, *80*, 1–9. https://doi.org/10.1016/j.biocel.2016.09.008.

Yun, J., & Lee, D. G. (2016b). Cecropin A-induced apoptosis is regulated by ion balance and glutathione antioxidant system in Candida albicans. *IUBMB Life*, *68*(8), 652–662. https://doi.org/10.1002/iub.1527.

Yun, J., Woo, E.-R., & Lee, D. G. (2016). Isoquercitrin, isolated from Aster yomena triggers ROS-mediated apoptosis in Candida albicans. *Journal of Functional Foods*, *22*, 347–357. https://doi.org/10.1016/j.jff.2016.01.041.

Zhang, M., Chang, W., Shi, H., Li, Y., Zheng, S., Li, W., et al. (2018). Floricolin C elicits intracellular reactive oxygen species accumulation and disrupts mitochondria to exert fungicidal action. *FEMS Yeast Research*, *18*(1). https://doi.org/10.1093/femsyr/foy002.

Zhang, L., & Gallo, R. L. (2016). Antimicrobial peptides. *Current Biology*, *26*(1), R14–R19. https://doi.org/10.1016/j.cub.2015.11.017.

CHAPTER THREE

TRPM2 channel in oxidative stress-induced mitochondrial dysfunction and apoptotic cell death

Philippa Malko[a], Ran Ding[b], and Lin-Hua Jiang[a,b,*]

[a]Faculty of Biological Sciences, School of Biomedical Sciences, University of Leeds, Leeds, United Kingdom
[b]Department of Physiology and Pathophysiology and Sino-UK Joint Laboratory of Brain Function and Injury of Henan Province, Xinxiang Medical University, Xinxiang, China
*Corresponding author: e-mail address: l.h.jiang@leeds.ac.uk

Contents

Abstract

Mitochondria, conserved intracellular organelles best known as the powerhouse of cells for generating ATP, play an important role in apoptosis. Oxidative stress can induce mitochondrial dysfunction and activate mitochondria-mediated apoptotic cell death. TRPM2 is a Ca^{2+}-permeable cation channel that is activated by pathologically relevant concentrations of reactive oxygen species (ROS) and one of its well-recognized roles is to confer susceptibility to ROS-induced cell death. Increasing evidence from recent studies supports TRPM2 channel-mediated cell death as an important cellular mechanism linking miscellaneous oxidative stress-inducing pathological factors to associated

Advances in Protein Chemistry and Structural Biology, Volume 125
ISSN 1876-1623
https://doi.org/10.1016/bs.apcsb.2020.12.001

diseased conditions. In this chapter, we will discuss the role of the TRPM2 channel in neurons in the brain and pancreatic β-cells in mediating mitochondrial dysfunction and cell death, focusing mainly on apoptotic cell death, that are induced by pathological stimuli implicated in the pathogenesis of neurodegenerative diseases, ischemic stroke and diabetes.

1. Introduction

Oxidative stress, which can result from both excessive generation of reactive oxygen species (ROS) and compromized antioxidant capacity, is a well-documented pathological feature of numerous diseased conditions, such as neurodegenerative diseases, ischemic stroke, cardiovascular disorders and diabetes (Bordoni & Gabbianelli, 2020; Bugger & Pfeil, 2020; Cheignon et al., 2018; Davis & Pennypacker, 2017; Förstermann, Xia, & Li, 2017; Granger & Kvietys, 2015; Jiang, Sun, & Chen, 2016; Newsholme, Cruzat, Keane, Carlessi, & de Bittencourt, 2016; Rochette, Zeller, Cottin, & Vergely, 2014; Yang, Huang, Hu, & Tang, 2019). Oxidative stress is also gaining increasing recognition as an important pathological factor for the development of these conditions, leading to growing interest in targeting oxidative stress as a potential therapeutic strategy (Bordoni & Gabbianelli, 2020; Jiang et al., 2016). ROS are highly reactive chemicals and persistent exposure to high level of ROS or oxidative stress can damage various functionally important macromolecules including nuclear DNA. It has long been known that poly(ADPR) polymerase (PARP) and poly(ADPR) glycohydrolase (PARG) in the nucleus participate in repairing DNA damage, a process generating ADP-ribose (ADPR) as a by-product (Ray Chaudhuri & Nussenzweig, 2017). However, ADPR is nowadays known as an intracellular signaling molecule, owing to the two seminal studies in 2001 that reported that intracellular ADPR specifically binds to and gates the mammalian transient receptor potential melastatin 2 (TRPM2) Ca^{2+}-permeable cation channel (Perraud et al., 2001; Sano et al., 2001). Several studies shortly demonstrated that the TRPM2 channel exhibits a salient sensitivity to activation by pathological levels of ROS and, furthermore, confers susceptibility to ROS-induced cell death (Hara et al., 2002; Wehage et al., 2002; Zhang et al., 2003), which depends on activation of PARP and PARG (Fig. 1). Over the past two decades, particularly in the recent few years, studies have accumulated a large body of evidence to suggest that the TRPM2 channel plays an important role in mediating Ca^{2+} signaling and the regulation of diverse cellular functions induced by a

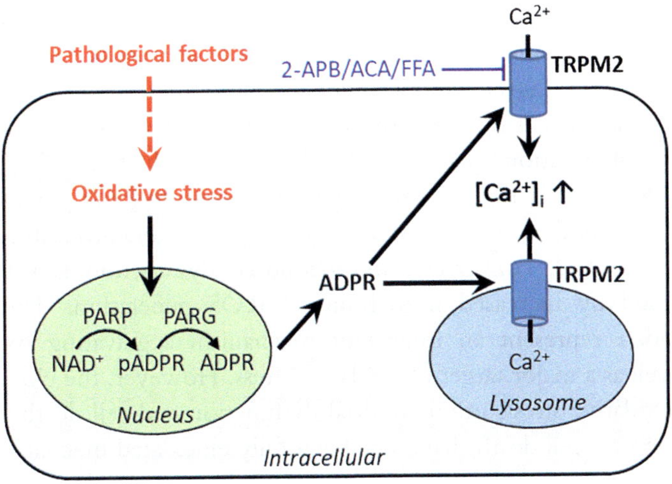

Fig. 1 TRPM2 channel activation by oxidative stress-induced pathological factors. Diverse oxidative stress-inducing pathological factors, such as H_2O_2, Aβ, MPTP/MPP^+ and reperfusion following transient ischemia, can induce generation of ADP-ribose (ADPR) *via* a nuclear DNA repair mechanism that catalyses formation of poly(ADPR) using NAD^+ by poly(ADPR) polymerase (PARP) and subsequent cleavage of poly(ADPR) into ADPR by poly(ADPR) glycohydrolase (PARG), and thereby activate the TRPM2 channel. The TRPM2 channel is located on the cell surface and functions as a Ca^{2+}-permeable channel mediating Ca^{2+} influx and is also present in lysosomes as a lysosomal Ca^{2+} release channel. TRPM2 channel activation results in an increase in intracellular Ca^{2+} concentration ($[Ca^{2+}]_i$). As discussed in the text and not depicted in the figure, activation of the lysosomal TRPM2 channel leads to lysosomal Zn^{2+} release in pancreatic β-cells and activation of the TRPM2 channel in mitochondria stimulates mitochondrial Zn^{2+} uptake in hippocampal neurons. 2-Aminoethoxyldiphenyl borate (2-APB), *N*-(*p*-amycinnamoyl) anthranilic acid (ACA) and flufenamic acid (FFA) are known inhibitors of the TRPM2 channel.

variety of pathological factors known to induce oxidative stress (Alim, Teves, Li, Mori, & Tymianski, 2013; Fanczal et al., 2020; Gao et al., 2014; Garcia-Rodriguez et al., 2018; Jia et al., 2011; Kheradpezhouh, Ma, Morphett, Barritt, & Rychkov, 2014; Manna et al., 2015; Miller, 2019; Ostapchenko et al., 2015; Sun et al., 2018; Wang, Li, et al., 2020; Xu et al., 2009; Ye et al., 2014; Zheng et al., 2020). In particular, TRPM2-mediated cell death has emerged as an important cellular mechanism that links miscellaneous oxidative stress-inducing pathological factors to associated diseased conditions (Malko & Jiang, 2020). It remains less well understood, though increasing efforts have been made to examine, how oxidative stress-induced TRPM2 channel activation leads to cell death.

Apoptosis, or programmed cell death, is a major self-eliminating cellular mechanism that is important in the maintenance of tissue homeostasis and function (Arnoult, 2007). Mitochondria, conserved intracellular organelles that are best known as the powerhouse of cells that generate ATP *via* oxidative phosphorylation, play a vital role in the induction of apoptosis (Jeong & Seol, 2008; Lin & Beal, 2006). Oxidative stress is a potent mitochondrial stressor that impairs mitochondrial function and prompts apoptotic cell death and, conversely, mitochondrial dysfunction is invariably accompanied by increased mitochondrial ROS generation. Therefore, mitochondria represent an important mechanism aggravating oxidative stress as well as a major target of oxidative stress. However, the close interplays of oxidative stress, mitochondrial dysfunction and cell death, particularly apoptotic cell death, have not been fully elucidated mechanistically. In this chapter, we will discuss the recent studies by ourselves and other groups that have examined the TRPM2 channel in neurons in the brain and pancreatic β-cells as a molecular mechanism mediating mitochondrial dysfunction and apoptotic cell death induced by oxidative stress-inducing pathological factors that are strongly implicated in Alzheimer's disease (AD), Parkinson's disease (PD), ischemic stroke and related ischemia/reperfusion (I/R) brain damage, and diabetes.

2. Mitochondrial dysfunction and the apoptotic cell death pathways

In order to facilitate the discussion of the role of the TRPM2 channel in oxidative stress-induced mitochondrial dysfunction and apoptotic cell death, we briefly introduce mitochondrial dynamics, mitochondrial dysfunction and its role in the apoptotic pathways, highlighting the proteins that are crucial in the regulation of these processes and often amenable to investigations. Readers can refer to reviews for detailed discussion of these topics (Detmer & Chan, 2007; Jeong & Seol, 2008; Karbowski & Neutzner, 2012). Mitochondria form an inter-connected network with a tubular morphology in healthy cells and such steady-state morphology is critical for mitochondrial function (Karbowski & Neutzner, 2012). However, mitochondria are highly dynamic, undergoing frequent fusion and fission, and thus the morphology, number and function of mitochondria are dictated by these two opposing processes. It is known that distinct mechanisms mediated by concerted actions of different sets of proteins direct or regulate mitochondrial fusion and fission and thereby shape the steady-state mitochondrial morphology. In mammalian cells, mitofusin 1, mitofusin 2 and optic atrophy protein 1 are important in mediating mitochondrial fusion, and dynamin related protein 1 (Drp-1) and several other proteins are critical in driving mitochondrial fission. Mitochondrial fragmentation due to increased mitochondrial fission can impair mitochondrial function and initiate apoptosis.

The so-called intrinsic and extrinsic pathways are the best-defined apoptotic cell death pathways in mammalian cells. Members of the B-cell lymphoma 2 (Bcl-2) protein family are the master regulators of the intrinsic pathway, with Bax and Bak being pro-apoptotic and Bcl-2 and Bcl-xL anti-apoptotic *via* inhibition of Bax and Bak. The Bcl-2 homology 3 (BH3)-only proteins, including Bim, Bid and Bad, are also pro-apoptotic, owing to their ability to interact with Bcl-2 and Bcl-xL and neutralize their inhibition of Bax and Bak. The intrinsic pathway is also referred to as the mitochondria-mediated programmed cell death pathway. Mitochondrial dysfunction induced by intracellular signals, such as oxidative stress, nuclear DNA damage and endoplasmic reticulum (ER) stress, activates Bax and Bak directly or indirectly through BH3-only proteins. As a result, Bax is recruited to the outer mitochondrial membrane, where it works together with Bak to induce mitochondrial membrane permeabilization (MMP), leading to the release of various pro-apoptotic factors from the mitochondrial intermembrane space, including cytochrome *c*, apoptosis-inducing factor and endonuclease G. In the cytosol, released cytochrome *c* interacts with apoptosis-activating factor-1 and procaspase-9 to form the apoptosome that drives the activation of the initiator caspase-9, which in turn activates caspase-3 and other effector caspases. Further activation of nucleases such as caspase-activated DNases causes DNA fragmentation and caspase-dependent apoptosis. In addition, released apoptosis-inducing factor and endonuclease G induce nuclear condensation and DNA fragmentation, acting as a caspase-independent apoptotic mechanism. The extrinsic pathway begins with ligand-induced activation of the death receptors on the cell surface that activates the initiator capase-8, which in turn activates the effector caspases (caspase-3, caspase-6 and/or caspase-7) and nucleases, mediating caspase-dependent apoptosis. Caspase-8 can also proteolytically cleave Bid and the resulting truncated Bid can translocate to the outer mitochondrial membrane and stimulate MMP, which links the extrinsic pathway to the intrinsic pathway. Mitochondria are therefore central to the intrinsic pathway and also critical in the extrinsic pathway by emanating various pro-apoptotic factors to induce caspase-dependent and caspase-independent apoptosis in response to the apoptotic signals. In addition, mitochondrial dysfunction induces the opening of the mitochondrial permeability transition pore (mPTP), a transmembrane channel formed at the contact sites between the inner and outer mitochondrial membranes, leading to dissipation of the mitochondrial membrane potential that impairs ATP generation and increases mitochondrial ROS generation.

3. TRPM2 channel in oxidative stress-induced mitochondrial dysfunction and cell death in neuronal cells

Neurons in the brain are highly susceptible to oxidative stress, and oxidative stress-induced neuronal cell death is well recognized in the pathogenesis of AD, PD and ischemic stroke brain damage. Studies provide evidence to implicate a significant role for the TRPM2 channel in mediating neuronal cell death due to exposure to oxidative stress-inducing factors that are associated with these diseased conditions and readers can consult recent reviews for further details (Belrose & Jackson, 2018; Jiang, Li, Syed Mortadza, Lovatt, & Yang, 2018; Li, Meng, Li, Li, & Jiang, 2015; Mai et al., 2020; Malko & Jiang, 2020; Sita, Hrelia, Graziosi, & Morroni, 2020; Turlova, Feng, & Sun, 2018). Our discussion will focus on the role of TRPM2 channel-dependent mitochondrial dysfunction in neuronal cell death induced by various oxidative stress-inducing stimuli.

3.1 H_2O_2-induced mitochondrial dysfunction and neuronal cell death

Exposure to H_2O_2 is widely used to induce oxidative stress *in vitro*. Several groups have studied the role of the TRPM2 channel in mediating H_2O_2-induced neuronal cell death using cultured cortical, hippocampal and striatal neurons, providing pharmacological and genetic evidence to support an important role for the TRPM2 channel and, more specifically, TRPM2-mediated Ca^{2+} signaling in H_2O_2-induced neuronal cell death (Fonfria et al., 2005; Jia et al., 2011; Kaneko et al., 2006; Li, Yang, & Jiang, 2017; Verma et al., 2012). These studies have drawn less attention to the signaling pathways underlying cell death and, nonetheless, some of these studies suggest that H_2O_2-induced mitochondrial dysfunction is critically involved in H_2O_2-induced TRPM2-mediated neuronal cell death. For example, in cultured hippocampal neurons (Li, Yang, et al., 2017), we showed that short exposure to H_2O_2 (300 μM for 30 min) induced loss of mitochondria and mitochondrial fragmentation revealed by confocal imaging of MitoTracker staining (Fig. 2A). In addition, extended exposure

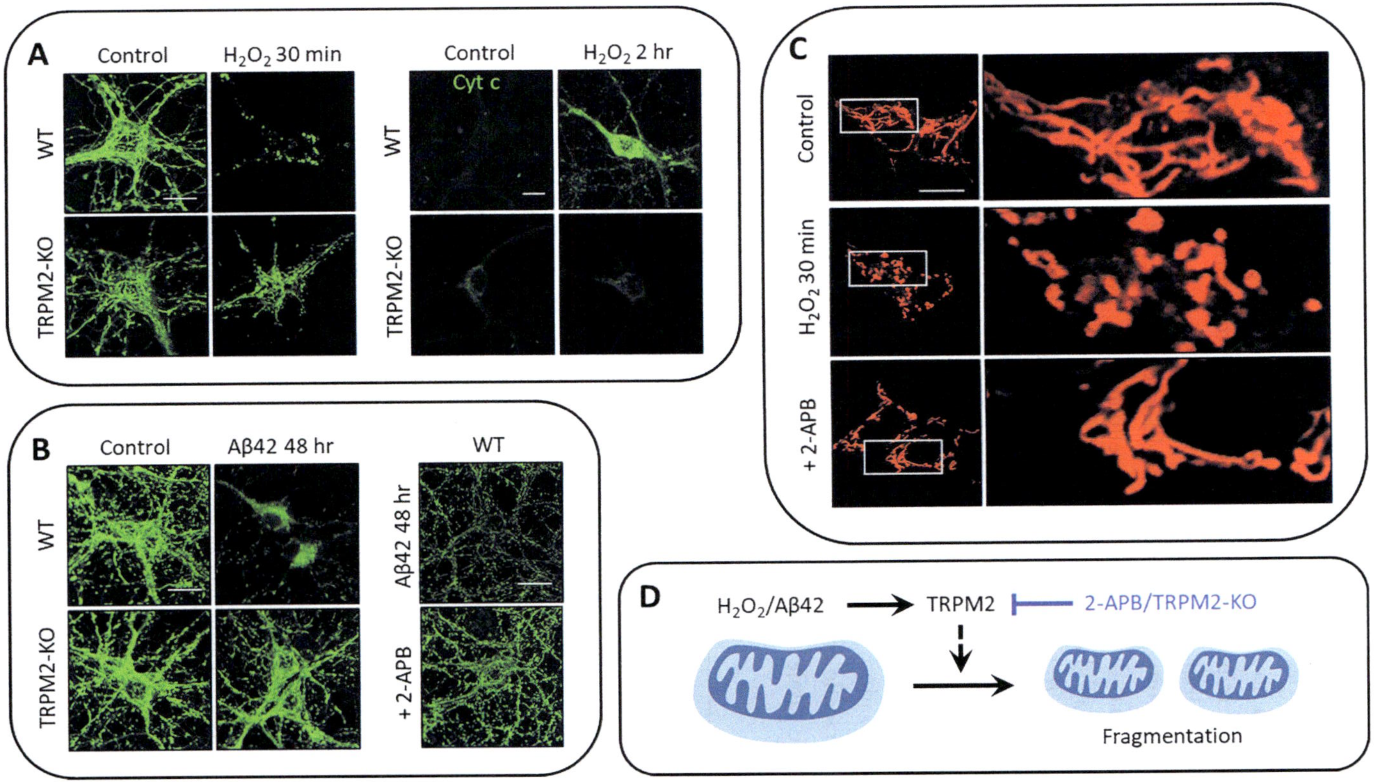

Fig. 2 See figure legend on next page.

H_2O_2 (100–300 μM for up to 24 h) resulted in release of cytochrome *c* into the cytosol shown by immunofluorescent staining (Fig. 2A) and mitochondrial ROS generation by staining with MitoTracker Red CM-H_2Xros, a fluorescent probe for mitochondrial ROS. H_2O_2-induced mitochondrial fragmentation and release of cytochrome *c* (Fig. 2A) as well as mitochondrial ROS generation were strongly attenuated in hippocampal neurons from transgenic TRPM2-knockout (TRPM2-KO) mice. It is known that disruption of intracellular Zn^{2+} homeostasis as well as intracellular Ca^{2+} homeostasis triggers neuronal cell death in the central nervous system (CNS) (Shuttleworth & Weiss, 2011). Indeed, we showed that H_2O_2-induced TRPM2-mediated neuronal cell death in cultured hippocampal neurons was attenuated by treatment with TPEN, a Zn^{2+} chelator, supporting that alteration in intracellular Zn^{2+} homeostasis is important in H_2O_2-induced TRPM2-mediated neuronal cell death. By examining the co-localization of staining with FluoZin-3, a fluorescent Zn^{2+} indicator, and intracellular organelle-specific fluorescent markers, we demonstrated that the labile Zn^{2+} in hippocampal neurons under control conditions

Fig. 2 TRPM2 channel in H_2O_2/Aβ42-induced mitochondrial fragmentation in neuronal cells. (A) MitoTracker Green staining showing loss and fragmentation of mitochondria (left) and immunostaining showing release of cytochrome (cyt) c (right) in hippocampal neurons from wild-type (WT) and TRPM2-KO mice, after exposure to 300 μM H_2O_2 for 30 min (MitoTracker staining) or 2 h (cyt c immunostaining). (B) MitoTracker Green staining showing loss and fragmentation of mitochondria in hippocampal neurons from wild-type (WT) and TRPM2-KO mice after exposure to 1 μM Aβ for 48 h (left) or WT neurons exposed to 1 μM Aβ for 48 h without or with prior treatment with 10 μM 2-APB (right). (C) MitoTracker Red staining showing mitochondrial fragmentation in human SH-SY5Y cells after exposure to 300 μM H_2O_2 for 2 h without or with prior treatment with 10 μM 2-APB. The areas outlined in the images on the left are enlarged and shown on the right. (D) A graphic summary of H_2O_2/Aβ42-induced activation of the TRPM2 channel that leads to mitochondrial fragmentation that can be inhibited by pharmacological inhibition of the TRPM2 channel with 2-APB or genetic deletion of TRPM2 expression as shown in A–C. *Panel A adapted from Li, X., Yang, W., & Jiang, L. H. (2017). Alteration in intracellular Zn^{2+} homeostasis as a result of TRPM2 channel activation contributes to ROS-induced hippocampal neuronal death. Frontiers in Molecular Neuroscience, 10, 414. doi:10.3389/fnmol.2017.00414. Panel B adapted from Li, X., & Jiang, L. H. (2018). Multiple molecular mechanisms form a positive feedback loop driving amyloid beta42 peptide-induced neurotoxicity via activation of the TRPM2 channel in hippocampal neurons. Cell Death & Disease, 9(2), 195. doi:10.1038/s41419-018-0270-1. Panel C adapted from Li, X., & Jiang, L. H. (2019). A critical role of the transient receptor potential melastatin 2 channel in a positive feedback mechanism for reactive oxygen species-induced delayed cell death. Journal of Cellular Physiology, 234(4), 3647–3660. doi:10.1002/jcp.27134.*

was mainly present in the lysosomes, but not in the mitochondria or ER. Short exposure to H_2O_2 (300 μM for 30 min) resulted in loss of lysosomes and lysosomal Zn^{2+} and intracellular Zn^{2+} increase and, in addition, mitochondrial Zn^{2+} accumulation shown using RhodZin3, a mitochondrial Zn^{2+} fluorescent indicator. These effects were observed in hippocampal neurons from wild-type (WT) mice, but not from TRPM2-KO mice. Furthermore, mitochondrial dysfunction indicated by mitochondrial ROS generation was also blunted by treatment with TPEN as well as by TRPM2-KO. Collectively, our results suggest that the TRPM2 channel mediates H_2O_2-induced neuronal cell death *via* induction of Ca^{2+} influx and lysosomal Zn^{2+} release to disturb the intracellular Ca^{2+} and Zn^{2+} homeostasis and mitochondrial functions (Li, Yang, et al., 2017). H_2O_2-induced TRPM2-mediated neuronal cell death occurs possibly *via* the intrinsic apoptotic pathway as a result of mitochondrial dysfunction, though the evidence to support this mechanism is still lacking.

3.2 Aβ42-induced mitochondrial dysfunction and neuronal cell death

In the most well-known amyloid cascade hypothesis of AD, generation and aggregation of amyloid β (Aβ) peptides, particularly Aβ42, represent a key pathological factor or event that can sequentially induce oxidative stress and neurotoxicity such as synapse loss and neuronal cell death (Hardy & Selkoe, 2002). Transgenic APP/PS1 mice that express the human mutant amyloid precursor protein (APP) and presenilin 1 (PS1) proteins are one of the most widely-used rodent models of AD. A study by Ostapchenko et al. showed that genetic deletion of TRPM2 expression in APP/PS1 mice almost completely rescued age-related memory deficit as well as synapse loss and microglial cell activation (Ostapchenko et al., 2015). Such striking effects strongly implicate a crucial role for the TRPM2 channel in Aβ-induced AD-related pathologies. In a recent study, we have further explored the role of the TRPM2 channel and the signaling mechanisms involved in Aβ42-induced neuronal cell death in cultured mouse hippocampal neurons (Li & Jiang, 2018). Exposure to Aβ42 at pathologically relevant concentrations ($\leq$100 nM to 1 μM for up to 96 h) induced substantial neuronal cell death. Such Aβ42-induced neuronal cell death was largely absent in hippocampal neurons from TRPM2-KO mice, providing compelling evidence to suggest a key role for the TRPM2 channel in mediating Aβ-induced neurotoxicity. There is substantial evidence to show that extensive mitochondrial dysfunction occurs in human AD brains and suggest that mitochondrial dysfunction is likely an early event leading to AD (Swerdlow, 2018; Wang, Zhao, Ma, Perry, & Zhu, 2020). In the aforementioned study (Li & Jiang, 2018), we have consistently shown by confocal imaging of MitoTracker staining that exposure to Aβ42 (1 μM for 24–48 h) resulted in substantial loss and fragmentation of mitochondria (Fig. 2B) and mitochondrial ROS generation in cultured hippocampal neurons. Aβ42-induced mitochondrial fragmentation and mitochondrial ROS generation as well as Aβ42-induced neuronal cell death was inhibited by TRPM2-KO or by treatment with 2-aminoethoxyldiphenyl borate (2-APB) (Fig. 2B), which is known to inhibit the TRPM2 channel (Fig. 1), as well as TPEN. As observed in hippocampal neurons after exposure to H_2O_2, exposure to Aβ42 induced lysosomal dysfunction, lysosomal Zn^{2+} release, mitochondrial Zn^{2+} accumulation, mitochondrial fragmentation and mitochondrial ROS generation, and each of these effects were prevented by treatment with 2-APB, by TRPM2-KO, or by treatment with TPEN. We further demonstrated that Aβ42-induced detrimental effects on lysosomal and mitochondrial functions and alteration in intracellular Zn^{2+} homeostasis in hippocampal neurons were strongly suppressed by treatment with Gö6983, a protein kinase C (PKC) inhibitor, apocynin, a generic inhibitor of NADPH oxidase (NOX) or GKT137831, a NOX1/4 inhibitor. Interestingly, bafilomycin-induced lysosomal dysfunction also led to lysosomal Zn^{2+} release, mitochondrial Zn^{2+} accumulation, mitochondrial fragmentation, and mitochondrial ROS generation, all of which were inhibited by treatment with 2-APB or by TRPM2-KO, and also by treatment with TPEN, and moreover, by treatment with Gö6983, apocynin or GKT137831. We detected, using western blotting, TRPM2 proteins in isolated mitochondria prepared from mouse cortex and hippocampus. Consistently, we showed that application of ADPR to isolated mitochondria induced mitochondrial Zn^{2+} accumulation in the presence but not in the absence of Ca^{2+}, and ADPR-induced Ca^{2+}-dependent Zn^{2+} accumulation was not observed in isolated mitochondria from the cortex and hippocampus of TRPM2-KO mice (Li & Jiang, 2018). Collectively, these results have led us to propose that TRPM2-dependent mitochondrial Zn^{2+} uptake is central in the induction of a positive feedback mechanism driving Aβ42-induced neuronal cell death. Such a mechanism is comprised of multiple steps, including activation of PKC and NOX, NOX-mediated ROS generation, lysosomal dysfunction, lysosomal Zn^{2+} release, mitochondrial Zn^{2+} accumulation, mitochondrial dysfunction and mitochondrial ROS generation (Fig. 3). In fact, identification

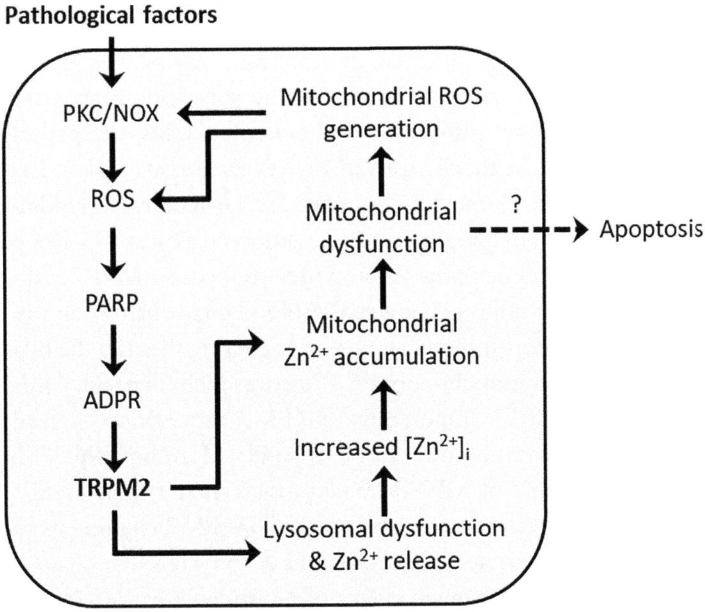

Fig. 3 TRPM2-dependent positive feedback mechanism for ROS/Aβ-induced neuronal cell death. The key steps or events in the proposed positive feedback mechanism that drives H_2O_2- and Aβ42-induced neuronal cell death and ischemia/reperfusion-related delayed neuronal cell death. The pathological factors initially activate protein kinase C (PKC) and NADPH oxidases (NOX) to generate ROS, which in turn induces ADP-ribose (ADPR) *via* poly(ADPR) polymerase (PARP) and poly(ADPR) glycohydrolase (not depicted) to activate the TRPM2 channel. ROS-induced, potentially *via* the TRPM2 channel, lysosomal dysfunction and lysosomal Zn^{2+} release results in intracellular Zn^{2+} increase ($[Zn^{2+}]_i$). Subsequent TRPM2-dependent mitochondrial Zn^{2+} accumulation leads to mitochondrial fragmentation, mitochondrial dysfunction, release of cytochrome *c* and mitochondrial ROS that feeds back to induction of oxidative stress. Mitochondrial fragmentation and dysfunction may trigger the apoptotic cell death pathways to drive cell death. *Adapted from Li, X., & Jiang, L. H. (2019). A critical role of the transient receptor potential melastatin 2 channel in a positive feedback mechanism for reactive oxygen species-induced delayed cell death. Journal of Cellular Physiology, 234(4), 3647–3660. doi:10.1002/jcp.27134.*

of such Aβ42-induced TRPM2-mediated neuronal cell death and multiple other positive feedbacks loops among the key AD-related events has led to the compelling proposal of the amyloid feedback hypothesis rather than the cascade hypothesis (Doig, 2018). Again, it is tempting to hypothesize that Aβ42-induced TRPM2-mediated mitochondrial dysfunction activates the intrinsic apoptotic pathway to prompt neuronal cell death.

3.3 MPP$^+$-induced apoptotic neuronal cell death

Oxidative stress and mitochondrial dysfunction also play a vital role in mediating loss of dopaminergic neurons in the substantia nigra pars compacta (SNpc) induced by misfolded α-synuclein proteins, the pathological hallmark and key cellular mechanism of PD (Poewe et al., 2017). Exposure to neurotoxins MPTP (1-methyl-4-phenyl-1,2,3,6-tetrahydropyridine) and MPP$^+$ (1-methyl-4-phenylpyridinium ion) is known to cause PD-like pathologies in humans, non-human primates and rodents, because MPP$^+$ as the toxic metabolite of MPTP readily penetrates the blood–brain barrier and is selectively accumulated in dopaminergic neurons, triggering them to die by inducing oxidative stress and mitochondrial dysfunction (Schildknecht, Di Monte, Pape, Tieu, & Leist, 2017). Importantly, TRPM2 expression was found in a recent study using western blotting to be upregulated in the SNpc of human PD brains and the SNpc of MPTP-treated mouse brains (Sun et al., 2018). The study also showed similar upregulation of TRPM2 channel expression in MPP$^+$-exposed human neuroblastoma SH-SY5Y cells, which are widely used to study the molecular mechanisms of neurodegeneration, particularly those related to PD pathogenesis (Xicoy, Wieringa, & Martens, 2017). In addition, it was shown that exposure to MPP$^+$ promoted ROS generation in SH-SY5Y cells and thereby induced apoptotic cell death assessed by co-staining with annexin V and propidium iodide (PI). Consistently, as shown by western blotting, exposure of SH-SY5Y cells to MPP$^+$ led to increases in the levels of pro-apoptotic proteins, Bak, Bid and Bad, release of cytochrome *c* to the cytosol, and the activity of calpain and caspase-3, suggesting MPP$^+$-induced activation of the intrinsic apoptotic pathway. Both MPP$^+$-induced cell death and increases or activation of proteins associated with the apoptotic pathway were suppressed by treatment with flufenamic acid (FFA), which is known to inhibit the TRPM2 channel (Fig. 1), or small interference RNA (siRNA) to knock down TRPM2 expression (Sun et al., 2018). In a recent study, we have shown that overexpression of the human TRPM2 channel rendered SH-SY5Y cells more sensitive to H_2O_2-induced cell death (An et al., 2019). Taken together, these studies support that oxidative stress-induced TRPM2 channel activation as well as upregulation of its expression leads to activation of the apoptotic pathway that mediates MPP$^+$-induced neuronal cell death. Such findings suggest TRPM2-mediated mitochondrial dysfunction and apoptotic cell death as an important mechanism mediating oxidative stress-induced loss of dopaminergic neurons in PD pathogenesis.

3.4 ROS-induced mitochondrial dysfunction and delayed neuronal cell death

Reperfusion, following transient ischemia that occurs in ischemic stroke or due to cardiac arrest, is a well-documented condition that promotes excessive ROS generation and exacerbates tissue damage by inducing neuronal cell death, which is often called delayed neuronal cell death. Numerous studies using *in vitro* and various *in vivo* I/R models support an important role for the TRPM2 channel in mediating delayed neuronal cell death in brain damage and cognitive deficits due to ischemic stroke and other I/R conditions (Belrose & Jackson, 2018; Li et al., 2015; Mai et al., 2020). TRPM2-mediated modulation of two different signaling pathways or mechanisms has been proposed to underpin I/R-induced neuronal cell death. Alim et al., by examining WT and TRPM2-KO mice subjected to middle cerebral artery occlusion followed by reperfusion (MCAO/R), an *in vivo* model of ischemic stroke, provided evidence to suggest engagement of the signaling pathways mediated by the NMDA-subtype glutamate receptors in neurons (Alim et al., 2013). More specifically, they proposed that TRPM2 channel activation promotes the GluN2B-dependent pro-death signaling pathway and inhibits the GluN2A-dependent pro-survival signaling pathway. In our study, we showed that TRPM2-KO prevented intracellular Zn^{2+} increase as well as delayed neuronal cell death in pyramidal neurons in mice exposed to transient bilateral common carotid artery occlusion and reperfusion (BCCAO/R), an *in vivo* model of global ischemia. Furthermore, using time-lapse imaging to monitor the intracellular Zn^{2+} level in cultured hippocampal neurons during oxygen-glucose deprivation (OGD) followed by reoxygenation (OGD-R), an *in vitro* model of I/R, we revealed that TRPM2-KO selectively abolished the increase in intracellular Zn^{2+} during reperfusion, consistent with reperfusion-induced oxidative stress. Our results support the notion that reperfusion-associated oxidative stress activates the TRPM2 channel and disrupts intracellular Zn^{2+} homeostasis to induce delayed neuronal cell death (Ye et al., 2014). It is worth noting that these two different mechanisms are not mutually exclusive.

In another recent study using SH-SY5Y cells, we have shown that exposure to H_2O_2 (100–300 μM for 2 h) induced negligible cell death immediately after exposure; however, cell death was readily detected when cells were subsequently maintained in normal and H_2O_2-free culture medium for a further 24 h, indicating that cell death occurred with a substantial delay (Li & Jiang, 2019). Such delayed cell death was suppressed in the absence of extracellular Ca^{2+} as well as by treatment with 2-APB and *N*-(*p*-amycinnamoyl)anthranilic acid (ACA) to inhibit the TRPM2 channel (Fig. 1), supporting the importance of TRPM2-mediated Ca^{2+} signaling. H_2O_2-induced delayed cell death was abolished by treatment with TPEN or clioquinol, another Zn^{2+} chelator. These results support that TRPM2-dependent alteration in intracellular Zn^{2+} homeostasis is critical in H_2O_2-induced delayed cell death, as shown in delayed neuronal death in hippocampal neurons induced by OGD/R *in vitro* and BCCAO/R *in vivo* (Ye et al., 2014). Interestingly, H_2O_2-induced delayed cell death in SH-SY5Y cells was also inhibited by treatment with Gö6983, apocynin, DPI, and GKT137831. In addition, exposure to H_2O_2 induced lysosomal dysfunction, lysosomal Zn^{2+} release, mitochondrial Zn^{2+} accumulation, mitochondrial fragmentation (Fig. 2C) and mitochondrial ROS generation. These effects were subdued by treatment with 2-APB (Fig. 2C), TPEN, Gö6983, apocynin, DPI or GKT137831. As described above in hippocampal neurons, bafilomycin-induced lysosomal dysfunction in SH-SY5Y cells resulted in intracellular Zn^{2+} increase, mitochondrial Zn^{2+} accumulation, mitochondrial fragmentation, and mitochondrial ROS generation, all of which were suppressed by treatment with 2-APB or TPEN. We further demonstrated TRPM2 protein and ADPR-induced Ca^{2+}-dependent mitochondrial Zn^{2+} accumulation in mitochondria isolated from SH-SY5Y cells. Collectively, these results suggest that a similar positive feedback mechanism that we propose for Aβ42-induced neuronal cell death drives I/R-related delayed neuronal cell death (Fig. 3) (Li & Jiang, 2019).

4. TRPM2 channel in oxidative stress-induced mitochondrial dysfunction and apoptosis in pancreatic β-cells

Pancreatic β-cells have a unique function in generating insulin in response to elevated glucose in the blood, and thus loss of pancreatic β-cells and their function is key to the pathogenesis of type 1 diabetes (T1D) and also contributes to type 2 diabetes (T2D) (Cerf, 2013). Pancreatic β-cells are known for their weak antioxidant capacity and are highly vulnerable to damage by exposure to oxidative stress-inducing factors, for example, the diabetogenic agent streptozotocin (STZ), and free fatty acids such as palmitate. It has been shown that the TRPM2 channel is expressed in pancreatic β-cells with two distinct functional roles, depending on the stimuli. On one hand, the TRPM2 channel plays a significant role in the regulation of glucose-induced insulin secretion (Uchida & Tominaga, 2014). On the other hand, accumulating evidence supports that the TRPM2 channel mediates oxidative stress-induced pancreatic β-cell death, implicating the importance of the TRPM2 channel in the pathogenesis of both T1D and T2D.

4.1 H_2O_2/STZ-induced mitochondrial dysfunction and apoptotic cell death

We investigated the role of the TRPM2 channel in H_2O_2/STZ-induced pancreatic β-cells in relation to T1D pathogenesis (Manna et al., 2015). Exposure to H_2O_2 (200 μM for 2 h) resulted in cell death in INS-1 cells, a rat insulin-secreting β-cell line. Such cell death was attenuated by treatment with 2-APB, and also by treatment with Ac-DEVD, an inhibitor of caspase-dependent apoptosis, but not IM-54, a necrosis inhibitor, or necrostatin-1, a necroptosis inhibitor, suggesting that H_2O_2-induced TRPM2-mediated cell death occurs primarily *via* apoptosis, as is known for pancreatic β-cell death in diabetes. H_2O_2-induced cell death in INS-1 cells was inhibited by treatment with TPEN or clioquinol. Similar to hippocampal neurons and SH-SY5Y cells, the labile Zn^{2+} in INS-1 cells was found in the lysosomes, but not in the mitochondria or ER under control conditions, and exposure to H_2O_2 induced lysosomal Zn^{2+} release and intracellular Zn^{2+} increase, which were inhibited by treatment with 2-APB. Administration with multiple low doses of STZ (40 mg/kg 6 times per day) led to substantial loss of pancreatic β-cells and hyperglycaemia in WT mice, but such STZ-induced detrimental effects were not observed in TRPM2-KO mice. Likewise, exposure to STZ (2 mM for 48 h) caused substantial damage to mouse pancreatic islets, which was prevented by treatment with TPEN as well as by TRPM2-KO. Taken together, these results support that H_2O_2/STZ-induced TRPM2 channel activation results in Ca^{2+} influx and release of lysosomal Zn^{2+}, leading to alterations in intracellular Ca^{2+} and Zn^{2+} homeostasis and apoptotic cell death in pancreatic β-cells. It remains unclear how TRPM2-mediated alterations in intracellular Ca^{2+} and Zn^{2+} homeostasis activate the apoptotic pathways.

4.2 Palmitate-induced mitochondrial dysfunction and apoptotic cell death

Accumulation of free fatty acids can promote oxidative stress and impair insulin secretion, and thus represents a risk factor for T2D (Oprescu et al., 2007). Li et al. have recently investigated the role of the TRPM2 channel in palmitate-induced cell death using INS-1832/13 cells and also mouse and human islets (Li, Munsey, et al., 2017). Exposure to palmitate (500 μM for 12 h) induced apoptotic cell death in INS-1832/13 cells shown by annexin V-GFP and PI staining, which was inhibited by treatment with TRPM2-specific siRNA or TPEN. Consistently, exposure to palmitate (500 μM for 5 days) induced cell death in pancreatic islets from WT mice, but not from TRPM2-KO mice. Exposure to a combination of palmitate (500 μM), IL-1β (5 ng/mL) and IFN-γ (5 ng/mL) for 7 days led to significant apoptotic cell death in human islets assessed by TUNEL staining, which was reduced by treatment with ACA or TRPM2-specific siRNA (Li, Munsey, et al., 2017). These results support that the TRPM2 channel mediates palmitate-induced alteration in intracellular Zn^{2+} homeostasis and apoptotic cell death in pancreatic β-cells, implicating TRPM2 channel activation as a significant mechanism in T2D pathogenesis. The study further showed that exposure to palmitate (500 μM for 4 h) induced mitochondrial fragmentation in INS-1832/13 cells revealed by MitoTracker staining, which was inhibited by treatment with apocynin and NOX-2 specific peptide gp91ds-tat. Palmitate-induced mitochondrial fragmentation was also inhibited by treatment with TRPM2-specific siRNA. Similarly, exposure to palmitate (500 μM for 5 days) also induced mitochondrial fragmentation in pancreatic islets from WT mice, but not from TRPM2-KO mice (Li, Munsey, et al., 2017). As discussed above for H_2O_2 and STZ, exposure to palmitate (500 μM for 4 h) induced mitochondrial Zn^{2+} accumulation in INS-1832/13 cells that was inhibited by treatment with TRPM2-specific siRNA. Prolonged exposure to palmitate triggered MMP, which was prevented by treatment with TPEN or TPA, another Zn^{2+} chelator. In addition, exposure to palmitate resulted in a reduction of mitochondrial membrane potential shown using JC-10 staining. Furthermore, exposure to palmitate promoted the recruitment of Drp-1 to mitochondria. Both palmitate-induced reduction of mitochondrial membrane potential and Drp-1 recruitment were attenuated by treatment with TRPM2-specific siRNA or TPEN. Taken together, these results suggest that fatty acids like palmitate induce pancreatic β-cell death *via* a multi-step mechanism, including activation of NOX, NOX-mediated ROS generation, ROS-induced activation of PARP and the TRPM2 channel, lysosomal Zn^{2+}, mitochondrial Zn^{2+} accumulation, Drp-1-mediated mitochondrial fission, mitochondrial dysfunction and activation of the apoptotic pathway. These findings provide new mechanistic insights into the role of the TRPM2 channel in mediating oxidative stress-induced pancreatic β-cell death in T2D pathogenesis.

5. Summary and perspective

Increasing *in vitro* and *in vivo* studies have gathered compelling evidence to support a significant role for the TRPM2 channel in mediating oxidative stress-induced cell death in CNS neurons and pancreatic β-cells. In particular, studies show that diverse pathological factors induce NOX-mediated generation of ROS and subsequent TRPM2 channel activation, leading to alterations in intracellular Ca^{2+} and Zn^{2+} homeostasis that induce mitochondrial dysfunction and activation of the apoptotic pathways. There is evidence to suggest that mitochondrial ROS generation as a result of mitochondrial dysfunction constitutes a vicious positive feedback loop in driving ROS/Aβ42-induced neuronal cell death (Fig. 3). Such a mechanism, once set in motion by the pathological factors, becomes self-amplified and self-drives the pathological process. It would be of interest to test whether a similar mechanism drives apoptotic cell death in pancreatic β-cells induced by STZ, palmitate or other diabetogenic agents. On the other hand, TRPM2-mediated alterations in mitochondrial Zn^{2+} homeostasis stimulate Drp1 recruitment to mediate mitochondrial fission and induce mitochondrial fragmentation and dysfunction to trigger pancreatic β-cell apoptosis. It would also be interesting to examine whether such a mechanism underpins TRPM2-mediated mitochondrial fragmentation and mitochondrial dysfunction in neurotoxicity induced by neurotoxins such as Aβ peptides, MPP^+ or α-synuclein proteins.

In summary, the TRPM2 channel has been identified as a crucial molecular mechanism that mediates mitochondrial dysfunction and cell death, particularly mitochondria-mediated apoptosis, and thereby links oxidative stress-inducing pathological factors to associated pathological conditions. Further investigations are needed to provide a better understanding of how TRPM2 channel activation induces mitochondrial dysfunction and apoptotic cell death and the contribution of such mechanisms to the pathogenesis of AD, PD and ischemic stroke, and evaluate the perspective of the TRPM2 channel as a therapeutic target for the treatment of these debilitating conditions with unmet clinical needs.

Acknowledgments

We are grateful to the Wellcome Trust (072275), Alzheimer's Research Trust (ART/PPG2009A/2), and National Science of China (31471118) for their support to our research described in this paper. P.M. is supported by a PhD studentship from the University of Leeds.

Author contributions

P.M. and L.H.J.: literature analysis, discussion, writing and revision of manuscript; R.D.: discussion and revision of manuscript.

Conflict of interest statement

We declare no conflict of interest.

References

Alim, I., Teves, L., Li, R., Mori, Y., & Tymianski, M. (2013). Modulation of NMDAR subunit expression by TRPM2 channels regulates neuronal vulnerability to ischemic cell death. *Journal of Neuroscience*, *33*(44), 17264–17277. https://doi.org/10.1523/JNEUROSCI.1729-13.2013.

An, X., Fu, Z., Mai, C., Wang, W., Wei, L., Li, D., et al. (2019). Increasing the TRPM2 channel expression in human neuroblastoma SH-SY5Y cells augments the susceptibility to ROS-induced cell death. *Cells*, *8*(1), 28. https://doi.org/10.3390/cells8010028.

Arnoult, D. (2007). Mitochondrial fragmentation in apoptosis. *Trends in Cell Biology*, *17*(1), 6–12. https://doi.org/10.1016/j.tcb.2006.11.001.

Belrose, J. C., & Jackson, M. F. (2018). TRPM2: A candidate therapeutic target for treating neurological diseases. *Acta Pharmacologica Sinica*, *39*(5), 722–732. https://doi.org/10.1038/aps.2018.31.

Bordoni, L., & Gabbianelli, R. (2020). Mitochondrial DNA and neurodegeneration: Any role for dietary antioxidants? *Antioxidants*, *9*(8), 764. https://doi.org/10.3390/antiox9080764.

Bugger, H., & Pfeil, K. (2020). Mitochondrial ROS in myocardial ischemia reperfusion and remodeling. *Biochimica et Biophysica Acta—Molecular Basis of Disease*, *1866*(7), 165768. https://doi.org/10.1016/j.bbadis.2020.165768.

Cerf, M. E. (2013). Beta cell dysfunction and insulin resistance. *Frontiers in Endocrinology*, *4*(37), 37. https://doi.org/10.3389/fendo.2013.00037.

Cheignon, C., Tomas, M., Bonnefont-Rousselot, D., Faller, P., Hureau, C., & Collin, F. (2018). Oxidative stress and the amyloid beta peptide in Alzheimer's disease. *Redox Biology*, *14*, 450–464. https://doi.org/10.1016/j.redox.2017.10.014.

Davis, S. M., & Pennypacker, K. R. (2017). Targeting antioxidant enzyme expression as a therapeutic strategy for ischemic stroke. *Neurochemistry International*, *107*, 23–32. https://doi.org/10.1016/j.neuint.2016.12.007.

Detmer, S. A., & Chan, D. C. (2007). Functions and dysfunctions of mitochondrial dynamics. *Nature Reviews. Molecular Cell Biology*, *8*(11), 870–879. https://doi.org/10.1038/nrm2275.

Doig, A. J. (2018). Positive feedback loops in Alzheimer's disease: The Alzheimer's feedback hypothesis. *Journal of Alzheimer's Disease*, *66*(1), 25–36. https://doi.org/10.3233/JAD-180583.

Fanczal, J., Pallagi, P., Gorog, M., Diszhazi, G., Almassy, J., Madacsy, T., et al. (2020). TRPM2-mediated extracellular Ca^{2+} entry promotes acinar cell necrosis in biliary acute pancreatitis. *Journal of Physiology*, *598*(6), 1253–1270. https://doi.org/10.1113/JP279047.

Fonfria, E., Marshall, I. C., Boyfield, I., Skaper, S. D., Hughes, J. P., Owen, D. E., et al. (2005). Amyloid beta-peptide (1-42) and hydrogen peroxide-induced toxicity are mediated by TRPM2 in rat primary striatal cultures. *Journal of Neurochemistry*, *95*(3), 715–723. https://doi.org/10.1111/j.1471-4159.2005.03396.x.

Förstermann, U., Xia, N., & Li, H. (2017). Roles of vascular oxidative stress and nitric oxide in the pathogenesis of atherosclerosis. *Circulation Research*, *120*(4), 713–735. https://doi.org/10.1161/circresaha.116.309326.

Gao, G., Wang, W., Tadagavadi, R. K., Briley, N. E., Love, M. I., Miller, B. A., et al. (2014). TRPM2 mediates ischemic kidney injury and oxidant stress through RAC1. *Journal of Clinical Investigation*, *124*(11), 4989–5001. https://doi.org/10.1172/JCI76042.

Garcia-Rodriguez, S., Rosal-Vela, A., Botta, D., Cumba Garcia, L. M., Zumaquero, E., Prados-Maniviesa, V., et al. (2018). CD38 promotes pristane-induced chronic inflammation and increases susceptibility to experimental lupus by an apoptosis-driven and TRPM2-dependent mechanism. *Scientific Reports*, *8*(1), 3357. https://doi.org/10.1038/s41598-018-21337-6.

Granger, D. N., & Kvietys, P. R. (2015). Reperfusion injury and reactive oxygen species: The evolution of a concept. *Redox Biology*, *6*, 524–551. https://doi.org/10.1016/j.redox.2015.08.020.

Hara, Y., Wakamori, M., Ishii, M., Maeno, E., Nishida, M., Yoshida, T., et al. (2002). LTRPC2 Ca^{2+}-permeable channel activated by changes in redox status confers susceptibility to cell death. *Molecular Cell*, *9*(1), 163–173. https://doi.org/10.1016/s1097-2765(01)00438-5.

Hardy, J., & Selkoe, D. J. (2002). The amyloid hypothesis of Alzheimer's disease: Progress and problems on the road to therapeutics. *Science*, *297*(5580), 353–356. https://doi.org/10.1126/science.1072994.

Jeong, S. Y., & Seol, D. W. (2008). The role of mitochondria in apoptosis. *BMB Reports*, *41*(1), 11–22. https://doi.org/10.5483/bmbrep.2008.41.1.011.

Jia, J., Verma, S., Nakayama, S., Quillinan, N., Grafe, M. R., Hurn, P. D., et al. (2011). Sex differences in neuroprotection provided by inhibition of TRPM2 channels following experimental stroke. *Journal of Cerebral Blood Flow and Metabolism*, *31*(11), 2160–2168. https://doi.org/10.1038/jcbfm.2011.77.

Jiang, L. H., Li, X., Syed Mortadza, S. A., Lovatt, M., & Yang, W. (2018). The TRPM2 channel nexus from oxidative damage to Alzheimer's pathologies: An emerging novel intervention target for age-related dementia. *Ageing Research Reviews*, *47*, 67–79. https://doi.org/10.1016/j.arr.2018.07.002.

Jiang, T., Sun, Q., & Chen, S. (2016). Oxidative stress: A major pathogenesis and potential therapeutic target of antioxidative agents in Parkinson's disease and Alzheimer's disease. *Progress in Neurobiology*, *147*, 1–19. https://doi.org/10.1016/j.pneurobio.2016.07.005.

Kaneko, S., Kawakami, S., Hara, Y., Wakamori, M., Itoh, E., Minami, T., et al. (2006). A critical role of TRPM2 in neuronal cell death by hydrogen peroxide. *Journal of Pharmacological Sciences*, *101*(1), 66–76. https://doi.org/10.1254/jphs.fp0060128.

Karbowski, M., & Neutzner, A. (2012). Neurodegeneration as a consequence of failed mitochondrial maintenance. *Acta Neuropathologica*, *123*(2), 157–171. https://doi.org/10.1007/s00401-011-0921-0.

Kheradpezhouh, E., Ma, L., Morphett, A., Barritt, G. J., & Rychkov, G. Y. (2014). TRPM2 channels mediate acetaminophen-induced liver damage. *Proceedings of the National Academy of Sciences of the United States of America*, *111*(8), 3176–3181. https://doi.org/10.1073/pnas.1322657111.

Li, X., & Jiang, L. H. (2018). Multiple molecular mechanisms form a positive feedback loop driving amyloid beta42 peptide-induced neurotoxicity via activation of the TRPM2 channel in hippocampal neurons. *Cell Death & Disease*, *9*(2), 195. https://doi.org/10.1038/s41419-018-0270-1.

Li, X., & Jiang, L. H. (2019). A critical role of the transient receptor potential melastatin 2 channel in a positive feedback mechanism for reactive oxygen species-induced delayed cell death. *Journal of Cellular Physiology*, *234*(4), 3647–3660. https://doi.org/10.1002/jcp.27134.

Li, C., Meng, L., Li, X., Li, D., & Jiang, L. H. (2015). Non-NMDAR neuronal Ca^{2+}-permeable channels in delayed neuronal death and as potential therapeutic targets for ischemic brain damage. *Expert Opinion on Therapeutic Targets*, *19*(7), 879–892. https://doi.org/10.1517/14728222.2015.1021781.

Li, F., Munsey, T. S., & Sivaprasadarao, A. (2017). TRPM2-mediated rise in mitochondrial Zn^{2+} promotes palmitate-induced mitochondrial fission and pancreatic beta-cell death in rodents. *Cell Death and Differentiation*, *24*(12), 1999–2012. https://doi.org/10.1038/cdd.2017.118.

Li, X., Yang, W., & Jiang, L. H. (2017). Alteration in intracellular Zn^{2+} homeostasis as a result of TRPM2 channel activation contributes to ROS-induced hippocampal neuronal death. *Frontiers in Molecular Neuroscience*, *10*, 414. https://doi.org/10.3389/fnmol.2017.00414.

Lin, M. T., & Beal, M. F. (2006). Mitochondrial dysfunction and oxidative stress in neurodegenerative diseases. *Nature*, *443*(7113), 787–795. https://doi.org/10.1038/nature05292.

Mai, C., Mankoo, H., Wei, L., An, X., Li, C., Li, D., et al. (2020). TRPM2 channel: A novel target for alleviating ischaemia-reperfusion, chronic cerebral hypo-perfusion and neonatal hypoxic-ischaemic brain damage. *Journal of Cellular and Molecular Medicine*, *24*(1), 4–12. https://doi.org/10.1111/jcmm.14679.

Malko, P., & Jiang, L. H. (2020). TRPM2 channel-mediated cell death: An important mechanism linking oxidative stress-inducing pathological factors to associated pathological conditions. *Redox Biology*, *37*, 101755. https://doi.org/10.3390/antiox9121253.

Manna, P. T., Munsey, T. S., Abuarab, N., Li, F., Asipu, A., Howell, G., et al. (2015). TRPM2-mediated intracellular Zn^{2+} release triggers pancreatic beta-cell death. *Biochemical Journal*, *466*(3), 537–546. https://doi.org/10.1042/BJ20140747.

Miller, B. A. (2019). TRPM2 in cancer. *Cell Calcium*, *80*, 8–17. https://doi.org/10.1016/j.ceca.2019.03.002.

Newsholme, P., Cruzat, V. F., Keane, K. N., Carlessi, R., & de Bittencourt, P. I., Jr. (2016). Molecular mechanisms of ROS production and oxidative stress in diabetes. *Biochemical Journal*, *473*(24), 4527–4550. https://doi.org/10.1042/bcj20160503c.

Oprescu, A. I., Bikopoulos, G., Naassan, A., Allister, E. M., Tang, C., Park, E., et al. (2007). Free fatty acid-induced reduction in glucose-stimulated insulin secretion: Evidence for a role of oxidative stress in vitro and in vivo. *Diabetes*, *56*(12), 2927–2937. https://doi.org/10.2337/db07-0075.

Ostapchenko, V. G., Chen, M., Guzman, M. S., Xie, Y. F., Lavine, N., Fan, J., et al. (2015). The transient receptor potential melastatin 2 (TRPM2) channel contributes to beta-amyloid oligomer-related neurotoxicity and memory impairment. *Journal of Neuroscience*, *35*(45), 15157–15169. https://doi.org/10.1523/JNEUROSCI.4081-14.2015.

Perraud, A. L., Fleig, A., Dunn, C. A., Bagley, L. A., Launay, P., Schmitz, C., et al. (2001). ADP-ribose gating of the calcium-permeable LTRPC2 channel revealed by Nudix motif homology. *Nature*, *411*(6837), 595–599. https://doi.org/10.1038/35079100.

Poewe, W., Seppi, K., Tanner, C. M., Halliday, G. M., Brundin, P., Volkmann, J., et al. (2017). Parkinson disease. *Nature Reviews. Disease Primers*, *3*, 17013. https://doi.org/10.1038/nrdp.2017.13.

Ray Chaudhuri, A., & Nussenzweig, A. (2017). The multifaceted roles of PARP1 in DNA repair and chromatin remodelling. *Nature Reviews. Molecular Cell Biology*, *18*(10), 610–621. https://doi.org/10.1038/nrm.2017.53.

Rochette, L., Zeller, M., Cottin, Y., & Vergely, C. (2014). Diabetes, oxidative stress and therapeutic strategies. *Biochimica et Biophysica Acta*, *1840*(9), 2709–2729. https://doi.org/10.1016/j.bbagen.2014.05.017.

Sano, Y., Inamura, K., Miyake, A., Mochizuki, S., Yokoi, H., Matsushime, H., et al. (2001). Immunocyte Ca^{2+} influx system mediated by LTRPC2. *Science*, *293*(5533), 1327–1330. https://doi.org/10.1126/science.1062473.

Schildknecht, S., Di Monte, D. A., Pape, R., Tieu, K., & Leist, M. (2017). Tipping points and endogenous determinants of nigrostriatal degeneration by MPTP. *Trends in Pharmacological Sciences*, *38*(6), 541–555. https://doi.org/10.1016/j.tips.2017.03.010.

Shuttleworth, C. W., & Weiss, J. H. (2011). Zinc: New clues to diverse roles in brain ischemia. *Trends in Pharmacological Sciences*, *32*(8), 480–486. https://doi.org/10.1016/j.tips.2011.04.001.

Sita, G., Hrelia, P., Graziosi, A., & Morroni, F. (2020). Back to the fusion: Mitofusin-2 in Alzheimer's disease. *Journal of Clinical Medicine*, *9*(1). https://doi.org/10.3390/jcm9010126.

Sun, Y., Sukumaran, P., Selvaraj, S., Cilz, N. I., Schaar, A., Lei, S., et al. (2018). TRPM2 promotes neurotoxin MPP^+/MPTP-induced cell death. *Molecular Neurobiology*, *55*(1), 409–420. https://doi.org/10.1007/s12035-016-0338-9.

Swerdlow, R. H. (2018). Mitochondria and mitochondrial cascades in Alzheimer's disease. *Journal of Alzheimer's Disease*, *62*(3), 1403–1416. https://doi.org/10.3233/jad-170585.

Turlova, E., Feng, Z. P., & Sun, H. S. (2018). The role of TRPM2 channels in neurons, glial cells and the blood-brain barrier in cerebral ischemia and hypoxia. *Acta Pharmacologica Sinica*, *39*(5), 713–721. https://doi.org/10.1038/aps.2017.194.

Uchida, K., & Tominaga, M. (2014). The role of TRPM2 in pancreatic beta-cells and the development of diabetes. *Cell Calcium*, *56*(5), 332–339. https://doi.org/10.1016/j.ceca.2014.07.001.

Verma, S., Quillinan, N., Yang, Y. F., Nakayama, S., Cheng, J., Kelley, M. H., et al. (2012). TRPM2 channel activation following in vitro ischemia contributes to male hippocampal cell death. *Neuroscience Letters*, *530*(1), 41–46. https://doi.org/10.1016/j.neulet.2012.09.044.

Wang, M., Li, J., Dong, S., Cai, X., Simaiti, A., Yang, X., et al. (2020). Silica nanoparticles induce lung inflammation in mice via ROS/PARP/TRPM2 signaling-mediated lysosome impairment and autophagy dysfunction. *Particle and Fibre Toxicology*, *17*(1), 23. https://doi.org/10.1186/s12989-020-00353-3.

Wang, W., Zhao, F., Ma, X., Perry, G., & Zhu, X. (2020). Mitochondria dysfunction in the pathogenesis of Alzheimer's disease: Recent advances. *Molecular Neurodegeneration*, *15*(1), 30. https://doi.org/10.1186/s13024-020-00376-6.

Wehage, E., Eisfeld, J., Heiner, I., Jungling, E., Zitt, C., & Luckhoff, A. (2002). Activation of the cation channel long transient receptor potential channel 2 (LTRPC2) by hydrogen peroxide. A splice variant reveals a mode of activation independent of ADP-ribose. *Journal of Biological Chemistry*, *277*(26), 23150–23156. https://doi.org/10.1074/jbc.M112096200.

Xicoy, H., Wieringa, B., & Martens, G. J. (2017). The SH-SY5Y cell line in Parkinson's disease research: A systematic review. *Molecular Neurodegeneration*, *12*(1), 10. https://doi.org/10.1186/s13024-017-0149-0.

Xu, C., Li, P. P., Cooke, R. G., Parikh, S. V., Wang, K., Kennedy, J. L., et al. (2009). TRPM2 variants and bipolar disorder risk: Confirmation in a family-based association study. *Bipolar Disorders*, *11*(1), 1–10. https://doi.org/10.1111/j.1399-5618.2008.00655.x.

Yang, Q., Huang, Q., Hu, Z., & Tang, X. (2019). Potential neuroprotective teatment of stroke: Targeting excitotoxicity, oxidative stress, and inflammation. *Frontiers in Neuroscience*, *13*, 1036. https://doi.org/10.3389/fnins.2019.01036.

Ye, M., Yang, W., Ainscough, J. F., Hu, X. P., Li, X., Sedo, A., et al. (2014). TRPM2 channel deficiency prevents delayed cytosolic Zn^{2+} accumulation and CA1 pyramidal neuronal death after transient global ischemia. *Cell Death & Disease*, *5*(11), e1541. https://doi.org/10.1038/cddis.2014.494.

Zhang, W., Chu, X., Tong, Q., Cheung, J. Y., Conrad, K., Masker, K., et al. (2003). A novel TRPM2 isoform inhibits calcium influx and susceptibility to cell death. *Journal of Biological Chemistry*, *278*(18), 16222–16229. https://doi.org/10.1074/jbc.M300298200.

Zheng, Q., Zhu, T., Hu, H., Zhao, Y., Ying, Y., Luo, X., et al. (2020). TRPM2 ion channel is involved in the aggravation of cognitive impairment and downregulation of epilepsy threshold in pentylenetetrazole-induced kindling mice. *Brain Research Bulletin*, *155*, 48–60. https://doi.org/10.1016/j.brainresbull.2019.11.018.

CHAPTER FOUR

Intrinsic and extrinsic pathways of apoptosis: Role in cancer development and prognosis

Dharambir Kashyap[a], Vivek Kumar Garg[b], and Neelam Goel[c,*]

[a]Department of Histopathology, Postgraduation Institute of Medical Education and Research (PGIMER), Chandigarh, India

[b]Punjab Biotechnology Incubator, Mohali, Punjab, India

[c]Department of Information Technology, UIET, Panjab University, Chandigarh, India

*Corresponding author: e-mail address: neelam.goyal@85gmail.com

Contents

Advances in Protein Chemistry and Structural Biology, Volume 125
ISSN 1876-1623
https://doi.org/10.1016/bs.apcsb.2021.01.003

Abstract

Apoptosis, also named programmed cell death, is a fundament process required for morphogenetic homeostasis during early development and in pathophysiological conditions. It is come into existence in 1972 by work of Kerr, Wyllie and Currie and later on investigated during the research on development of the *C. elegans*. Trigger by several stimuli, apoptosis is necessary during the embryonic development and aging as homeostatic mechanism to control the cell population and also play a key role as defense mechanism against the immune responses and elimination of damaged cells. Cancer, a genetic disease, is a growing burden on the health and economy of both developing and developed countries. Every year there is tremendously increasing in the number of new cancer cases and mortality rate. Although, there is a significant improvement have been made in biotechnological and bioinformatic fields however, the therapeutic advantages and cancer etiology is still under explored. Several studies determined the deregulation of different apoptotic components during the cancer development and progression. Apoptosis relies on activation of distinct signaling pathways that are often deregulated in cancer. Thus, exploring the single or more than one apoptotic component underlying their expression in carcinogenesis could help to track the disease progression. Current book chapter will provide the several evidences supporting the use of different apoptotic components as prognosis and prediction markers in various human cancer types.

1. Introduction

Apoptosis, also named programmed cell death, is a fundament process required for morphogenetic homeostasis during early development and in pathophysiological conditions at any stage during individual's life(Aggarwal, Kashyap, Sak, et al., 2019; Elmore, 2007; Kashyap, Garg, Tuli, et al., 2019; Kashyap, Sharma, Sak, et al., 2018). Apoptosis firstly come into existence in 1972 by work of Kerr, Wyllie and Currie and further investigated and described during the research on development of the *C. elegans* (Fadeel & Orrenius, 2005). In this organism ~1000 cells are generated for the formation of adult and 131 of these cells later on undergone programmed cell death for morphogenesis (Elmore, 2007). In 2002 year, the Nobel Prize in medicine was awarded to the Sydney Brenner, Horvitz and John E Suston, for contribution in field of apoptosis research (Bruce, 2002).

Since that apoptosis is recognized as remarkable mode of cell death involve genetically determined elimination of unwanted cells in the body. However, other forms of programmed cell death are also described (Debnath, Baehrecke, & Kroemer, 2005; Garg, Kashyap, & Tuli, 2018; Kashyap, Kumar, Sharma, et al., 2017; Kashyap, Sharma, Garg, et al., 2016; Kashyap, Tuli, Garg, et al., 2017; Sperandio, De Belle, & Bredesen, 2000).

Normally, apoptosis is necessary during the embryonic development and aging as homeostatic mechanism to control the cell population (Garg et al., 2018; Goel & Garg, 2018; Goel, Karir, & Garg, 2017; Kashyap, Kumar, et al., 2017; Kashyap, Mittal, Sak, et al., 2016; Kashyap, Mondal, Tuli, et al., 2016; Kashyap, Sharma, Garg, et al., 2016; Kashyap, Tuli, et al., 2017; Renehan, Booth, & Potten, 2001; Tuli, Kashyap, Bedi, et al., 2015). In addition, apoptosis is also playing a key role as defense mechanism against the immune responses and elimination of damaged cells after disease course (Favaloro, Allocati, Graziano, et al., 2012; Kashyap, Sharma, Tuli, et al., 2018a; Kashyap, Tuli, Yerer, et al., 2019; Yadav, Jaswal, Sharma, et al., 2018). There are various conditions or stimuli that can trigger the apoptosis in different cells (Vermeulen, Berneman, & Van Bockstaele, 2003). For example, radiation and chemotherapy for cancer treatment damage the DNA and induce the apoptosis in p53 dependent manner (Williams & Schumacher, 2016). Similarly, certain hormones such as corticosteroids can also leads to apoptosis in thymocytes (Herold, McPherson, & Reichardt, 2006).

Apoptosis need to be distinguished from another very similar process called necrosis (Kashyap & Kaur, 2020; Kashyap, Sharma, S Tuli, et al., 2016; Mitra, Bal, Kashyap, et al., 2020; Tuli, Kashyap, Sharma, & Sandhu, 2015). Both of these processes are independent to each other, works sequentially, can occur simultaneously and leads to the cell death (Elmore, 2007; Kashyap, Sharma, Tuli, et al., 2018b; Tuli, Sharma, Sandhu, & Kashyap, 2013). It is actually the types of stimuli and degree of intensity of stimuli which determined whether cell has to die by apoptosis or necrosis (Fink & Cookson, 2005). Stimuli such as hypoxia, radiation, heat, and cytotoxic chemicals at low or control dose can induce the apoptosis. And all of these stimuli at high/intense dose activate the necrosis process (Elmore, 2007).

Cancer, a genetic disease, is a growing burden on the health and economy of both developing and developed countries. Every year there is tremendously increase in the number of new cancer cases and mortality rate due to cancer (Ma & Yu, 2006). Although, there is a significant improvement has been made in biotechnological and bioinformatics fields however, the therapeutic advantages and cancer biology under explore (Bayat, 2002). It is really creating the need for the knowledge of specific biomarker that can be used to monitor the cancer progress (Goossens, Nakagawa, Sun, & Hoshida, 2015).

After cancer initiation, apoptosis is a first process that counters the tumor cells proliferation (Elmore, 2007; Kashyap, Tuli, Garg, et al. 2018; Sharma, Kashyap, Sak, et al., 2018). Several studies determined the deregulation of different apoptotic components during the cancer development and progression (Pistritto, Trisciuoglio, Ceci, et al., 2016). Apoptosis relies on activation of distinct signaling pathways that are often deregulated in cancer (Kashyap, Tuli, & Sharma, 2016; Plati, Octavian, & Roya, 2011). Thus, exploring the single or more than one apoptotic component underlying their expression in carcinogenesis could to help to track the disease progression (Gerl & Vaux, 2005). Current book chapter will provide the several evidences for use of different apoptotic components as prognosis and prediction markers in various human cancer types.

2. Components of extrinsic/death receptor mediated pathway and their role in cancer prognosis

2.1 Overview of extrinsic/death receptor mediated pathway

The extrinsic pathway of apoptosis is mediated by external stimuli through death receptors including Fas receptors (Fas receptor), DR4/DR5, tumor necrosis factor receptors (TNF-R), and TNF-related apoptosis-inducing ligand receptors (TRAIL-R) present at the surface of different cells (Zhou, Jiang, Liu, et al., 2017). At cell surface, death receptor interacts with their specific ligands, recruit the adaptor proteins Fas-associated protein with death domain (FADD), tumor necrosis factor receptor type 1 associated death domain protein (TRADD) and then downstream factors caspase-8 (Guicciardi & Gores, 2009). This recruitment eventually leads in cell apoptosis activation. To date, the best-characterized ligands and corresponding death receptors are CD95/FasR, TNF-α/TNF-R1, Apo2L/DR4 and Apo2L/DR5. The TNF receptor family is characterized by the presence of a cysteine-rich extracellular domains and a cytoplasmic domain (death domain) that transmits the death signal from outside to inside the cell leading to apoptosis (Papenfuss, Cordier, & Walczak, 2008). Basically, ligand binding caused recruitment and dimerization of the death receptors which

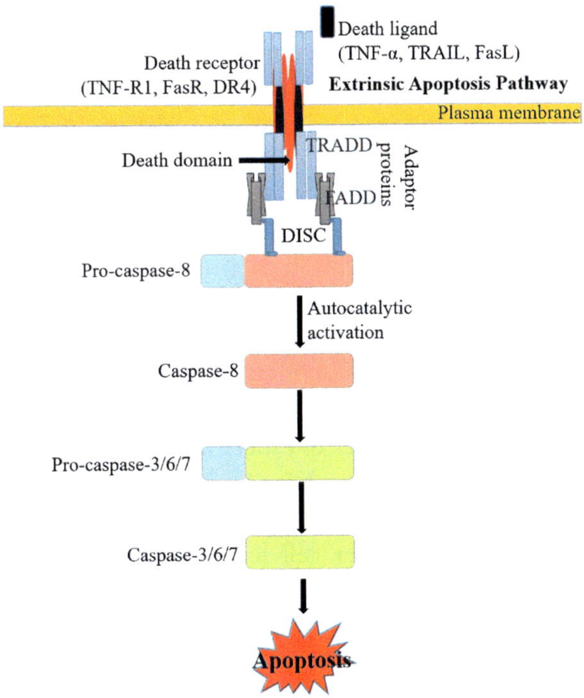

Fig. 1 Diagram representing the extrinsic pathways of apoptosis. Death ligands (TNF-α, TRAIL and FasL) recognize and binds to the death receptors (THN-R1, FasR, and DR4) causing the recruitment of adaptor proteins (FADD, TRADD) and form the DICS complex towards the cytoplasmic side. Formed DICS complex induces the autocatalysis of inactive pro-caspase-8 into active caspase-8. Subsequently, activated caspase-8 cleaved and activate the caspase-3. Activated caspase-3 then induce the apoptosis of target cell.

results in recruitment of death domains and activation of caspase-8 and caspase-10 (Wang, Chun, Wong, et al., 2001). Eventually, activated caspase-8 and caspase-10 initiate apoptosis by cleaving and activating executioner caspase-3, caspase-6 and caspase-7 (Brentnall, Rodriguez-Menocal, De Guevara, et al., 2013) (Fig. 1).

2.2 FasL/FasR (APO-1/CD95)

Fas receptor (FasR) also called CD95 or tumor necrosis factor receptor superfamily member 6 (TNFRSF6), is a member of TNF-R death receptor subfamily localized on the surface of various cells and triggers a signal transduction pathway leading to apoptosis. Mechanistically, FasL binds to FasR, form trimers, recruits FADD toward the cytoplasmic side and initiate an apoptotic cascade (Guicciardi & Gores, 2009). FasL (CD95L) is expressed on the surface of specific immune cells for example T-cells. FasL is a 40 kDa protein encoded by gene located on chromosome 1, composed of 280 aa (amino acid), comprises of an intracellular proline-rich domain, a single transmembrane domain, and an extracellular domain required for oligomerization (Yamada, Arakaki, Saito, et al., 2017). FasR is 48 kDa protein encode by the gene located on chromosome 10, composed of 319 aa and consists of three domains, an extracellular domain (157 aa), a transmembrane domain (17 aa), and a cytoplasmic domain (145 aa) (Guicciardi & Gores, 2009). Expression of FasL on the immune cells protects them from the autoreactive lymphocytes attack. Signal transduction mediated by FasL/FasR regulates several physiological and pathological processes including aging carcinogenesis (Elmore, 2007). Dysregulated expression of FasL/FasR was observed in several patients with different types of cancer including lung, breast, ovary, prostate, colon, and gastric and have been correlated with the disease prognosis (Bebenek, Duś, & Koźlak, 2013). For instance, a study found expression of FasL in pancreatic cancer cell lines and surgical specimens and correlated it with evading immune surveillance and developed resistant to apoptosis (Bernstorff, Glickman, Odze, et al., 2002). In particular the FasR, was detectable in a remarkable proportion of urinary bladder cancer surgical specimens and was associated with clinicopathological parameter (Giannopoulou, Nakopoulou, Zervas, et al., 2002). Another study investigated FasL and FasR expression in colorectal carcinoma found that FasR was expressed in normal glandular epithelium whereas FasL was strongly expressed in cancer tissues. Since expression differs distinctly between normal intestinal epithelium and cancer cells, it suggested as potential therapeutic targets for colorectal carcinoma (Pryczynicz, Guzińska-Ustymowicz, & Kemona, 2010). Similarly, decreased expression of FasR was also noted in formalin-fixed paraffin-embedded (FFPE) sections from epithelial ovarian cancer compared with benign ovarian epithelial lesions. Further correlation showed that low FasR expression was associated with decreased sensitivity to Fas-mediated apoptosis (Reed, Hakam, Nicosia, & Coppola, 2005). Using immunohistochemistry (IHC) and in fluorescent situ hybridization (FISH) techniques, FasL was detected in gastric tumors at protein and mRNA respectively. Study described that FasL expression might be used as prognosis indicator in stomach cancer (Bennett, O'Connell, O'Sullivan, et al., 1999). Further, the overexpression of Fas decoy receptors (DcR3) found in human pancreatic cancer resistant cell lines (Capan-1, AsPC-1, BxPC-3, and PANC-1) as well as in surgical specimens. Subsequently, an abrogate FasR function was noted at different levels of Fas signaling cascade, to be resulted in resistance to Fas-mediated apoptosis and may promoted the progression of this malignancy (Elnemr, Ohta, Yachie, et al., 2001). Similarly, in-vivo and in-vitro model of colon cancer using SW620 and colonic resected tumor respectively showed that expressed FasL might kill lymphoid cells by Fas-mediated apoptosis called "Fas counterattack" and thus, suggested that FasL is a common mediator of immune privilege in colon cancer (O'Connell, Bennett, O'Sullivan, et al., 1998). Moreover, IHC was used to analyze the FasR and its ligand FasL in 164 non-small cell lung carcinomas (NSCLC). Patients with Fas-positive tumors exhibited significantly longer survival times and had greater sensitivity to doxorubicin than patients with Fas-negative. Further, multivariate analysis indicated that lymph node status and Fas expression might be considered as prognostic factors in NSCLC (Koomägi & Volm, 1999). In another study, breast cancer surgical samples collected during cohort analysis had higher expression for FasR in benign lesions. The expression of FasL was also positively up-regulated in invasive carcinomas and significant associated with the presence of lymph nodes metastasis and larger tumor size (Mottolese, Buglioni, Bracalenti, et al., 2000). During IHC examinations of Fas-related molecules in urothelium surgical specimens, a low expression of FasR found associated with a higher pathological grade ($P < 0.0001$), a more advanced tumor stage ($P = 0.023$) and poorer prognosis ($P = 0.010$) (Yamana, Bilim, Hara, et al., 2005). In an experiment using bladder cancer cell lines RT4, RT112, T24, and SU of different tumor grade, a grade specific expression of FasR and FasL was observed. Real-Time PCR (RT-PCR) analyses found FasR similar expression levels in all bladder carcinoma cell lines whereas, FasL was highly expressed only in T24 (G3) and SUP (G4) cell lines, and thus, proposed as biomarker with prognostic importance (Perabo, Mattes, Wirger, et al., 2001). Similarly, a high expression of FasL, and reduced expression FasR was observed in stage D colorectal cancer which highlighting the therapeutic advantage of FasL in carcinogenesis (Kykalos, Mathaiou, Karayiannakis, et al., 2012). In addition, in esophageal carcinomas, a strong FasL expression was correlated with ($P = 0.0011$) decrease in tumor nest CD8+ cells and FasR expression was correlated with depth of invasion ($P = 0.0308$), histological differentiation ($P = 0.0347$), and lymph node metastasis ($P = 0.0129$) thus, study proposing these as an independent favorable prognostic factor for survival risk determination (Ohno, Tachibana, Shibakita, et al., 2000). In addition, body fluid majorly serum also showed presence of secreted FasR (sFasR). The measurement of sFas levels in serum of colon patients showed its correlation with cancer progression.

Study found that sFas levels were higher in the patients with lymph node positive status and liver and lung metastases. Therefore, sFasR could be used as a marker for predicting clinical course and outcome of colon cancer (Kushlinskii, Britvin, Abbasova, et al., 2001). Moreover, invasive breast cancer patients evaluated for sFasR showed 794.2 ± 183.0 pg/mL in serum than normal control 582.1 ± 62.8 pg/mL. Further, the levels of sFasR were significantly higher in the older patients (age ≥ 50) and in those with a more advanced TNM stage (Sheen-Chen, Chen, Eng, & Chen, 2003). Similar observations were also found in the bladder cancer where patients had three-fold higher mean sFasR in the serum than the normal control with less disease-specific survival rate after 5-year follow-up (Mizutanil, Yoshida, & Bonavida, 1998). In another study, breast carcinomas and fibroadenomas surgical species used for expression analysis of FasL also reflected significant observations. All breast carcinomas samples had higher FasL mRNA expression with positive correlation with histological grading ($P < 0.0001$). Further, a ratio of FasL/FasR mRNA transcripts >1 was found to be significantly associated with decreased patient's disease-free survival ($P < 0.03$) and increased mortality ($P = 0.19$) (Reimer, Herrnring, Koczan, et al., 2000). Epithelial-mesenchymal transition (EMT) which is an important hallmark in tumor progression and promote metastasis was found to be influenced by the FasL signaling. The FasL signaling through ERK1/2 pathway downregulated epithelial markers and upregulated mesenchymal markers in gastrointestinal (GI) cancer cells and underlying its contribution in tumorigenesis (Zheng, Cai, Wang, et al., 2013). Several studies also explored the FasR gene mutations in different cancer types. For example, human NSCLCs found to have four different FasR gene mutations with high mutational frequency in metastasis lesions samples (23%) than in the primary lesions of the NSCLCs without metastasis (5.4%) (Min, Hong, Sug, et al., 2002). A pilot study also associated different FasR SNPs such as rs1800682, FasL rs5030772, and FasL rs763110 with the high risk of breast cancer development in the examined population (Hashemi, Fazaeli, Ghavami, et al., 2013). Identification of "AG" or "GG" genotypes in FasR gene by restriction fragment-length polymorphism (RFLP) highlighted that presence of "G" allele may reduce sFasR levels and prevented the apoptotic inhibition (Lima, Morais, Lobo, et al., 2008). Another study examined FasR (1377G/A and 670A/G) and FasL (844T/C and 7896G/C) polymorphisms in breast cancer and proposed that moderately increased breast cancer risk was associated with FasR 1377AG and 1377AA genotypes compared with 1377GG genotype. On the other hand, decreased breast cancer risk was associated with FasL 844CT and 844TT genotypes compared with the 844CC genotype (Zhang, Sun, Xue, et al., 2007). Patients with cervical squamous cell carcinoma with advanced disease stage also showed overproduction of FasL and represent an attempt to destroy the host's lymphocytic reaction (Lerma, Romero, Gallardo, et al., 2008). Further reported that primary pancreatic ductal cancers and hepatic metastases samples expressed FasL and killed lymphoid cells by Fas-mediated apoptosis. The IHC results showed that tumor infiltrating lymphoid cells were frequently undergone apoptosis in FasL positive tumor and had significantly shorter survival times than FasL negative tumors (Ohta, Elnemr, Kitagawa, et al., 2004). Using three different techniques i.e. RT-PCR, western blotting, and IHC, FasL expression was analyzed in normal breast tissues and breast cancer tissues (ductal carcinoma in situ; invasive ductal, lobular, mucinous and medullary carcinomas). The result suggested that increased expression of FasL may confer an advantage on breast cancer cells, possibly by eliminating tumor infiltrating immune cells or by facilitating tissue destruction during invasion (Müllauer, Mosberger, Grusch, et al., 2000). Thus, it is cleared that expression of FasR/FasL had significant therapeutic association with the different cancer types and accurate knowledge of their expression can be utilized to monitored the disease out come and prognosis.

2.3 TNF-α/TNFR1

Tumor necrosis factor-α (TNF-α), a homotrimer, 17 kDa protein consisting of 157 aa and encoded by gene mapped on chromosome 6 (Kany, Vollrath, & Relja, 2019). TNF-α is produced from the activated macrophages, T-lymphocytes, and natural killer (NK) cells (Kany et al., 2019). TNF-α binds to the extracellular domain of TNF-R1, which then binds by an adaptor protein TRADD (Macewan, 2002). The binding of the ligand TNF-α to the TNF-R1 initiates the pro-inflammatory and pro-apoptotic signaling cascades and integrate the nuclear factor of kappa B (NF-kappaβ) family or mitogen-activated protein (MAP) kinase family signaling pathways (Macewan, 2002). Similarly, TNF-R1 also called tumor necrosis factor receptor superfamily member 1A (TNFRSF1A) or CD20a or death receptor (DR), is a 455 aa protein encoded by gene located on chromosome 12 and contains death domain (DD) at its cytoplasmic side (Morgan, Thorburn, Pandolfi, & Thorburn, 2002). On the other hand, TNF-R2 has short aa motif near the cytoplasmic terminus which enables recruitment of the adapter protein TRAF2 and TRAF2 associated proteins such as TRAF1 and cellular inhibitor of apoptosis protein 1 (cIAP1) and cIAP2 (Wajant & Siegmund, 2019). Therefore, TRAF2 do not execute the apoptosis but can induce the NF-κB family or MAP kinase family signaling pathways (Wajant & Siegmund, 2019). Expression of TNF-α/TNF-R1 have been determined in several human cancer types and correlated with the disease prognosis. For example, expression levels of TNF-α were found to be detectable in atypical hyperplasia or hyperplasia, compared to the normal breast tissues and hence this study proposed TNF-α a promising candidate to treat TNF-α positive tumors (Yu, Zhou, Niu, et al., 2013). In an in-vivo study, it was reported that chronic production of TNF-α in the tumor microenvironment (TME) enhanced myeloid cell recruitment in an IL-17 dependent manner and further contributed to the tumor-promoting action of this proinflammatory cytokine (Charles, Kulbe, Soper, et al., 2009). Mechanistically, this action is associated with EMT through NF-kappaβ mediated upregulation of Twist1 and CD44. Therefore, targeting TNF-α may offer an effective therapeutic strategy for cancer treatment (Li et al., 2012; Mikami, Mizuno, Kosaka, et al., 2015). According to an observation an increased level of TNF-α and its soluble receptors (TNF-R1, TNF-R2) were identified in endometrial cancer and related with poor prognosis (Dossus, Becker, Rinaldi, et al., 2011). Further, two variants of TNF-α gene, 308G/A (rs1800629) and 238G/A (rs361525) were identified and associated with risk of cervical cancer development (Liu, Yang, Chen, et al., 2012). Another study evaluated the significance of circulating TNF-α levels and its soluble receptors (sTNF-Rs) in the plasma of ovarian cancer patients. High plasma levels of TNF-α and sTNF Rs in epithelial ovarian cancer were correlated with high tumor stage, worst cancer progression and reduced mean survival time (Dobrzycka, Terlikowski, Kowalczuk, & Kinalski, 2009). Thus, TNF-α/TNFR1 has role in carcinogenesis and their expression measurement might be utilized to calculate the patient's prognosis.

2.4 Apo3L/DR3

Death receptor 3 (DR3) also known as TNFRSF25/Apo3 TRAMP (TNFR-related apoptosis- mediating protein)/APO-3/WSP-1/LARD and (lymphocyte-associated receptor of death)/TR3 (TNF receptor-like-3), is a TNFR superfamily member (Wang, 2012). It is 53.5 kDa protein encoded by the gene located on chromosome 1 and produced by lymphocytes and contains cysteine-rich motifs as well as a death domain (Kiraz, Adan, Kartal Yandim, & Baran, 2016). Death receptor interact with Apo3L activates the apoptosis and transcription NF-kappaβ (Heyninck & Beyaert, 2001). The Apo3L, a 249 aa protein encoded by the gene present on the chromosome 17, is a type II transmembrane protein. Role of Apo3L/DR3 in prognosis of various human cancer types have been observed in various studies (Méndez-Barbero, Gutiérrez-Muñoz, Blázquez-Serra, et al., 2020). For example, a decreased expression of DR3 in clinical cohort of breast cancer patients and breast cancer cell lines (MCF-7 and MDA-MB-231) was noted compared to normal breast tissues which were associated with a poorer prognosis as well as a significantly shorter long-term survival (Ge, Sanders, Ye, et al., 2013). Treatment with anti-mitotic chemotherapeutic agents such as taxol and vinblastine, arrest the cell cycle progression through the DR3/TL1A-mediated pathway therefore, suggested DR3 a biomarker for predicting clinical response to antimitotic therapeutics (Qi, Wang, Shen, et al., 2018). Further, high molecular weight isoforms of DR3 are found expressed in samples of primary human colon carcinoma. Findings identified novel signaling and functional roles of DR3 activated in response to *E*-selectin and highlighted the potential link between DR3 and metastasis (Gout, Morin, Houle, & Huot, 2006). Thus, data provided the evidences for Apo3L/DR3 as promising markers in the tumorigenesis.

2.5 Apo2L/DR4

Death receptor (DR4) also known as TRAILR 1/Apo-2/TNFRSF10A, is a N-glycosylated protein, composed of 468 aa and encoded by the gene located on chromosome 8 (Park, 2011). DR4 induces apoptosis in DD-dependent association with the proximal adaptor FADD after binding with ligand Apo2 ligand (Apop2L) (Park, Lee, Kim, et al., 2012). Apo2L also known as TRAIL, is a type-II transmembrane protein, composed of 281 aa and has highest homology to CD95L (Wajant, 2019). The monomer Apo2L/TRAIL polypeptide is of 32.5 kDa and its mature, fully glycosylated form, has a molecular mass of around 41 kDa (Naval, de Miguel, Gallego-Lleyda, et al., 2019). A dysregulated expression of Apo2L/DR4 have been measured in the different human caner types that might be used for prognostic assessment (Mehta, Shelling, Muthukaruppan, et al., 2010). A study confirmed that positive staining of either DR4 or DR5 receptors were associated with increased risk of death (odds ratio, 5.76; 95% confidence interval, 1.04–31.93; $P=0.045$) in stage III NSCLC. Therefore, might be the targeted for the use of rhTRAIL as a new adjunct in the treatment of stage III NSCLC (Spierings, De Vries, Timens, et al., 2003). Another study observed the cytoplasmic expression of DR4 and DR5 in bladder cancer patients and showed that higher expression of both the receptors was present in patients with longer postoperative recurrence-free rate. Further, multivariate analysis revealed that the expression of DR4 ($P<0.001$), DR5 ($P<0.001$) were acted as independent prognostic indicators of bladder cancer (Li, Jin, Li, et al., 2012). The genetic profiling of DR4 gene revelated homozygous C626G and G422A alterations in head and neck squamous and gastric adenocarcinoma. This allelic alteration when further evaluated shows amino acid changes near the ligand-binding domain of DR4 that affect TRAIL binding to DR4 and contributed to the genetic predisposition to lung cancer and head and neck cancer (Fisher, Virmani, Wu, et al., 2001). Another study provided the molecular epidemiological evidence for DR4 polymorphism particularly in exon 4 and promoted that G/G genotype of DR4 gene was associated with an overall decreased risk of bladder cancer in Caucasians (odds ratio (OR) = 0.58; 95% confidence interval (CI), 0.38–0.88) (Hazra, Chamberlain, Grossman, et al., 2003).

2.6 FADD/TRADD

Fas-associated protein with death domain (FADD) is a 23 kDa protein, encoded by the gene present on chromosome 11, and a main adaptor for transmitting apoptotic signals mediated through all known death receptors (He, Ren, Zheng, et al., 2016). Similarly, TRADD, a 34-kDa protein, encoded by the gene present on the chromosome 16, is an adaptor protein with two structurally distinct domains connected by a long linker peptide of 37 aa, participates in different signaling pathways, including apoptosis, necrosis, NF-kappaβ, and MAP kinase (LaCasse, Mahoney, Cheung, et al., 2008). It can also simultaneously bind to the death domain of FADD to recruit FADD. Similarly, TRADD contains an N-terminal region of unknown function and a C-terminal death domain that is 23% identical to the death domain of TNFR1 (Pobezinskaya & Liu, 2012). TRADD overexpression activates TNFR1-like signaling pathways for both apoptosis and activation of the transcription factor NF-kappaβ (Hsu, Huang, Shu, et al., 1996). Similarly, overexpression of FADD mimics FasR activation and leading to apoptosis. However, the mechanisms of signaling by TRADD and FADD are not identical. The induction of cell death by TRADD requires only their death domains, apoptosis induced by FADD overexpression occurs independently of its C-terminal death domain and require death effector domain at its N-terminus (Morgan et al., 2002).

In the several studies the expression of FADD/TRADD have been noted and corelated with risk of cancer initiation and out of the disease. For example, The Cancer Genomic Atlas (TCGA) data reveled that expression of FADD was associated with enhanced overall and disease-free survival in head and neck cancer (HNSCC) and established their clinical relevance (Reddy, Bhat, James, et al., 2016). Study using invasive ductal carcinoma of breast evaluated the association of FADD with clinicopathological parameters of the patients and concluded that, FADD expression was associated disease-free survival in breast cancer (Choi, Yun, Jabeen, et al., 2014). Another study using breast cancer with or without lymph node metastasis evaluated the copy number by TaqMan assays. Findings revealed that increase copy number and mRNA overexpression of FADD genes was observed in the breast tumors therefore may be used a prognostic biomarker (Li, Jin, et al., 2012). Similarly, study also detected gene amplification, expression, and mutation of FADD in NSCLC using FISH, single-strand conformation polymorphism (PCR-SSCP), and IHC techniques. Results showed that FADD gene mutation present in NSCLC and might play a crucial role in carcinogenesis of NSCLC. Further, gene amplification and protein level of FADD were closely correlates with cell apoptosis of NSCLC (Dobrzycka et al., 2009). Another study found that loss of FADD expression plays an important role in primary hepatocellular carcinogenesis (HCC) and restorage of FADD expression resulted in cell apoptosis in HCC (Sun, Zhao, Wang, et al., 2000). Further, in oral squamous cell carcinomas (OSCCs) samples FADD protein expression was measured by IHC. There was a strong correlation between protein expression and FADD gene copy number amplification. Both gene copy number and high protein expression were significantly associated with lymph node metastasis ($P < 0.001$), shortest disease-free survival (DFS; $P = 0.074$ and $P = 0.002$) and overall survival (OS; $P = 0.011$ and $P = 0.027$) and thus, could be considered as potential prognostic markers (Chien, De Cheng, Chuang, et al., 2016). Consequently, data presented here supporting the therapeutic advantages of FADD gene and proposing FADD as potential candidate in cancer risk management.

2.7 Caspase-8

Caspase-8 also known as FLICE (FADD-like IL-1β converting enzyme), is a monomer protein, member of the cysteine-aspartic acid protease family, encoded by the CASP8 gene that is located on chromosome 2 (McIlwain, Berger, & Mak, 2013). Once activated, caspase-8 acts as a crucial initiator in the death receptor-mediated apoptotic pathway. It involves formation of death-inducing signaling complex (DISC) comprises of death receptors, adaptor proteins and procaspase-8 leading to procaspase-8 dimerizes and activation. Eventually, activated caspase-8 cleave the downstream effector caspases which resulting in apoptosis (Danial & Korsmeyer, 2004; Hengartner, 2000). Activated caspase-8 can also cleave Bid to t-Bid which then translocate to the mitochondria and promotes cytochrome *c* release. The significance of genomic and genetic profile of caspase-8 have been explored in different human cancer types. For instance, data suggested that caspase-8 play a remarkable role in cancer development and sever as potential therapeutic or prognostic marker. A study investigated the expression of caspase-8 in rectal adenocarcinoma and the corresponding normal mucosa tissue, found caspase-8 upregulation in the cancer samples compared to the normal mucosa. These findings suggesting that caspase-8 deregulated expression was involved in the pathogenesis of rectal cancer (Xu, Zhou, Li, et al., 2008). Further, restorage of caspase-8 expression in neuroblastoma cells (NB) after treatment with 5-aza-2′-deoxycytidine increased the therapeutic sensitivity and induced apoptosis. Results reflecting that downregulation of caspase-8 could be an important mechanism of resistance to chemotherapy in neuroblastoma (Eggert, Grotzer, Zuzak, et al., 2000). Similar results were also reported in another study on ovarian carcinogenesis where downregulation of caspase-8 induced resistance to apoptosis (Borhani, Manoochehri, Saleh Gargari, et al., 2014). Moreover, downregulation of caspase-8 was significantly correlated with poorly differentiated HNSCC tumors compared to moderately and well differentiated HNSCC. Further, univariate analysis indicated that higher expression of caspase-8 had correlation with no metastasis, better disease-free survival and overall survival (Elrod, Fan, Muller, et al., 2010). A study by enrolled cases of gastric carcinoma diffuse-type, intestinal-type, and mixed-type gastric carcinomas characterized by Lauren's classification with stage I to IV and 293T, 293, and HT1080 cells, analyzed the genetic profile of caspase-8 gene. A spectrum of mutations was observed in caspase-8 gene (492_493delTG, 698delG, and 969_972delCTAT) which affects the pathogenesis of gastric cancers at the late stage of gastric carcinogenesis (Young, Jong, Su, et al., 2005). Further, genotyping analyses in Korean population showed that −678del−/IVS12-19A haplotype not −678del−/IVS12-19G haplotype of caspase-8 gene was associated with a significantly increased risk of lung cancer (adjusted OR = 0.58, 95% CI = 0.36–0.93,

$P = 0.023$, and $Pc = 0.046$), suggested that the CASP-8 gene may contribute to an inherited predisposition to squamous cell carcinoma of the lung (Son, Kang, Chae, et al., 2006).

3. Components of intrinsic/mitochondrial mediated pathway and their role in cancer prognosis

3.1 Overview of intrinsic/mitochondrial mediated pathway

In addition of external stimuli, apoptosis can also be regulated or activated by the mode of internal stimuli such as DNA damage and oxidative stress (Fulda, Gorman, Hori, & Samali, 2010). This pathway of apoptosis is called intrinsic/mitochondrial. Members of the Bcl family including Bax and Bcl-2, knows as pro- or antiapoptotic proteins respectively, present on mitochondrial membrane, are the key mediators for intrinsic apoptotic pathway (Kashyap, Tuli, Sak, et al., 2019). Their activity release cyt. *c* (cytochrome *c*) in the cytoplasm after mitochondrial membrane perturbation. Released cyt. *c* in the cytoplasm forms a complex called apoptosome with apoptotic protease activating factor 1 (APAF1) and procaspase 9. This complex cleaves procaspase-9 to caspase 9 which then cleaves and activates the caspases of executioner pathway such as caspase-3, -6, and -7 and induce cell apoptosis eventually (Wang & Youle, 2009) (Fig. 2).

3.2 Bcl-2 family

Based on their primary function Bcl-2 (B-cell leukemia/lymphoma-2) family is divided into three groups (1) anti-apoptotic (Bcl-2, Bcl-xl, Bcl—W, Mcl-1, and BFL-1/A1), (2) pro-apoptotic (Bax, Bak, and Bok) and (3) pro-apoptotic BH3-only (Bad, Bid, Bik, Bim, BMF, HRK, Noxa, and PUMA). Apoptosis activation is partly depending upon the balance between anti-apoptotic and pro-apoptotic proteins (Shamas-Din, Kale, Leber, & Andrews, 2013).

3.2.1 Anti-apoptotic

Family members of anti-apoptotic protein majorly Bcl-2 and Bcl-xl works against programmed cell death and inhibit it by the mechanism involve mitochondria (Tsujimoto, 1998). The Bcl-2 family proteins are found on outer mitochondrial member and suppress the apoptosis process by inhibiting the release of cyt. *c* from membrane permeabilization (Kale, Osterlund, & Andrews, 2018). Till date five different anti-apoptotic or

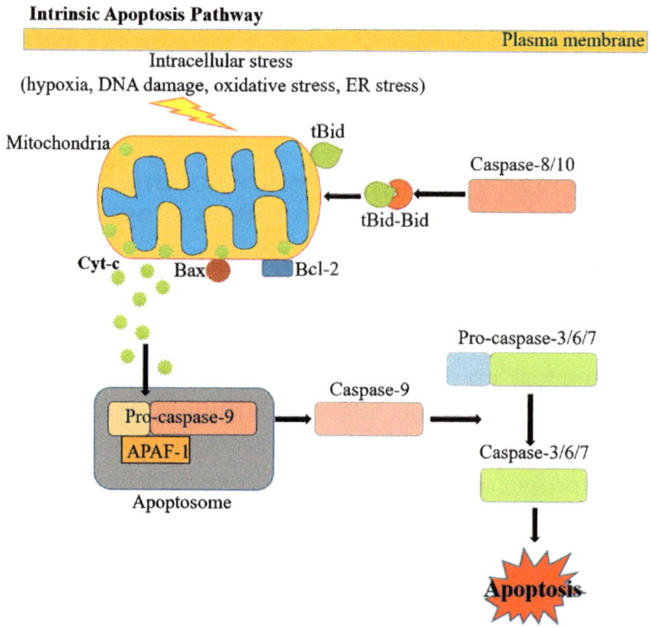

Fig. 2 Diagram representing the intrinsic pathway of apoptosis. Different cellular stress including hypoxia, DNA damage, and endoplasmic reticulum stress (ER) etc. signal for intrinsic pathway. Following the signal, mitochondrial releases Cyc. c into the cytoplasm which form apoptosome complex. Formed apoptosome further recruit and activate the executioner caspase-3.

pro-survival members have been discovered: Bcl-2, Bcl-xL, myeloid cell leukemia sequence 1 (Mcl-1), Bcl-w, and BFL-1/A1. All of these members share four Bcl-2 homology domains (BH1–BH4) (Shamas-Din et al., 2013). These pro-survival proteins expresses in different body tissues i.e. Bcl-2 in renal epithelial stem cells during embryogenesis melanocyte progenitors and mature B and T lymphocytes (Bouillet, Cory, Zhang, et al., 2001; Veis, Sorenson, Shutter, & Korsmeyer, 1993) Bcl-xL in erythroid progenitors and neuronal cells during embryogenesis (Motoyama, Wang, Roth, et al., 1995), Bcl-w in spermatogenesis (Print, Loveland, Gibson, et al., 1998) and Mcl-1 is essential for implantation during early development (Rinkenberger, Horning, Klocke, et al., 2000). A deregulated expression of anti-apoptotic proteins has been found in human cancers and there is significant evidence that deregulated expression of anti-apoptotic proteins is critical in tumor development and maintenance (Fernald & Kurokawa, 2013). For example, renal cell carcinomas patients with high Bcl-2 and Bcl-xl expression showed absence of apoptosis when compared with proteins those had low Bcl-2 and Bcl-xl expression (Fernald & Kurokawa, 2013). A study analyzed the expression of Bcl-2 anti-apoptotic family members including Bcl-2, Bcl-X, and Mcl-1 in prostate adenocarcinomas patients using IHC and observed their higher expression in high grade tumors and nodal metastases only and not in low grade primary tumors. Findings of this study suggesting that Bcl-2 gene family had elevated expression during progression of prostate cancers which might be the reason for therapy failure (Krajewska, Krajewski, Epstein, et al., 1996). In another study surgical specimens of localized prostate cancer and biopsy specimens of hormone refractory prostate cancer (HRPC) were used for immunostaining of Bcl-2, Bak, and Bax. Mutations screening in Bak and Bax genes was also done using SSCP, and confirmed by direct DNA sequencing. Multivariate analysis revealed that the level of Bcl-2 expression before treatment with taxane-based chemotherapy was an independent predictor for cause-specific survival ($P < 0.01$) and baseline prostate-specific antigen level was an independent predictor for progression-free survival ($P < 0.01$). Bax gene mutation was found in only one HRPC specimen (Yoshino, Shiina, Urakami, et al., 2006). The IHC staining for Bcl-2 was evaluated using tissue microarray and correlated with clinicopathologic parameters of the breast cancer patients. Positive expression Bcl-2 were observed in 38.4% breast cancer cases. Results suggest that breast cancer might be due to deregulated Bcl-2 protein expression (Won, Kim, Kim, et al., 2010). Another study confirmed that positive staining of Bcl-2, and Bcl-XL and negative staining of Bax is related with radioresistant laryngeal carcinoma development and therefore, could be a potential mechanism by which cancer cells avoid the destructive effects of radiotherapy (Nix, Cawkwell, Patmore, et al., 2005). Further, another multivariate analysis found a p53 dependent predictive role of Bcl-2 expression on 6-year relapse-free and overall survival in lymph node negative breast cancer (Silvestrini, Veneroni, Daidone, et al., 1994). Moreover, using male nude mice a study found that altered expression of Bcl-2 was related with tumor progression and had protective effect against apoptosis resulted in refractory to the growth-inhibiting effects of castration (Raffo, Perlman, Chen, et al., 1995). A study tried to identified the relationship between Bcl-2 and p53 protein expression in normal human ovaries and different types of human ovarian epithelial tumors. A significant inverse correlation was found between Bcl-2 and p53 expression in the malignant ovarian tumors. Statistically, a positive correlation between apoptosis and p53 expression, but negative correlation between apoptosis and Bcl-2 expression was determined (Chan, Cheung, Schorge, et al., 2000). Similarly, colorectal carcinomas patients showed downregulation of Bcl-2 than normal control. This report showed that loss of Bcl-2 expression is associated with high stage tumor and poor prognosis. Phenotype of Bcl-2 positive and p53 negative samples was very much similar to that of normal mucosa, confirming the role of Bcl-2 in tumor progression (Kaklamanis, Savage, Whitehouse, et al., 1998). In another study, on gastric cancer patients with positive Bcl-2 staining showed a significant negative correlation with depth of invasion, lymph node metastasis and better 5-year survival rate than those with Bcl-2 negative status and described Bcl-2 a prognostic factor in gastric cancer (Lee, Lee, Yang, et al., 2003). Moreover, during an investigation on NSCLC the Bcl-2 expression was significantly lower in patients who developed metastasis and survival probability was higher in patients with high Bcl-2 positivity. Hence, Bcl-2 proteins and their expression may have prognostic importance (Fontaninil, Vignatil, Bigini, et al., 1995). Another study found that imbalance between anti-apoptosis molecules such as Bcl-2, Bcl-XL, and Mcl-1 and proapoptotic molecules such as Bax and Bcl-Xs predominantly might be involved in the carcinogenesis of pancreas adenocarcinomas (Miyamoto, Hosotani, Wada, et al., 1999). Similarly, in a study using IHC on gastric carcinomas, Bcl-2 expression positivity was observed in well and moderately differentiated tumors than in poorly differentiated tumors and proposing Bcl-2 a prognostic marker (Müller, Schneiders, Hommel, & Gabbert, 1998). Further, a multivariate analysis of squamous cell lung carcinomas patients with Bcl-2 negative and p53 positive staining showed negative Bcl-2 status as an independent prognostic factor for shorter survival (Apolinario, Van Der Valk, De Jong, et al., 2017). Likewise, Bcl-2 was found to be independent prognostic indicators of survival in ovarian cancer where patients with high expression of Bcl-2 had improved survival (Herod, Eliopoulos, Warwick, et al., 1996). Moreover, multivariate logistic regression analysis found that Bcl-2 strongly associated with histologic dedifferentiation and increasing N stage in the patients with neck cancer. In univariate analyses Bcl-2 overexpressed patients had a lower locoregional relapse rate and had improved survival compared to patients with under expression of Bcl-2 (Wilson, Saunders, Dische, et al., 2001). Similarly, in colorectal tumors with high Bcl-2 expression had longer overall survival and therefore, described Bcl-2 an independent prognostic factor for out in colorectal cancer (Manne, Myers, Moron, et al., 1997). Further, another study found that the loss of Bcl-2 expression during the tumor course was correlated with high tumor stage and increased chance of relapses (Poincloux, Durando, Seitz, et al., 2009). Univariate analysis of positive Bcl-2 immunostaining proved it as prognostic factor in squamous cell carcinomas, adenocarcinomas, and large cell lung carcinomas (Kaiser, Schilli, Haag, et al., 1996). Western blotting analysis of PANC-1/Rad and AsPC-1/Rad cells showed that overexpression of Bcl-xl had related to abnormal apoptosis induction and served as important factors in acquired radio-resistance (Lee, Hosotani, Wada, et al., 1999). Similarly, high Bcl-2 expression was associated with improved survival in node-positive estrogen receptor (ER) and progesterone receptor (PR) positive breast cancer and might be benefited from Bcl-2 targeting agents in addition to anti-hormonal therapy (Nadler, Camp, Giltnane, et al., 2008). Moreover, study analyzed Bcl-2 expression in breast infiltrating ductal carcinoma and found that Bcl-2 was associated with ER/PR positive cases and tumor diploid status (Linjawi, Kontogiannea, Halwani, et al., 2004). Also, a retrospectively study found that Bcl-2 was widely expressed in ER/PR positive breast cancer, had inverse correlation histological grade and undifferentiated tumor (Sierra, Lloveras, Castellsague, et al., 1995). In addition, survival analysis showed that NSCLC patients with high Bcl-2 expression had a longer median survival compared to patients with low Bcl-2 expressers (22 vs 17.5 months, log rank $P=0.014$). Further, multivariate analysis revealed an independent lower risk for all patients with Bcl-2 expressing tumors (HR $=0.53$, 95% CI 0.37–0.75, $P=0.0003$) (Anagnostou, Lowery, Zolota, et al., 2010). IHC performed on FFPE rectal carcinomas tissues with UICC stages I–III, found that p53 positive/Bcl-2 negative subgroup had poorest prognosis in contrast to the patients with p53 negative/Bcl-2 positive. Hence, assessment of these makers may predict the recurrence and survival rate for rectal cancer (Schwandner, Schiedeck, Bruch, et al., 2000). Evidences provided here confirmed that Bcl-2 could be used as prognostic marker, but accurate measurement and validation of Bcl-2 expression is required before proposing it as prognostic factor.

3.2.2 Pro-apoptotic

Anti-apoptotic proteins family members maintain cell survival by regulating or maintaining the pro-apoptotic proteins levels (Carrington, Zhan, Brady, et al., 2017). Pro-apoptotic proteins are multi-BH domain Bcl-2 family members (Reed, 2006). Member of this family contain and share extensive structural similarity with pro-survival relatives (Boise, Gonzblez-garcia, Postema, et al., 1993). During apoptotic event Bax and Bak undergo homodimerization and oligomerization in outer mitochondrial membrane and leads to release cyt. *c* in the cytoplasm (Park, Lee, Kim, et al., 1994). The prognostic role of pro-apoptotic proteins has been described in several studies. For example, in colorectal cancer patients, mRNA expression of Bax and Bcl-2 was examined. Results suggested that tumor cell apoptosis is strongly induced by Bax and influenced by Bcl-2, which inhibits tumor cell apoptosis (Katsumata, Sumi, Tomioka, et al., 2003). A study noted that ovarian cancer tumor with high Bax protein expression achieved a complete response (CR) to first-line chemotherapy of paclitaxel and platinum compared to the tumor with low Bax protein expression and had increased median disease-free survival (Tai, Lee, Niloff, et al., 1998). Similarly, another study concluded that colorectal cancer patients with high Bax protein expression had a median survival of 53.6 months compared with 35.4 months for patients with low Bax expression. Further, low Bax expression was served as an independent negative prognostic marker in multivariate regression analysis (Sturm, Köhne, Wolff, et al., 1999). The IHC of primary colorectal adenocarcinoma patients showed stronger Bax expression than normal tissues which might be benefited from therapy (Jansson & Sun, 2002). Another study reported that pro-apoptotic marker Bax expression was associated with both progression-free and overall survival in ovarian cancer patients and hence represented a prognostic indicator (Schuyer, Van der Burg, Henzen-Logmans, et al., 2001). Moreover, the expression of Bax was found to be altered in NSCLC and SCLC tissues compared to histologically normal bronchial epithelium and could be used to reflect the pathogenesis of these two lung pathologies (Porebska, Wyrodek, Kosacka, et al., 2006). In esophageal squamous cell carcinoma those patients presented chemoradio-resistant, the loss of Bax expression has been reported to be associated with poor treatment response and thus, suggested Bax a good prognosis marker for chemoradio-sensitivity (Ikeguchi, Maeta, & Kaibara, 2001). In another study, the relative mRNA expression of Bax was determined using RT-qPCR technique in adenocarcinoma colorectal cancer. Results revealed a significant correlation between Bax expression and clinicopathologic parameters of colorectal carcinoma, including tumor size, tumor stage, and differentiation status which suggesting Bax as a potential molecular prognostic marker of colorectal cancer (Khodapasand, Jafarzadeh, Farrokhi, et al., 2015). Using fresh frozen oral squamous cell carcinomas samples, the expression of Bcl-2 and Bax was analyzed by IHC method. The report suggested that a lowered Bcl-2/Bax ratio was related with increased apoptosis and expression of Bax correlated with histological grading in oral squamous cell carcinoma (Loro, Vintermyr, Liavaag, et al., 1999).

The in-vitro study using MKN-45, LoVo, and PANC-1 cancer cell lines treated with cisplatin (10 μg/mL), reported an increased and decreased expression of Bax and Bcl-2 proteins and confirmed these proteins as prognosis therapeutic markers in cancer treatment (Ikeguchi & Nakamura, 2002). In another study, examined bladder cancer biopsy using immunostain for Bax and Bcl-2 to assess the Bax/Bcl-2 ratio for apoptosis index, a significant association with the CR rate is observed that confirming the prognostic index of these two apoptosis markers in bladder cancer (Matsumoto, Wada, Fukunaga, et al., 2004). Infiltrating ductal breast carcinoma with stages pT1-pT2 were analyzed for protein expression of Bcl-2 and Bax and their relation with prognosis, tumor recurrence, and overall survival through univariate and multivariate analysis revealed the prognostic significance of Bax and Bcl-2. (Sirvent, Aguilar, Olona, et al., 2004). Further, gastric cancer patients with metastatic or recurrent disease after treatment with FOLFOX (Folinic acid "FOL," Fluorouracil "F," and Oxaliplatin "OX") regimen showed high expression of Bax and its association with good median overall survival. Univariate and multivariate analysis reported that low expression of Bax was found to associated with poor overall survival (Jeong, Han, Kim, et al., 2011). In resected NSCLC tissues with non-squamous histology, stage II or III, low expression of Bax was significantly correlated with poor disease-free survival and hence, suggested Bax a most independent prognostic factor in etiology of NSCLC (Jeong, Lee, Han, et al., 2008). Data presented in several studies here confirming the prognostic importance of the Bax protein and proposing that it could be used as biomarker for the assessment of cancer patient's outcome.

3.3 Cytochrome *c*

Cytochrome *c* (cyc. *c*) is a small, water-soluble protein of molecular weight about 12 kDa and a key component of intrinsic pathway, releases from mitochondrial membrane in cytoplasm and initiate the executive apoptotic pathways (Wang & Youle, 2009). In the oxidized form created by ROS, cyt. *c* can readily form apoptosome assembly (Garrido, Galluzzi, Brunet, et al., 2006). In the cytoplasm cyt. *c* binds with Apaf-1 caused its oligomerizes which then binds with procaspase-9 and results in proteolytic activation of procaspase-9 and caspase-3 (Beere, Wolf, Cain, et al., 2000). Data accumulated so far on cyt. *c* and its relation with cancer, supporting Cyc. *c* as independent apoptotic prognostic marker in several cancer types. For examples, data noted that induction of apoptosis by NS398 drugs associated with inhibition of COX-2 expression, occurred through the cyt. *c* dependent pathway in esophageal cancer cell (Li, Wu, & Xu, 2001). Exposure of the HER2/neu-overexpressing breast cancer cells to apigenin resulted in apoptosis activation through cyt. *c* release (Chaudhary, Madhukrishna, Adhya, et al., 2016). Similarly, another study, reported that teniposide induced a dose-dependent increase in the expression of cyt. *c* and which played a role in the early cellular defense against anticancer drugs and induced the programmed cell death in cancer cell (Sánchez-Alcázar, Khodjakov, & Schneider, 2001).

3.4 Caspase-9

Caspase-9 enzyme is encoded by CASP9 gene located on the chromosome 1 (Li, Zhou, Zhao, et al., 2017). Procaspase-9 which acts as an initiator caspase for intrinsic apoptosis pathway exists as monomer and consist of three domains: N-terminal pro-domain, large subunit, and a small subunit pi (Qin, Srinivasula, Wu, et al., 1999). The N-terminal pro-domain called long pro-domain, contain the caspase activation domain (CARD) motif (Huber, Serrano, & Hardy, 2018). Pro-domain is linked to the catalytic domain by a linker loop. There are two proposed hypotheses for activation of caspase-9, First "induced conformation model," which involve Apaf-1 binding to the caspase-9 through respective CARD domains (Zou, Li, Liu, & Wang, 1999). Second, "induced proximity model," which involve dimerization caspase-9 supported by apoptosome (Hakem, Hakem, Duncan, et al., 1998; Hu, Benedict, Wu, et al., 1998). A deep understanding on the roles as well as regulators of caspase-9 is required to find more effective ways to conquer apoptosis-related diseases especially cancer (Kim, Srivastava, & Kim, 2015). There are many studies supporting the role of caspase-9 in cancer and its knowledge genetic and genomics may help in the patient's outcome. For instance, Genotype study done in pancreatic cancer patients and healthy controls found a 1263A/G polymorphism in caspase-9 gene and proposed that presence of "A" allele can increase the susceptibility of pancreatic cancer development (Theodoropoulos, Michalopoulos, Panoussopoulos, et al., 2010). Similarly, another study also found 1263A/G and 1263G/G genotype in the caspase-9 gene and confirmed that "GG" homozygous condition may confer a better prognosis in colorectal cancer patients (Theodoropoulos, Gazouli, Vaiopoulou, et al., 2011). Further, lung cancer tissue examined the polymorphisms 1263A > G, 905T > G, 712C > T, and 293del in caspase-9 promoter and found that patients with 263G/G, 712C genotype had significantly decreased risk of lung cancer (Park, Park, Jang, et al., 2006).

The susceptibility of caspase-91263G/G genotype gene was determined by a PCR-RFLP observed a significantly association with a reduced risk of gastric cancer and increased overall survival in gastric cancer patients and suggested to play a protective role in gastric cancer predisposition (Liamarkopoulos, Gazouli, Aravantinos, et al., 2011). Further, prognostic influence of caspase-9 was determined in colon carcinoma with stage II/III showed that decreased expression of caspase-9 was significantly associated with poor survival and may acted as an independent prognosticator in colon carcinoma (Sträter, Herter, Merkel, et al., 2010). In prostate cancer PC-3 cell line, higher caspase-9 expression was associated with higher apoptosis rate, decrease in migration and invasion, and low VEGF-C protein expression (Sun, Wang, Cui, et al., 2014). RT-PCR was used to analyze caspase-9 mRNA expression in stage II colorectal cancer tissue and corresponding normal mucosa found its downregulation in cancer tissues. Further, poorly differentiated cancer showed lower caspase-9 mRNA expression than cancer with greater differentiation and had worst prognosis as per Kaplan–Meier survival analysis (Shen, Wang, Li, et al., 2010). In the study which used transduced U-373MG cells harboring mutant p53 found that expression of Apaf-1 and caspase-9 in tumors may be an important determinant of the therapeutic effect of irradiation in cancer treatment (Shinoura, Sakurai, Asai, et al., 2001). Another study analyzed the expression of caspase-9 protein gastric adenocarcinomas by IHC approach detected positive staining in early and advanced gastric carcinomas and not in normal epithelial cells, showed that caspase-9 may sever as biomarker in the gastric cancer (Shen et al., 2010).

3.5 Apoptosis inducing factor (AIF)

The apoptosis inducing factor (AIF) also known as AIFM1 (apoptosis inducing factor mitochondrial 1), encoded by the gene located on X chromosome, is a 67 kDa precursor molecule having 613 aa long, and consisting of a N-terminal mitochondrial leading sequence, two nuclear leading sequences, and a NAD (nicotinamide adenine dinucleotide) and FAD (flavin adenine dinucleotide) binding motifs (Sevrioukova, 2011). It releases form mitochondria and induced the apoptosis in caspase independent manner (Wang & Youle, 2009). After imported into mitochondria N-terminal mitochondrial localization signal cleaved off resulting in a mature protein of 57 kDa (Otera, Ohsakaya, Nagaura, & Ishihara, 2005). Under normal physiological conditions, AIF function as a ROS scavenger and under apoptotic stimuli it released from mitochondria and translocate to the nucleus (Sinha, Das, Pal, & Sil, 2013). AIF also believed to mediate caspase-independent apoptosis confirmed by inhibition of caspase activation or caspase activity that does not abolish the proapoptotic action of AIF (Candé, Cohen, Daugas, et al., 2002). Different studies confirmed the association of AIF with cancer and their prognostic role in cancer patients. For example, in the transduction and short interference RNA using breast cancer cell line MDA-MB-231 and female nude mic, study demonstrated that up-regulation and down-regulation of AIF-1 promoted and inhibited the proliferation of cancer cells respectively (Liu, Tan, Chen, et al., 2008). Similarly, another study found that elevated expression of AIF in MDA-MB-231 and MCF-7 cells enhanced that migration of both the cell lines cells (Li, Feng, Jia, et al., 2012). Another study aimed to explored the role of AIF in gastric cancer analyzed the high expression of AIF in gastric adenocarcinomas in early and advanced stage but not in mucosal cells, suggest that AIF expression may play a role in the function of the gastric parietal cells (Lee, Jeong, Soung, et al., 2006). Study analyzed the expression and mutation of AIF gene in exons 10–15 using colorectal, gastric, breast, hepatocellular carcinomas, and leukemias demonstrated that all the tumor types had higher intensities of AIF immunostaining. Further, a two-base deletion mutation was also detected in intron 15 of AIF gene in the colorectal carcinomas. Finding of the study posing AIF as an independent prognostic factor in different cancer types (Jeong, Lee, Soung, et al., 2006). In addition, in non-small cell lung cancer, high AIF expression was associated with poor prognosis and drive the progression of lung cancer (Rao, Mondragón, Pranjic, et al., 2019). Thus, evidences confirming that accurate measured AIF expression and mutation screening could be utilized for assessment of cancer patients.

3.6 Endonuclease B

Endonuclease G (EndoG) acts as mitochondrial peptidase, a 28 kDa protein, encoded by the gene located on chromosome 9, and initially synthesized as inactive precursor of 33 kDa (Li, Luo, & Wang, 2001). It is a nuclear-encoded mitochondrial protein that activates the apoptosis by transporting from mitochondria to the nucleus (Wiehe, Gole, Chatre, et al., 2018). EndoG is releases from mitochondria and induce cell death in caspase-independent pathways via DNA fragmentation (Li, Luo, & Wang, 2001). Role of EndoG has been implicated in cancer, aging, and neurodegenerative diseases. Findings of study suggested that the presence of EndoG in non-invasive breast cancer cells determines their sensitivity to apoptosis and might be taken into consideration for developing the chemotherapeutic strategy for cancer treatment (Basnakian, Apostolov, Yin, et al., 2006). A mechanistic analysis demonstrated that CTX III-induced apoptotic cell death accompanied by up-regulation of both Bax and Endo G in human leukemia K562 cells (Yang, Chien, Lu, et al., 2006).

4. Components of execution pathway and their role in cancer prognosis

4.1 Overview of execution pathway

Execution phase is the final step of apoptosis induction characterized by cell shrinkage, chromosomal condensation, DNA fragmentation, and cell membrane blebbing (Elmore, 2007). Caspase-3 a main component of execution pathway cleaves the DNAase inhibitor, cytoskeleton proteins and caused the cellular fragmentation (Porter & Ja, 1999). Microtubule disassembly is main event in execution pathways occur through depolymerization of actomyosin filaments (Porter & Ja, 1999). This phase helps to disperse cellular fragments, keep the plasma membrane intact to the package the cellular toxic components and protect from inflammatory event (Yang, Jiang, Zhang, & Fan, 2015). Finally, phagocytosis removes the apoptotic bodies to avoiding any adverse reactions may happen through inflammation (Yang et al., 2015).

4.2 Caspase-3

Caspase-3 (CPP32, apopain, or YAMA) an endoproteases, member of asparate-specific cysteinyl proteases family, is a small protein consists of two subunits of 12 kDa and 17 kDa with 3 and 5 thiol functions respectively (Miller, 1997). The role of caspase-3 in apoptosis is to cleave and activate caspases-6, -7, and -9 in order to break down the apoptotic cells before removal (Miller, 1997). After this process, the caspase-3 protein is cleaved and broken down itself by caspase-8 and 10 (Miller, 1997). The sequential cleaving and activation of these proteins is crucial to the execution stage of programmed cellular death (Elmore, 2007). The expression of caspase-3 becomes deregulated in the carcinogenesis therefore may sever as good indicator of disease progression. For example, Study examined the expression of caspase-3 in normal and malignant human prostates (PC-3, DU-145, TSU-Pr1m and LNCaP) at the mRNA and the protein level and found that lack of caspase-3 immunoreactivity in cancer cells implicated in the loss of apoptotic in tumorigenesis may have prognostic significance in disease progression (Winter, Kramer, Borkowski, & Kyprianou, 2001). Study using MCF-7 cells demonstrated that overexpression of caspase-3 enhances the chemosensitivity against the acquired drug resistance (Friedrich, Wieder, Von Haefen, et al., 2001). Moreover, down-regulated expression of caspase-3 plays an important role in gastric carcinogenesis and considered an effective marker to reveal differentiation, growth, invasion and metastasis of gastric cancer by modulating apoptosis of infiltrating lymphocytes. (Zheng, Sun, Wei, et al., 2003). Further study using A2780 and COC1 cancer cells after treatment with cisplatin concluded that down-regulation of caspase-3 activity may be associated with cisplatin resistance in human ovarian cancer (Yang, Zheng, Xing, et al., 2004). In univariate and multivariate analysis, patients with high expression of cleaved caspase-3 had a significant shorter overall survival time compared with those with low cleaved caspase-3 expression in gastric, ovarian, cervical cancer, and colorectal cancer and well correlated to disease progression and aggressive behaviors. Thus, implicated it as a potential predictive factor prognosis factor of the four cancer types (Hu, Peng, Liu, et al., 2014). Another study also concluded that caspase-3 enhanced apoptosis of anaplastic thyroid cancer cells (ARO and KAT-4 cells) after treatment with manumycin plus paclitaxel and highlighted the importance of caspase-3 in carcinogenesis (Pan, Xu, & Yeung, 2001). Further, study examined relationship between expression of caspase-3 and clinical outcome in NSCLC by IHC and western blot analyses found that caspase-3 expression had correlation with lower incidence of lymph node involvement (Koomägi & Volm, 2000). In a knockout experiment using HCT116 cells a reduced EMT phenotypes was examined in absence of caspase-3. Therefore, therapeutic targeting of caspase-3 may not only increase the sensitivity of cancer cell to chemotherapy and radiotherapy, but also inhibit cancer cell invasion and metastasis (Zhou, Liu, Li, et al., 2018). Similarly, gastric cancer patients showed enhanced 5-year overall survival and disease-free survival rates with caspase-3 expression. However, lack of caspase-3 expression was related with more advanced, and aggressive stromal reaction and lymph node involvement, suggesting the importance of caspase-3 as a good prognostic marker in gastric cancer (Huang, Fang, Li, et al., 2018). Also, a meta-analysis and systematic review on more than 2000 patients with esophageal, gastric, and colorectal cancers established a relation between low caspase-3 expression poor prognosis and recurrence-free survival (HR = 1.65, 95% CI (1.12–2.41), I2 = 43.9%, P = 0.148) (Chen, Yang, Feng, et al., 2015). In the prostate cancer tissues the expression of caspase-3 related to pathological grading, (Gleason scores), clinical stages, and development of prostatic cancer and proposed it a new target for diagnosis and treatment in future (Gu, Ren, Wang, et al., 2015). In addition, certain studies also identified some contrast results. For instance, buccal mucosa squamous cell carcinoma (BMSCC) showed an evaluated expression of caspase-3 associated with tumorigenesis compared to the normal adjacent normal tissue. Moreover, the low expression of caspase-3 was associated with better disease-specific survival in patients with stage I/II without lymph node invasion ($P=0.043$) compared to the high expression (Huang, Yang, Wang, et al., 2017). Similarly, meta-analysis revealed that increased caspase-3 expression was significantly associated with worse prognosis in breast cancer (Yang, Zhong, Qin, et al., 2018). Moreover, higher caspase-3 expression is involved in the progression or differentiation of gastric carcinoma and may be employed as an indicator of the pathobiological behavior and prognosis of gastric carcinomas (Xiao, Zhao, Zhao, et al., 2013). In colorectal cancer patient caspase-3 were stained in comparison with adjacent normal tissues and correlated with decreased overall survival ($P<0.05$), and unfavorable clinicopathologic characteristics. This study proposed caspase-3 expressions a novel candidate prognostic marker for colorectal cancer patients (Yao, Wang, Jin, et al., 2018). More prospective and validation studies are urgently needed to define the prognostic value of caspase-3 expression cancer.

4.3 Caspase-6

Caspase-6 expresses is a dimeric zymogen and consist of a short pro-domain, a large subunit, an inter-subunit linker, and a small subunit and encoded by the gene mapped on chromosome 6 (Julien & Wells, 2017). Caspase-6 first identified in 1995 in human Jurkat T lymphocytes named "mammalian Ced-3 homologue" or Mch2 caspase-6 (Julien & Wells, 2017). It activated by proteolytic with or without the activity of caspase-3 and usually found engaged in neuropathological conditions (Albrecht, Bourdeau, Bennett, et al., 2007; Guo, Albrecht, Bourdeau, et al., 2004; Klaiman, Petzke, Hammond, & LeBlanc, 2008). A limited gene expression data is available on caspase-6 and their association with cancer behavior. For instance, study analyzed the expression of caspase-6 and -7 in gastric carcinomas indicated that decreased expression of caspase-6 and -7 expression might be involved in the mechanisms of gastric cancer development (Yoo, Lee, Kim, et al., 2004). Further, in oral squamous cell carcinoma caspase-6 acted as tumor suppressor gene, as expression level of caspase-6 is decreased in primary tumors as compared with that matched normal samples. Failure of caspase-6 response in tumor samples suggest caspase-6 as a candidate tumor suppressor gene in a subset of oral cancer and measuring its levels might help to predict the disease progression (Shinno, Gunduz, Gunduz, et al., 2005). Treatment of A172 and U251MG cells with e23sFv-Fdt-casp6 proteins induced tumor cell apoptosis and inhibited the proliferation of HER-2 overexpressing and therefore, considered caspase-6 a promising HER2-targeted treatment option for human gliomas (Zhang, Ren, Zhang, et al., 2016). So, certain evidences provided here supporting the role of caspase-6 in cancer prognosis. Further, validation at large scale is required before confirming it as prognostic factor in cancer.

4.4 Caspase-7

Caspase-7, a 303 aa long, 34 kDa protein, encoded by the gene mapped on chromosome 10, is acts as effector enzyme in apoptosis activation after cleaved and activated by initiator caspases (caspase-1) (Julien & Wells, 2017). Caspase-7 contains a catalytic cysteine residue in its active site that helps to cleave various substrates such as PARP (Boucher, Blais, & Denault, 2012). The precursors of caspase-7 comprise a pro-domain, and a large and a small subunit separated by a short linker (Julien & Wells, 2017). Main function of caspase-7 involves ROS production and aids in cell detachment during apoptosis (Brentnall et al., 2013). The activation of procaspase-7 is initiated by spontaneous anti-parallel dimerization of two precursor forms. Few studies implicating the function of caspase in apoptosis regulation and their relation with the clinicopathological features in cancer. For instance, a downregulation of caspase-7 was determined in the colonic cancer samples by the combination of IHC and Western blot analysis and described that caspase-7 appeared as a new immunohistochemical marker of colonic neoplasia (Palmerini, Devilard, Jarry, et al., 2001). Similarly, another study found that low cleaved caspase-7 protein levels proved as highly prognostic factor of unfavorable outcome across all breast cancers (Lindner, Lucantoni, Varešlija, et al., 2018). Findings also suggested that caspase-7 is aberrantly expressed in breast cancer and contributed to cell growth and proliferation therefore, considered a target in breast cancer therapeutic strategies (Chaudhary et al., 2016).

5. Components of perforin/granzyme pathway and their role in cancer

5.1 Overview of perforin/granzyme pathway

Perforin/granzyme is another additional apoptotic pathway consists of two components i.e., granzyme A (GzmA) and granzyme B (GzmB) (Osińska, Popko, & Demkow, 2014). Granzymes itself are serine proteases released form granules present within cytoplasm of CTLs and NKs (Trapani, 2001). This pathways function in caspase independent/dependent manner by inducing DNA damage against tumor cells and virus infected cells (Trapani, 2001). Perforin protein creates pore in the membrane of targets cells and facilitate the entry of granzymes into that cells (Osińska et al., 2014). Once inside the target cell GzmB activates caspase-10 and -9 leading to the activation of caspase-3 downstream. Activated caspase-3 then recruits CAD resulting in the DNA fragmentation and cell death via apoptosis (Fig. 3).

5.2 Granzyme B and perforin

As mentioned above GzmB is member of serine proteases family expressed in the granules of cytotoxic T-lymphocyte and NK cells (Chowdhury & Lieberman, 2008). It is encoded by gene GZMB located on chromosome 14 and can exist in two glycosylated form i.e., high mannose (32 kDa) and complex (35 kDa) (Chowdhury & Lieberman, 2008). Initially, GzmB synthesized as an inactive precursor contain additional amino terminal peptide sequence which subsequently cleaved by cathepsin C/H resulting in a two amino acids short activated enzyme (Fu, Thorpe, Akula, & Hellman, 2016). Prognostic impact of GzmB has been described in many studies. For example, in breast and lung cancer patients, the percentage of cells positive for GzmB and perforin expression showed a inversely correlated with regional node metastasis and confirmed that these enzyme are important for apoptotic cell death of cancer (Kontani, Sawai, Hanaoka, et al., 2001). Similarly, in lung cancer patients a decreased percentage of GzmB and perforin expressing T-lymphocytes and NKs cells had observed. This study proposing the some released soluble factors inhibited the GzmB and perforin as a part of immune evasion mechanism (Hodge, Barnawi, Jurisevic, et al., 2014). Further, increased radiosensitivity of prostate cancer cells PC-3 and DU145 after resveratrol treatment correlated with up-regulation of GzmB and perforin, demonstrating a possible mechanism for tumor apoptosis

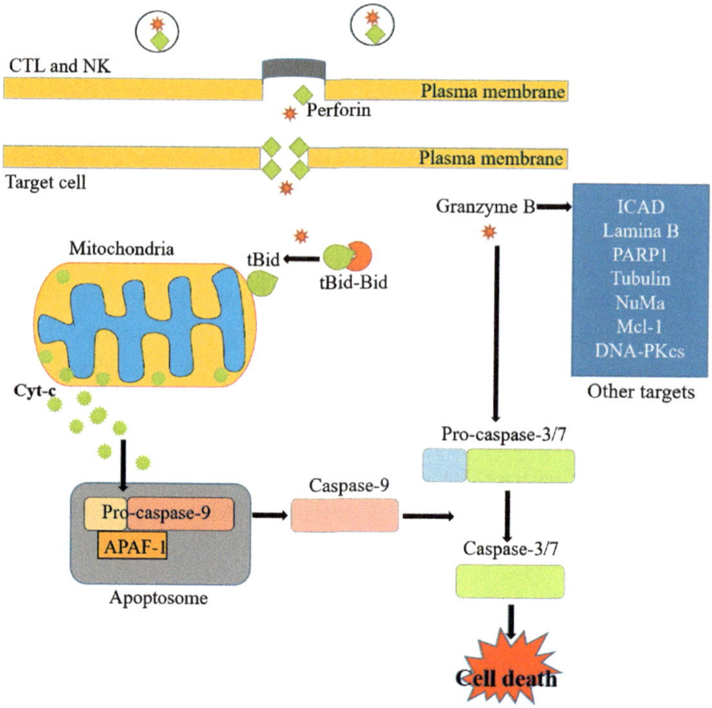

Fig. 3 Representing the perforin/granzyme pathway of apoptosis. Granzyme B synthesize in CTLs and NK cells transport in target cell through the pore made by perforin protein. Inside the target cell GzmB recruit the caspase 3 and 7 and induce the apoptosis. At the same GzmB can also activate the apoptosis through the mitochondrial mediated pathway.

(Kontani et al., 2001). In colorectal carcinoma an enhanced GzmB expression was found in tumors with a dense lymphocytic infiltrate but decreased GzmB expression in tumors with vascular invasion and positive nodal status. Univariate analysis confirmed that low GzmB was associated with early signs of metastasis (Salama, Phillips, Platell, & Iacopetta, 2011). Moreover, a significant increase in the concentrations of GzmB was noted after chemotherapy in stage II and stage III of breast cancer patients. Thus, elevation in the level of GzmB after chemotherapy may serve as a indicator of good prognosis (Kadam & Abhang, 2015). Another study done on metastatic and non-metastatic lymph nodes OCSCC showed a longer survival time in patients with high GzmB levels compared with patients having low GzmB levels and suggested that the increased expression of GzmB in the TME of OCSCC and in lymph nodes may have beneficial effect against neoplastic cells, contributing to apoptosis and increased survival of patient (Costa, Alencar, Valadares, et al., 2010; Costa, Gonçalves, Souza-Lima, et al., 2011). Study also examined the expression of GzmB and perforin in Treg cells and suggested that both these enzymes are crucial of Treg cells mediated suppression of tumor clearance (Cao, Cai, Fehniger, et al., 2007). Another, study tested the anti-cancer function of GrmB in AR-independent or androgen-insensitive (LNCaP cells. LNCaP), suggested the GrmB a potential prodrug for cancer treatment (Chuang, Wang, Li, et al., 2014).

5.3 Perforin

Perforin is a 70-kDa glycoprotein, composed of 555 aa, encoded by the gene present on chromosome 10 and identified in the dense cytoplasmic granules of both NK and CTL (Hwang, Ryu, Park, et al., 2004). It has four domains, a N-terminal containing Ca2+ binding site, a C-terminal typical for biological functions and other two domains located in the center of perforin (Osińska et al., 2014). The N and C terminal domains together comprises of 163–368 aa form a membrane attack complex called Complement/Perforin (MACPF) domain (Osińska et al., 2014). The MACPF domain made is made of C6, C7, C8α, C8β and C9 pore forming proteins form complement system and sea anemone toxins AvTX-60A (Hadders, Beringer, & Gros, 2007). In addition, perforin also contains a cysteine-rich domain which is homologous to low density lipoprotein (LDL) receptor type B and epithelial growth factor (EGF) precursor (Hadders et al., 2007). Perforin protein form a cylindrical, hydrophobic pore of 5–20 nm diameter in the cancerous target cells and allow non-selective and passive transport, water, small-molecule substances and enzymes (Osińska et al., 2014). Like other Apoptosis components perforin also found to have significance in carcinogenesis. Its deregulated expression has been analyzed in many cancer related studies. For example, in colorectal cancer tissues, the decrease in the number of perforin positive cells in accordance with tumor progression may reflect the suppression of the hosts local immunity (Nakanishi, Monden, Morimoto, et al., 1991). Another study revealed that co-expression of GrmB and perforin leads to the focus formation inhibition and apoptosis activation in Hep-2 cells and provided evidence for potential therapeutic applications in laryngeal cancer (Li, Li, An, et al., 2014). Role of perforin induced autoimmunity also confirmed in the study on lip and oral cavity squamous cell carcinoma where increased production of perforin leads to direct killing of tumor cells (Costa et al., 2011). Several studies provide the evidence that both GrmB and perforin positive tumors usually has good prognosis compared to the tumors with their low levels. Therefore, these two enzymes may be considered as prognostic factor and could be used to measure the survival rate.

5.4 Caspase-activated DNase

Caspase-activated DNase (CAD), an endonuclease, also known as DNA fragmentation factor (DFF) subunit beta or caspase activated nuclease (CPAN), is a protein encoded by gene mapped on chromosome 1 (Cao, Pei, Lan, et al., 2001). DFF is a heterodimeric protein and composed of two subunits 40-kD (DFFB) and 45-kD (DFFA) (Cao et al., 2001). Subunit DFFA is acts as a substrate for caspase-3, which cleaved and activates FFFA leading to DNA fragmentation during apoptosis (Tang & Kidd, 1998). On the other hand, subunit DFFB triggers both DNA fragmentation and chromatin condensation during apoptosis (Tang & Kidd, 1998). In the inactive state CAD remain binds with its inhibitor called ICAD and its dissociation form ICAD allow it to form oligomerize, an active DNase (Cao et al., 2001). Few studies implicating the function of caspase in apoptosis regulation and their relation with the clinicopathological features in cancer. For example, a study determined the role of CAD in apoptosis by using CAD-null mice model of skin cancer. Treatment of mouse skin with dimethylbenz [a] anthracene (DMBA) and 12-O-tetradecanoylphorbol-13-acetate (TPA) caused papillomas formation in the CAD knockout mice compared with wild-type littermates. Data indicated that loss of CAD function enhances tumorigenesis and therefore, could be used as an indicator for carcinogenesis (Yan, Wang, Xie, et al., 2009). Further, the apoptotic function of CAD was also demonstrated in Ehrlich Ascites Tumor (EAT) cells after treatment with curcumin. Curcumin treatment induced nuclear condensation, DNA fragmentation and translocation of caspase-activated DNase (CAD) to nucleus. Monitoring the high CAD expression in the tumor cells indicates tumor inhibition and good prognosis (Belakavadi & Salimath, 2005). Similarly, role of CAD in apoptosis induction and cancer progression is also described using acute myelogenous leukemia (t-AML), where CAD expression inhibit the tumor cells (Hars, Lyu, Lin, & Liu, 2006). Thus, data presented the apoptotic activation function of CAD in tumor cells which might be used as signature for disease progression.

6. Conclusion and future perspective

Burden of the cancer in increasing every year and affecting the tens of thousands of people around the world. Besides, improvement and evolution of high throughput molecular techniques, lack of specific and sensitive molecular markers made the cancer diagnosis and treatment onerous. All developed and developing countries investing a huge amount for the management of cancer but due to under acknowledge of cancer etiology the success rate in far away. This situation really creating an enthusiasm in the researchers and clinicians to identify markers with good paradigm in cancer diagnosis and therapeutics. Apoptosis is such process with different activation modes which could open the way to trace the cancer biology. Evidenced by several studies, different components of apoptosis can be observed in cancer patients and may be used to understand the cancer prognosis. Understanding of cancer behaviors can help to manage the patients and reduced exposer of cytotoxic drug overdose. Present book chapter summarized the several studies and the provided the research evidence in support to use different apoptosis components as prognosis markers. The genetic studies of different apoptosis component tell the predisposition for cancer risk. Further, validation of these signature use with available technologies for is must before put them in clinical. A good maker is one which has accurate and significant association with disease. Different components of apoptosis have a better potential being a prognostic biomarker for cancer patients.

Acknowledgment

Author(s) would like to acknowledge the Department of Histopathology, Post Graduate Institute of Medical Education and Research (PGIMER), Chandigarh for providing flatform to fulfill the requirements for this manuscript writing.

Conflict of interest

Author(s) declared that there is no conflict of interest.

Funding

Author(s) declared that this study in not funded by any resource.

References

Aggarwal, V., Kashyap, D., Sak, K., et al. (2019). Molecular mechanisms of action of tocotrienols in cancer: Recent trends and advancements. *International Journal of Molecular Sciences*, *20*, 656.

Albrecht, S., Bourdeau, M., Bennett, D., et al. (2007). Activation of caspase-6 in aging and mild cognitive impairment. *American Journal of Pathology*, *170*, 1200–1209. https://doi.org/10.2353/ajpath.2007.060974.

Anagnostou, V. K., Lowery, F. J., Zolota, V., et al. (2010). High expression of BCL-2 predicts favorable outcome in non-small cell lung cancer patients with non squamous histology. *BMC Cancer*, *10*, 1–11. https://doi.org/10.1186/1471-2407-10-186.

Apolinario, B. R. M., Van Der Valk, P., De Jong, J. S., et al. (2017). Prognostic value of the expression of p 53, bcl-2, and bax oncoproteins, and neovascularization in patients with radically resected non-small-cell lung cancer. *Journal of Clinical Oncology*, *15*, 2456–2466.

Basnakian, A. G., Apostolov, E. O., Yin, X., et al. (2006). Endonuclease G promotes cell death of non-invasive human breast cancer cells. *Experimental Cell Research*, *312*, 4139–4149. https://doi.org/10.1016/j.yexcr.2006.09.012.

Bayat, A. (2002). Science, medicine, and the future bioinformatics. *British Medical Journal*, *324*, 1018–1022.

Bebenek, M., Duś, D., & Koźlak, J. (2013). Prognostic value of the Fas/Fas ligand system in breast cancer. *Wspolczesna Onkologia*, *17*, 120–122. https://doi.org/10.5114/wo.2013.34612.

Beere, H. M., Wolf, B. B., Cain, K., et al. (2000). Heat-shock protein 70 inhibits apoptosis by preventing recruitment of procaspase-9 to the Apaf-1 apoptosome. *Nature Cell Biology*, *2*, 469–475.

Belakavadi, M., & Salimath, B. P. (2005). Mechanism of inhibition of ascites tumor growth in mice by curcumin is mediated by NF-kB and caspase activated DNase. *Molecular and Cellular Biochemistry*, *273*, 57–67. https://doi.org/10.1007/s11010-005-7717-2.

Bennett, M. W., O'Connell, J., O'Sullivan, G. C., et al. (1999). Expression of Fas ligand by human gastric adenocarcinomas: A potential mechanism of immune escape in stomach cancer. *Gut*, *44*, 156–162. https://doi.org/10.1136/gut.44.2.156.

Bernstorff, W. V., Glickman, J. N., Odze, R. D., et al. (2002). Fas (CD95/APO-1) and Fas ligand expression in normal pancreas and pancreatic tumors: Implications for immune privilege and immune escape. *Cancer*, *94*, 2552–2560. https://doi.org/10.1002/cncr.10549.

Boise, L. H., Gonzblez-garcia, M., Postema, C. E., et al. (1993). bcl-x, a bcl-2-related gene that functions as a dominant regulator of apoptotic cell death. *Cell*, *74*, 597–606.

Borhani, N., Manoochehri, M., Saleh Gargari, S., et al. (2014). Decreased expression of proapoptotic genes caspase-8- and BCL2-associated agonist of cell death (BAD) in ovarian cancer. *Clinical Ovarian and Other Gynecologic Cancer*, 7, 18–23. https://doi.org/10.1016/j.cogc.2014.12.004.

Boucher, D., Blais, V., & Denault, J. B. (2012). Caspase-7 uses an exosite to promote poly(ADP ribose) polymerase 1 proteolysis. *Proceedings of the National Academy of Sciences of the United States of America*, *109*, 5669–5674. https://doi.org/10.1073/pnas.1200934109.

Bouillet, P., Cory, S., Zhang, L. C., et al. (2001). Degenerative disorders caused by Bcl-2 deficiency prevented by loss of its BH3-only antagonist Bim. *Developmental Cell*, *1*, 645–653. https://doi.org/10.1016/S1534-5807(01)00083-1.

Brentnall, M., Rodriguez-Menocal, L., De Guevara, R. L., et al. (2013). Caspase-9, caspase-3 and caspase-7 have distinct roles during intrinsic apoptosis. *BMC Cell Biology*, *14*, 1–9. https://doi.org/10.1186/1471-2121-14-32.

Bruce, D. (2002). The nobel prize for physiology or medicine 2002. *Genome Biology*, *3*, 1–4. https://doi.org/10.1186/gb-spotlight-20021008-01.

Candé, C., Cohen, I., Daugas, E., et al. (2002). Apoptosis-inducing factor (AIF): A novel caspase-independent death effector released from mitochondria. *Biochimie*, *84*, 215–222.

Cao, X., Cai, S. F., Fehniger, T. A., et al. (2007). Granzyme B and perforin are important for regulatory T cell-mediated suppression of tumor clearance. *Immunity*, *27*, 635–646. https://doi.org/10.1016/j.immuni.2007.08.014.

Cao, G., Pei, W., Lan, J., et al. (2001). Caspase-activated DNase/DNA fragmentation factor 40 mediates apoptotic DNA fragmentation in transient cerebral ischemia and in neuronal cultures. *Journal of Neuroscience*, *21*, 4678–4690. https://doi.org/10.1523/jneurosci.21-13-04678.2001.

Carrington, E. M., Zhan, Y., Brady, J. L., et al. (2017). Anti-apoptotic proteins BCL-2, MCL-1 and A1 summate collectively to maintain survival of immune cell populations both in vitro and in vivo. *Cell Death and Differentiation*, *24*, 1–11. https://doi.org/10.1038/cdd.2017.30.

Chan, W.-Y., Cheung, K.-K., Schorge, J. O., et al. (2000). Bcl-2 and p53 protein expression, apoptosis, and p53 mutation in human epithelial ovarian cancers. *American Journal of Pathology*, *156*(2), 409–417. February 2000.

Charles, K. A., Kulbe, H., Soper, R., et al. (2009). The tumor-promoting actions of TNF-α involve TNFR1 and IL-17 in ovarian cancer in mice and humans. *Journal of Clinical Investigation*, *119*, 3011–3023. https://doi.org/10.1172/JCI39065.

Chaudhary, S., Madhukrishna, B., Adhya, A., et al. (2016). Overexpression of caspase 7 is ERα dependent to affect proliferation and cell growth in breast cancer cells by targeting p21Cip. *Oncogenesis*, *5*, e219. 5:1–12. https://doi.org/10.1016/j.febslet.2004.11.061.

Chen, H., Yang, X., Feng, Z., et al. (2015). Prognostic value of Caspase-3 expression in cancers of digestive tract: A meta-analysis and systematic review. *International Journal of Clinical and Experimental Medicine*, *8*, 10225–10234.

Chien, H. T., De Cheng, S., Chuang, W. Y., et al. (2016). Clinical implications of FADD gene amplification and protein overexpression in taiwanese oral cavity squamous cell carcinomas. *PLoS One*, *11*, 1–13. https://doi.org/10.1371/journal.pone.0164870.

Choi, E. J., Yun, J. A., Jabeen, S., et al. (2014). Prognostic significance of TMEM16A, PPFIA1, and FADD expression in invasive ductal carcinoma of the breast. *World Journal of Surgical Oncology*, *12*, 1–7. https://doi.org/10.1186/1477-7819-12-137.

Chowdhury, D., & Lieberman, J. (2008). Death by a thousand cuts: Granzyme pathways of programmed cell death. *Annual Review of Immunology*, *26*, 389–420. https://doi.org/10.1146/annurev.immunol.26.021607.090404.

Chuang, C. H., Wang, W. J., Li, C. F., et al. (2014). The combination of the prodrugs perforin-CEBPD and perforin-granzyme B efficiently enhances the activation of caspase signaling and kills prostate cancer. *Cell Death & Disease*, *5*, 1–11. https://doi.org/10.1038/cddis.2014.106.

Costa, N. L., Alencar, R. D., Valadares, M. C., et al. (2010). The clinicopathological significance of the expression of Granzyme B in oral squamous cell carcinoma. *Oral Oncology*, *46*, 185–189. https://doi.org/10.1016/j.oraloncology.2009.11.016.

Costa, N. L., Gonçalves, A. S., Souza-Lima, N. C., et al. (2011). Distinct expression of perforin and granzyme B in lip and oral cavity squamous cell carcinoma. *Journal of Oral Pathology and Medicine*, *40*, 380–384. https://doi.org/10.1111/j.1600-0714.2011.01014.x.

Danial, N. N., & Korsmeyer, S. J. (2004). Cell death: Critical control points. *Cell*, *116*, 205–219. https://doi.org/10.1016/S0092-8674(04)00046-7.

Debnath, J., Baehrecke, E. H., & Kroemer, G. (2005). Does autophagy contribute to cell death? *Autophagy*, *1*, 66–74. https://doi.org/10.4161/auto.1.2.1738.

Dobrzycka, B., Terlikowski, S. J., Kowalczuk, O., & Kinalski, M. (2009). Circulating levels of TNF-α and its soluble receptors in the plasma of patients with epithelial ovarian cancer. *European Cytokine Network*, *20*, 131–134. https://doi.org/10.1684/ecn.2009.0161.

Dossus, L., Becker, S., Rinaldi, S., et al. (2011). Tumor necrosis factor (TNF)-α, soluble TNF receptors and endometrial cancer risk: The EPIC study. *International Journal of Cancer*, *129*, 2032–2037. https://doi.org/10.1002/ijc.25840.

Eggert, A., Grotzer, M. A., Zuzak, T. J., et al. (2000). Resistance to TRAIL-induced apoptosis in neuroblastoma cells correlates with a loss of caspase-8 expression. *Medical and Pediatric Oncology*, *35*, 603–607. https://doi.org/10.1002/1096-911X(20001201)35:6<603::AID-MPO24>3.0.CO;2-1.

Elmore, S. (2007). Apoptosis: A review of programmed cell death. *Toxicologic Pathology*, *35*, 495–516. https://doi.org/10.1080/01926230701320337.

Elnemr, A., Ohta, T., Yachie, A., et al. (2001). Human pancreatic cancer cells disable function of Fas receptors at several levels in Fas signal transduction pathway. *International Journal of Oncology*, *18*, 311–316. https://doi.org/10.3892/ijo.18.2.311.

Elrod, H. A., Fan, S., Muller, S., et al. (2010). Analysis of death receptor 5 and caspase-8 expression in primary and metastatic head and neck squamous cell carcinoma and their prognostic impact. *PLoS One*, *5*, e12178. https://doi.org/10.1371/journal.pone.0012178.

Fadeel, B., & Orrenius, S. (2005). Apoptosis: A basic biological phenomenon with wide-ranging implications in human disease. *Journal of Internal Medicine*, *258*, 479–517. https://doi.org/10.1111/j.1365-2796.2005.01570.x.

Favaloro, B., Allocati, N., Graziano, V., et al. (2012). Role of apoptosis in disease. *Aging*, *4*, 330–349.

Fernald, K., & Kurokawa, M. (2013). Evading apoptosis in cancer. *Trends in Cell Biology*, *23*, 620633. https://doi.org/10.1016/j.tcb.2013.07.006.Evading.

Fink, S. L., & Cookson, B. T. (2005). Apoptosis, pyroptosis, and necrosis: Mechanistic description of dead and dying eukaryotic cells. *Infection and Immunity*, *73*, 1907–1916. https://doi.org/10.1128/IAI.73.4.1907-1916.2005.

Fisher, M. J., Virmani, A. K., Wu, L., et al. (2001). Nucleotide substitution in the ectodomain of TRAIL receptor DR4 is associated with lung cancer and head and neck cancer. *Clinical Cancer Research*, *7*, 1688–1697.

Fontaninil, G., Vignatil, S., Bigini, D., et al. (1995). Bcl-2 protein: A prognostic factor inversely correlated to p53 in non-small-cell lung cancer. *British Journal of Cancer*, *71*, 1003–1007. https://doi.org/10.1038/bjc.1995.193.

Friedrich, K., Wieder, T., Von Haefen, C., et al. (2001). Overexpression of caspase-3 restores sensitivity for drug-induced apoptosis in breast cancer cell lines with acquired drug resistance. *Oncogene*, *20*, 2749–2760. https://doi.org/10.1038/sj.onc.1204342.

Fu, Z., Thorpe, M., Akula, S., & Hellman, L. (2016). Asp-ase activity of the opossum granzyme B supports the role of granzyme B as part of anti-viral immunity already during early mammalian evolution. *PLoS One*, *11*, 1–20. https://doi.org/10.1371/journal.pone.0154886.

Fulda, S., Gorman, A. M., Hori, O., & Samali, A. (2010). Cellular stress responses: Cell survival and cell death. *International Journal of Cell Biology*, *2010*, 1–23. https://doi.org/10.1155/2010/214074.

Garg, V. K., Kashyap, D., & Tuli, H. S. (2018). Targeting telomerase and topoisomerase-II by natural moieties: An anti-cancer approach. *Novel Approaches in Cancer Study*, *1*, 3–4. https://doi.org/10.31031/nacs.2018.01.000520.

Garrido, C., Galluzzi, L., Brunet, M., et al. (2006). Mechanisms of cytochrome c release from mitochondria. *Cell Death and Differentiation*, *13*, 1423–1433. https://doi.org/10.1038/sj.cdd.4401950.

Ge, Z., Sanders, A. J., Ye, L., et al. (2013). Expression of death receptor-3 in human breast cancer and its functional effects on breast cancer cells in vitro. *Oncology Reports*, *29*, 1356–1364. https://doi.org/10.3892/or.2013.2259.

Gerl, R., & Vaux, D. L. (2005). Apoptosis in the development and treatment of cancer. *Carcinogenesis*, *26*, 263–270. 2005. https://doi.org/10.1101/gad.14.1.23.

Giannopoulou, I., Nakopoulou, L., Zervas, A., et al. (2002). Immunohistochemical study of pro-apoptotic factors Bax, Fas and CPP32 in urinary bladder cancer: Prognostic implications. *Urological Research*, *30*, 342–345. https://doi.org/10.1007/s00240-002-0274-1.

Goel, N., & Garg, V. K. (2018). Aging in humans and role of DNA methylation. *EC Pharmacology and Toxicology*, *6*, 891–892.

Goel, N., Karir, P., & Garg, V. K. (2017). Role of DNA methylation in human age prediction. *Mechanisms of Ageing and Development*, *166*, 33–41. https://doi.org/10.1016/j.mad.2017.08.012.

Goossens, N., Nakagawa, S., Sun, X., & Hoshida, Y. (2015). Cancer biomarker discovery and validation. *Translational Cancer Research*, *4*, 256–269. https://doi.org/10.3978/j.issn.2218-676X.2015.06.04.

Gout, S., Morin, C., Houle, F., & Huot, J. (2006). Death receptor-3, a new E-selectin counter-receptor that confers migration and survival advantages to colon carcinoma cells by triggering p38 and ERK MAPK activation. *Cancer Research*, *66*, 9117–9124. https://doi.org/10.1158/0008-5472.CAN-05-4605.

Gu, J., Ren, L., Wang, X., et al. (2015). Expression of livin, survivin and caspase-3 in prostatic cancer and their clinical significance. *International Journal of Clinical and Experimental Pathology*, *8*, 14034–14039.

Guicciardi, M. E., & Gores, G. J. (2009). Life and death by death receptors. *The FASEB Journal*, *23*, 1625–1637. https://doi.org/10.1096/fj.08-111005.

Guo, H., Albrecht, S., Bourdeau, M., et al. (2004). Active caspase-6 and caspase-6-cleaved tau in neuropil threads, neuritic plaques, and neurofibrillary tangles of Alzheimer's disease. *American Journal of Pathology*, *165*, 523–531. https://doi.org/10.1016/S0002-9440(10)63317-2.

Hadders, M. A., Beringer, D. X., & Gros, P. (2007). Structure of C8α-MACPF reveals mechanism of membrane attack in complement immune defense. *Science*, *317*, 1552–1554. https://doi.org/10.1126/science.1147103.

Hakem, R., Hakem, A., Duncan, G. S., et al. (1998). Differential requirement for caspase 9 in apoptotic pathways in vivo. *Cell*, *94*, 339–352. https://doi.org/10.1016/S0092-8674(00)81477-4.

Hars, E. S., Lyu, Y. L., Lin, C. P., & Liu, L. F. (2006). Role of apoptotic nuclease caspase-activated DNase in etoposide-induced treatment-related acute myelogenous leukemia. *Cancer Research*, *66*, 8975–8979. https://doi.org/10.1158/0008-5472.CAN-06-1724.

Hashemi, M., Fazaeli, A., Ghavami, S., et al. (2013). Functional polymorphisms of FAS and FASL gene and risk of breast cancer—pilot study of 134 cases. *PLoS One*, *8*, 1–9. https://doi.org/10.1371/journal.pone.0053075.

Hazra, A., Chamberlain, R. M., Grossman, H. B., et al. (2003). Death receptor 4 and bladder cancer risk. *Cancer Research*, *63*, 1157–1159.

He, L., Ren, Y., Zheng, Q., et al. (2016). Fas-associated protein with death domain (FADD) regulates autophagy through promoting the expression of Ras homolog enriched in brain (Rheb) in human breast adenocarcinoma cells. *Oncotarget*, 7, 24572–24584. https://doi.org/10.18632/oncotarget.8249.

Hengartner, M. O. (2000). The biochemistry of apoptosis. Abstract: Nature *Nature*, *407*, 770–776.

Herod, J. J. O., Eliopoulos, A. G., Warwick, J., et al. (1996). The prognostic significance of Bcl-2 and p53 expression in ovarian carcinoma. *Cancer Research*, *56*, 2178–2184.

Herold, M. J., McPherson, K. G., & Reichardt, H. M. (2006). Glucocorticoids in T cell apoptosis and function. *Cellular and Molecular Life Sciences*, *63*, 60–72. https://doi.org/10.1007/s00018-005-5390-y.

Heyninck, K., & Beyaert, R. (2001). Crosstalk between NF- kappaB -activating and apoptosis- inducing proteins of the TNF-receptor complex. *Molecular Cell Biology Research Communications*, *4*, 259–265. https://doi.org/10.1006/mcbr.2001.0295.

Hodge, G., Barnawi, J., Jurisevic, C., et al. (2014). Lung cancer is associated with decreased expression of perforin, granzyme B and interferon (IFN)-γ by infiltrating lung tissue T cells, natural killer (NK) T-like and NK cells. *Clinical and Experimental Immunology*, *178*, 79–85. https://doi.org/10.1111/cei.12392.

Hsu, H., Huang, J., Shu, H. B., et al. (1996). TNF-dependent recruitment of the protein kinase RIP to the TNF receptor-1 signaling complex. *Immunity*, *4*, 387–396. https://doi.org/10.1016/S1074-7613(00)80252-6.

Hu, Y., Benedict, M. A., Wu, D., et al. (1998). Bcl-XL interacts with Apaf-1 and inhibits Apaf-1-dependent caspase-9 activation. *Proceedings of the National Academy of Sciences of the United States of America*, *95*, 4386–4391. https://doi.org/10.1073/pnas.95.8.4386.

Hu, Q., Peng, J., Liu, W., et al. (2014). Elevated cleaved caspase-3 is associated with shortened overall survival in several cancer types. *International Journal of Clinical and Experimental Pathology*, 7, 5057–5070.

Huang, K. H., Fang, W. L., Li, A. F. Y., et al. (2018). Caspase-3, a key apoptotic protein, as a prognostic marker in gastric cancer after curative surgery. *International Journal of Surgery*, *52*, 258–263. https://doi.org/10.1016/j.ijsu.2018.02.055.

Huang, J. S., Yang, C. M., Wang, J. S., et al. (2017). Caspase-3 expression in tumorigenesis and prognosis of buccal mucosa squamous cell carcinoma. *Oncotarget*, *8*, 84237–84247. https://doi.org/10.18632/oncotarget.20494.

Huber, K. L., Serrano, B. P., & Hardy, J. A. (2018). CASPASE-9 card: Core domain interactions require a properly-formed active site. *The Biochemical Journal*, *475*, 1177–1196. https://doi.org/10.1042/BCJ20170913.CASPASE-9.

Hwang, W. S., Ryu, Y. J., Park, J. H., et al. (2004). Evidence of a pluripotent human embryonic stem cell line derived from a cloned blastocyst. *Science*, *303*, 1669–1674. https://doi.org/10.1126/science.1094515.

Ikeguchi, M., Maeta, M., & Kaibara, N. (2001). Bax expression as a prognostic marker of postoperative chemoradiotherapy for patients with esophageal cancer. *International Journal of Molecular Medicine*, 7, 413–417. https://doi.org/10.3892/ijmm.7.4.413.

Ikeguchi, M., & Nakamura, S. (2002). Quantitative analysis of expression levels of bax, bcl-2, and survivin in cancer cells during cisplatin treatment. *Oncology Reports*, *9*, 1121–1126.

Jansson, B. A., & Sun, X. (2002). Bax expression decreases significantly from primary tumor to metastasis in colorectal cancer. *Journal of Clinical Oncology*, *20*, 811–816.

Jeong, S. H., Han, J. H., Kim, J. H., et al. (2011). Bax predicts outcome in gastric cancer patients treated with 5-fluorouracil, leucovorin, and oxaliplatin palliative chemotherapy. *Digestive Diseases and Sciences*, *56*, 131–138. https://doi.org/10.1007/s10620-010-1280-8.

Jeong, S. H., Lee, H. W., Han, J. H., et al. (2008). Low expression of bax predicts poor prognosis in resected non-small cell lung cancer patients with non-squamous histology. *Japanese Journal of Clinical Oncology*, *38*, 661–669. https://doi.org/10.1093/jjco/hyn089.

Jeong, E. G., Lee, J. W., Soung, Y. H., et al. (2006). Immunohistochemical and mutational analysis of apoptosis-inducing factor (AIF) in colorectal carcinomas. *APMIS*, *114*, 867–873. https://doi.org/10.1111/j.1600-0463.2006.apm_502.x.

Julien, O., & Wells, J. A. (2017). Caspases and their substrates. *Cell Death and Differentiation*, *24*, 1380–1389. https://doi.org/10.1038/cdd.2017.44.

Kadam, C. Y., & Abhang, S. A. (2015). Serum levels of soluble Fas ligand, granzyme B and cytochrome c during adjuvant chemotherapy of breast cancer. *Clinica Chimica Acta*, *438*, 98–102. https://doi.org/10.1016/j.cca.2014.08.012.

Kaiser, U., Schilli, M., Haag, U., et al. (1996). Expression of bcl-2—Protein in small cell lung cancer. *Lung Cancer, 15*, 31–40. https://doi.org/10.1016/0169-5002(96)00568-5.

Kaklamanis, L., Savage, A., Whitehouse, R., et al. (1998). Bcl-2 protein expression: Association with p53 and prognosis in colorectal cancer. *British Journal of Cancer, 77*, 1864–1869. https://doi.org/10.1038/bjc.1998.310.

Kale, J., Osterlund, E. J., & Andrews, D. W. (2018). BCL-2 family proteins: Changing partners in the dance towards death. *Cell Death and Differentiation, 25*, 65–80. https://doi.org/10.1038/cdd.2017.186.

Kany, S., Vollrath, J. T., & Relja, B. (2019). Cytokines in inflammatory diseases. *International Journal of Molecular Sciences, 20*, 1–31.

Kashyap, D., Garg, V. K., Tuli, H. S., et al. (2019). Fisetin and quercetin: Promising flavonoids with chemopreventive potential. *Biomolecules, 9*, 174. https://doi.org/10.3390/biom9050174.

Kashyap, D., & Kaur, H. (2020). Cell-free miRNAs as non-invasive biomarkers in breast cancer: Significance in early diagnosis and metastasis prediction. *Life Sciences, 246*, 117417.

Kashyap, D., Kumar, G., Sharma, A., et al. (2017). Mechanistic insight into carnosol-mediated pharmacological effects: Recent trends and advancements. *Life Sciences, 169*, 27–36.

Kashyap, D., Mittal, S., Sak, K., et al. (2016). Molecular mechanisms of action of quercetin in cancer: Recent advances. *Tumor Biology, 37*, 12927–12939. https://doi.org/10.1007/s13277-016-5184-x.

Kashyap, D., Mondal, R., Tuli, H. S., et al. (2016). Molecular targets of gambogic acid in cancer: Recent trends and advancements. *Tumor Biology, 37*, 12915–12925.

Kashyap, D., Sharma, A., Garg, V., et al. (2016). Reactive oxygen species (ROS): An activator of apoptosis and autophagy in cancer. *Journal of Biological and Chemical Sciences, 3*, 256–264.

Kashyap, D., Sharma, A., S Tuli, H., et al. (2016). Ursolic acid and oleanolic acid: Pentacyclic terpenoids with promising anti-inflammatory activities. *Recent Patents on Inflammation & Allergy Drug Discovery, 10*, 21–33. https://doi.org/10.2174/1872213x10666160711143904.

Kashyap, D., Sharma, A., Sak, K., et al. (2018). Fisetin: A bioactive phytochemical with potential for cancer prevention and pharmacotherapy. *Life Sciences, 194*, 75–87.

Kashyap, D., Sharma, A., Tuli, H. S., et al. (2018a). Molecular targets of celastrol in cancer: Recent trends and advancements. *Critical Reviews in Oncology/Hematology, 128*, 70–81.

Kashyap, D., Sharma, A., Tuli, H. S., et al. (2018b). Apigenin: A natural bioactive flavone-type molecule with promising therapeutic function. *Journal of Functional Foods, 48*, 457–471.

Kashyap, D., Tuli, H. S., Garg, V. K., et al. (2017). Ursolic acid and quercetin: Promising anticancer phytochemicals with antimetastatic and antiangiogenic potential. *Tumor and Microenvironment, 1*, 1–7. https://doi.org/10.4103/tme.tme.

Kashyap, D., Tuli, H. S., Garg, V. K., et al. (2018). Oncogenic and tumor-suppressive roles of microRNAs with special reference to apoptosis: Molecular mechanisms and therapeutic potential. *Molecular Diagnosis & Therapy, 22*, 179–201.

Kashyap, D., Tuli, H. S., Sak, K., et al. (2019). Role of reactive oxygen species in cancer progression. *Current Pharmacology Reports, 5*, 79–86.

Kashyap, D., Tuli, H. S., & Sharma, A. K. (2016). Ursolic acid (UA): A metabolite with promising therapeutic potential. *Life Sciences, 146*, 201–213. https://doi.org/10.1016/j.lfs.2016.01.017.

Kashyap, D., Tuli, H. S., Yerer, M. B., et al. (2019). Natural product-based nanoformulations for cancer therapy: Opportunities and challenges. In *Seminars in cancer biology*.

Katsumata, K., Sumi, T., Tomioka, H., et al. (2003). Induction of apoptosis by p53, bax, bcl-2, and p21 expressed in colorectal cancer. *International Journal of Clinical Oncology, 8*, 352–356. https://doi.org/10.1007/s10147-003-0352-6.

Khodapasand, E., Jafarzadeh, N., Farrokhi, F., et al. (2015). Is Bax/Bcl-2 ratio considered as a prognostic marker with age and tumor location in colorectal cancer? *Iranian Biomedical Journal, 19*, 69–75. https://doi.org/10.6091/ibj.1366.2015.

Kim, B., Srivastava, S. K., & Kim, S. H. (2015). Caspase-9 as a therapeutic target for treating cancer. *Expert Opinion on Therapeutic Targets, 19*, 113–127. https://doi.org/10.1517/14728222.2014.961425.

Kiraz, Y., Adan, A., Kartal Yandim, M., & Baran, Y. (2016). Major apoptotic mechanisms and genes involved in apoptosis. *Tumor Biology, 37*, 8471–8486. https://doi.org/10.1007/s13277-016-5035-9.

Klaiman, G., Petzke, T. L., Hammond, J., & LeBlanc, A. C. (2008). Targets of caspase-6 activity in human neurons and Alzheimer disease. *Molecular and Cellular Proteomics, 7*, 1541–1555. https://doi.org/10.1074/mcp.M800007-MCP200.

Kontani, K., Sawai, S., Hanaoka, J., et al. (2001). Involvement of granzyme B and perforin in suppressing nodal metastasis of cancer cells in breast and lung cancers. *European Journal of Surgical Oncology, 27*, 180–186. https://doi.org/10.1053/ejso.2000.1060.

Koomägi, R., & Volm, M. (1999). Expression of Fas (CD95/APO-1) and Fas ligand in lung cancer, its prognostic and predictive relevance. *International Journal of Cancer, 84*, 239–243. https://doi.org/10.1002/(SICI)1097-0215(19990621)84:3<239::AID-IJC7>3.0.CO;2-S.

Koomägi, R., & Volm, M. (2000). Relationship between the expression of caspase-3 and the clinical outcome of patients with non-small cell lung cancer. *Anticancer Research, 20*, 493–496.

Krajewska, M., Krajewski, S., Epstein, J. I., et al. (1996). Immunohistochemical analysis of bcl-2, bax, bcl-X, and mcl-1 expression in prostate cancers. *American Journal of Pathology, 148*, 1567–1576.

Kushlinskii, N. E., Britvin, T. A., Abbasova, S. G., et al. (2001). Soluble Fas antigen in the serum of patients with colon cancer. *Bulletin of Experimental Biology and Medicine, 131*, 361–363. https://doi.org/10.1023/A:1017908320634.

Kykalos, S., Mathaiou, S., Karayiannakis, A. J., et al. (2012). Tissue expression of the proteins Fas and Fas ligand in colorectal cancer and liver metastases. *Journal of Gastrointestinal Cancer, 43*, 224–228. https://doi.org/10.1007/s12029-011-9252-6.

LaCasse, E. C., Mahoney, D. J., Cheung, H. H., et al. (2008). IAP-targeted therapies for cancer. *Oncogene, 27*, 6252–6275. https://doi.org/10.1038/onc.2008.302.

Lee, J. U., Hosotani, R., Wada, M., et al. (1999). Role of Bcl-2 family proteins (Bax, Bcl-2 and Bcl-X) on cellular susceptibility to radiation in pancreatic cancer cells. *European Journal of Cancer, 35*, 1374–1380. https://doi.org/10.1016/S0959-8049(99)00134-3.

Lee, J. W., Jeong, E. G., Soung, Y. H., et al. (2006). Immunohistochemical analysis of apoptosis-inducing factor (AIF) expression in gastric carcinomas. *Pathology Research and Practice, 202*, 497–501. https://doi.org/10.1016/j.prp.2006.03.004.

Lee, H. K., Lee, H. S., Yang, H. K., et al. (2003). Prognostic significance of Bcl-2 and p53 expression in gastric cancer. *International Journal of Colorectal Disease, 18*, 518–525. https://doi.org/10.1007/s00384-003-0491-2.

Lerma, E., Romero, M., Gallardo, A., et al. (2008). Prognostic significance of the Fas-receptor/Fas-ligand system in cervical squamous cell carcinoma. *Virchows Archiv, 452*, 65–74. https://doi.org/10.1007/s00428-007-0535-z.

Li, T., Feng, Z., Jia, S., et al. (2012). Daintain/AIF-1 promotes breast cancer cell migration by up-regulated TNF-α via activate p38 MAPK signaling pathway. *Breast Cancer Research and Treatment, 131*, 891–898. https://doi.org/10.1007/s10549-011-1519-x.

Li, Y., Jin, X., Li, J., et al. (2012). Expression of TRAIL, DR4, and DR5 in bladder cancer: Correlation with response to adjuvant therapy and implications of prognosis. *Urology, 79*, 968.e7–968.e15. https://doi.org/10.1016/j.urology.2011.11.011.

Li, X. Y., Li, Z., An, G. J., et al. (2014). Co-expression of perforin and granzyme B genes induces apoptosis and inhibits the tumorigenicity of laryngeal cancer cell line Hep-2. *International Journal of Clinical and Experimental Pathology*, *7*, 978–986.

Li, L. Y., Luo, X., & Wang, X. (2001). Endonuclease G is an apoptotic DNase when released from mitochondria. *Nature*, *412*, 95–99.

Li, M., Wu, X., & Xu, X. C. (2001). Induction of apoptosis by cyclo-oxygenase-2 inhibitor NS398 through a cytochrome C-dependent pathway in esophageal cancer cells. *International Journal of Cancer*, *93*, 218–223. https://doi.org/10.1002/ijc.1322.

Li, C.-W., Xia, W., Huo, L., Lim, S.-O., Wu, Y., Hsu, J. L., et al. (2012). Epithelial-mesenchyme transition induced by TNF-α requires NF- κB-mediated transcriptional upregulation of Twist1. *Cancer Research*, *72*, 1290–1300. https://doi.org/10.1158/0008-5472.CAN-11-3123.Epithelial-mesenchyme.

Li, P., Zhou, L., Zhao, T., et al. (2017). Caspase-9: structure, mechanisms and clinical application. *Oncotarget*, *8*, 23996–24008.

Liamarkopoulos, E., Gazouli, M., Aravantinos, G., et al. (2011). Caspase 8 and caspase 9 gene polymorphisms and susceptibility to gastric cancer. *Gastric Cancer*, *14*, 317–321. https://doi.org/10.1007/s10120-011-0045-1.

Lima, L., Morais, A., Lobo, F., et al. (2008). Association between FAS polymorphism and prostate cancer development. *Prostate Cancer and Prostatic Diseases*, *11*, 94–98. https://doi.org/10.1038/sj.pcan.4501002.

Lindner, A. U., Lucantoni, F., Varešlija, D., et al. (2018). Low cleaved caspase-7 levels indicate unfavourable outcome across all breast cancers. *Journal of Molecular Medicine*, *96*, 1025–1037. https://doi.org/10.1007/s00109-018-1675-0.

Linjawi, A., Kontogiannea, M., Halwani, F., et al. (2004). Prognostic significance of p53, bcl-2, and Bax expression in early breast cancer. *Journal of the American College of Surgeons*, *198*, 83–90. https://doi.org/10.1016/j.jamcollsurg.2003.08.008.

Liu, S., Tan, W. Y., Chen, Q. R., et al. (2008). Daintain/AIF-1 promotes breast cancer proliferation via activation of the NF-κB/cyclin D1 pathway and facilitates tumor growth. *Cancer Science*, *99*, 952–957. https://doi.org/10.1111/j.1349-7006.2008.00787.x.

Liu, L., Yang, X., Chen, X., et al. (2012). Association between TNF-α polymorphisms and cervical cancer risk: A meta-analysis. *Molecular Biology Reports*, *39*, 2683–2688. https://doi.org/10.1007/s11033-011-1022-9.

Loro, L. L., Vintermyr, O. K., Liavaag, P. G., et al. (1999). Oral squamous cell carcinoma is associated with decreased bcl-2/bax expression ratio and increased apoptosis. *Human Pathology*, *30*, 1097–1105. https://doi.org/10.1016/S0046-8177(99)90229-0.

Ma, X., & Yu, H. (2006). Global burden of cancer. *Yale Journal of Biology and Medicine*, *79*, 85–94.

Macewan, D. J. (2002). TNF ligands and receptors—A matter of life and death. *British Journal of Pharmacology*, *135*, 855–975.

Manne, U., Myers, R. B., Moron, C., et al. (1997). Prognostic significance of Bcl-2 expression and p53 nuclear accumulation in colorectal adenocarcinoma. *International Journal of Cancer*, *74*, 346–358. https://doi.org/10.1002/(SICI)1097-0215(19970620)74:3<346::AID-IJC19>3.0.CO;2-9.

Matsumoto, H., Wada, T., Fukunaga, K., et al. (2004). Bax to Bcl-2 ratio and Ki-67 index are useful predictors of neoadjuvant chemoradiation therapy in bladder cancer. *Japanese Journal of Clinical Oncology*, *34*, 124–130. https://doi.org/10.1093/jjco/hyh026.

McIlwain, D. R., Berger, T., & Mak, T. W. (2013). Caspase functions in cell death and disease. *Cold Spring Harbor Perspectives in Biology*, *5*, 1–28. https://doi.org/10.1101/cshperspect.a008656.

Mehta, S., Shelling, A., Muthukaruppan, A., et al. (2010). Predictive and prognostic molecular markers for cancer medicine. *Therapeutic Advances in Medical Oncology*, *2*, 125–148. https://doi.org/10.1177/1758834009360519.

Méndez-Barbero, N., Gutiérrez-Muñoz, C., Blázquez-Serra, R., et al. (2020). Tumor necrosis factor-like weak inducer of apoptosis (TWEAK)/fibroblast growth factor-inducible 14 (Fn14) axis in cardiovascular diseases: Progress and challenges. *Cell, 9*, 405.

Mikami, S., Mizuno, R., Kosaka, T., et al. (2015). Expression of TNF-α and CD44 is implicated in poor prognosis, cancer cell invasion, metastasis and resistance to the sunitinib treatment in clear cell renal cell carcinomas. *International Journal of Cancer, 136*, 1504–1514. https://doi.org/10.1002/ijc.29137.

Miller, D. K. (1997). The role of the caspase family of cysteine proteases in apoptosis. In *Vol. 9. Seminars in immunology* (pp. 35–49).

Min, S. S., Hong, S. K., Sug, H. L., et al. (2002). Alterations of Fas-pathway genes associated with nodal metastasis in non-small cell lung cancer. *Oncogene, 21*, 4129–4136. https://doi.org/10.1038/sj.onc.1205527.

Mitra, S., Bal, A., Kashyap, D., et al. (2020). Tumour angiogenesis and c-Met pathway activation—Implications in breast cancer. *APMIS, 128*, 316–325. https://doi.org/10.1111/apm.13031.

Miyamoto, Y., Hosotani, R., Wada, M., et al. (1999). Immunohistochemical analysis of Bcl-2, Bax, Bcl-X, and Mcl-1 expression in pancreatic cancers. *Oncology, 56*, 73–82. https://doi.org/10.1159/000011933.

Mizutanil, Y., Yoshida, O., & Bonavida, B. (1998). Prognostic significance of soluble Fas in the serum of patients with bladder cancer. *Journal of Urology, 160*, 571–576. https://doi.org/10.1016/S0022-5347(01)62960-4.

Morgan, M., Thorburn, J., Pandolfi, P. P., & Thorburn, A. (2002). Nuclear and cytoplasmic shuttling of TRADD induces apoptosis via different mechanisms. *Journal of Cell Biology, 157*, 975–984. https://doi.org/10.1083/jcb.200204039.

Motoyama, N., Wang, F., Roth, K. A., et al. (1995). Massive cell death of immature hematopoietic cells and neurons in Bcl-x-deficient mice. *Science, 267*, 1506–1510. https://doi.org/10.1126/science.7878471.

Mottolese, M., Buglioni, S., Bracalenti, C., et al. (2000). Prognostic relevance of altered Fas (CD95)-system in human breast cancer. *International Journal of Cancer, 86*, 127–132. https://doi.org/10.1002/(sici)1097-0215(20000320)89:2<127::aid-ijc5>3.0.co;2-4.

Müllauer, L., Mosberger, I., Grusch, M., et al. (2000). Fas ligand is expressed in normal breast epithelial cells and is frequently up-regulated in breast cancer. *Journal of Pathology, 190*, 20–30. https://doi.org/10.1002/(SICI)1096-9896(200001)190:1<20::AID-PATH497>3.0.CO;2-S.

Müller, W., Schneiders, A., Hommel, G., & Gabbert, H. E. (1998). Prognostic value of bcl-2 expression in gastric cancer. *Anticancer Research, 18*, 4699–4704.

Nadler, Y., Camp, R. L., Giltnane, J. M., et al. (2008). Expression patterns and prognostic value of Bag-1 and Bcl-2 in breast cancer. *Breast Cancer Research, 10*, 1–12. https://doi.org/10.1186/bcr1998.

Nakanishi, H., Monden, T., Morimoto, H., et al. (1991). Perforin expression in lymphocytes infiltrated to human colorectal cancer. *British Journal of Cancer, 64*, 239–242. https://doi.org/10.1038/bjc.1991.283.

Naval, J., de Miguel, D., Gallego-Lleyda, A., et al. (2019). Importance of TRAIL molecular anatomy in receptor oligomerization and signaling. Implications for cancer therapy. *Cancers, 11*, 444. https://doi.org/10.3390/cancers11040444.

Nix, P., Cawkwell, L., Patmore, H., et al. (2005). Bcl-2 expression predicts radiotherapy failure in laryngeal cancer. *British Journal of Cancer, 92*, 2185–2189. https://doi.org/10.1038/sj.bjc.6602647.

O'Connell, J., Bennett, M. W., O'Sullivan, G. C., et al. (1998). Fas ligand expression in primary colon adenocarcinomas: Evidence that the Fas counterattack is a prevalent mechanism of immune evasion in human colon cancer. *Journal of Pathology, 186*, 240–246. https://doi.org/10.1002/(SICI)1096-9896(199811)186:3<240::AID-PATH173>3.0.CO;2-L.

Ohno, S., Tachibana, M., Shibakita, M., et al. (2000). Prognostic significance of Fas and Fas ligand system-associated apoptosis in gastric cancer. *Annals of Surgical Oncology*, *7*, 750–757. https://doi.org/10.1007/s10434-000-0750-1.

Ohta, T., Elnemr, A., Kitagawa, H., et al. (2004). Fas ligand expression in human pancreatic cancer. *Oncology Reports*, *12*, 749–754. https://doi.org/10.3892/or.12.4.749.

Osińska, I., Popko, K., & Demkow, U. (2014). Perforin: An important player in immune response. *Central European Journal of Immunology*, *39*, 109–115. https://doi.org/10.5114/ceji.2014.42135.

Otera, H., Ohsakaya, S., Nagaura, Z., & Ishihara, N. (2005). Export of mitochondrial AIF in response to proapoptotic stimuli depends on processing at the intermembrane space. *The EMBO Journal*, *24*, 1375–1386. https://doi.org/10.1038/sj.emboj.7600614.

Palmerini, F., Devilard, E., Jarry, A., et al. (2001). Caspase 7 downregulation as an immunohistochemical marker of colonic carcinoma. *Human Pathology*, *32*, 461–467. https://doi.org/10.1053/hupa.2001.24328.

Pan, J., Xu, G., & Yeung, S. C. J. (2001). Cytochrome c release is upstream to activation of caspase-9, caspase-8, and caspase-3 in the enhanced apoptosis of anaplastic thyroid cancer cells induced by manumycin and paclitaxel. *Journal of Clinical Endocrinology and Metabolism*, *86*, 4731–4740. https://doi.org/10.1210/jcem.86.10.7860.

Papenfuss, K., Cordier, S. M., & Walczak, H. (2008). Death receptors as targets for anti-cancer therapy. *Journal of Cellular and Molecular Medicine*, *12*, 2566–2585. https://doi.org/10.1111/j.1582-4934.2008.00514.x.

Park, H. H. (2011). Structural analyses of death domains and their interactions. *Apoptosis*, *16*, 209–220. https://doi.org/10.1007/s10495-010-0571-z.

Park, K.-J., Lee, C.-H., Kim, A., et al. (1994). BH1 and BH2 domains of Bcl-2 are required for inhibition of apoptosis and heterodimerization with bax. *Nature*, *369*, 321–333.

Park, K., Lee, C., Kim, A., et al. (2012). Death receptors 4 and 5 activate Nox1 NADPH oxidase through riboflavin kinase to induce reactive oxygen species-mediated apoptotic cell death. *The Journal of Biological Chemistry*, *287*, 3313–3325. https://doi.org/10.1074/jbc.M111.309021.

Park, J. Y., Park, J. M., Jang, J. S., et al. (2006). Caspase 9 promoter polymorphisms and risk of primary lung cancer. *Human Molecular Genetics*, *15*, 1963–1971. https://doi.org/10.1093/hmg/ddl119.

Perabo, F. G. E., Mattes, R. H., Wirger, A., et al. (2001). Soluble Fas and Fas-ligand in bladder cancer in vitro and in vivo. *Urologic Oncology*, *6*, 163–169. https://doi.org/10.1016/S1078-1439(00)00130-7.

Pistritto, G., Trisciuoglio, D., Ceci, C., et al. (2016). Apoptosis as anticancer mechanism: Function and dysfunction of its modulators and targeted therapeutic strategies. *Aging*, *8*, 603–619. https://doi.org/10.18632/aging.100934.

Plati, J., Octavian, B., & Roya, K.-F. (2011). Apoptotic cell signaling in cancer progression and therapy. *Integrative Biology*, *3*, 279–296. https://doi.org/10.1039/c0ib00144a.Apoptotic.

Pobezinskaya, Y. L., & Liu, Z. (2012). The role of TRADD in death receptor signaling. *Cell Cycle*, *11*, 871–876.

Poincloux, L., Durando, X., Seitz, J. F., et al. (2009). Loss of Bcl-2 expression in colon cancer: A prognostic factor for recurrence in stage II colon cancer. *Surgical Oncology*, *18*, 357–365. https://doi.org/10.1016/j.suronc.2008.09.003.

Porebska, I., Wyrodek, E., Kosacka, M., et al. (2006). Apoptotic markers p53, Bcl-2 and bax in primary lung cancer. *In Vivo*, *20*, 599–604.

Porter, A. G., & Ja, R. U. (1999). Emerging roles of caspase-3 in apoptosis. *Cell Death and Differentiation*, *6*, 99–104. 1999.

Print, C. G., Loveland, K. L., Gibson, L., et al. (1998). Apoptosis regulator Bcl-w is essential for spermatogenesis but appears otherwise redundant. *Proceedings of the National Academy of Sciences of the United States of America*, *95*, 12424–12431. https://doi.org/10.1073/pnas.95.21.12424.

Pryczynicz, A., Guzińska-Ustymowicz, K., & Kemona, A. (2010). Fas/fasL expression in colorectal cancer. An immunohistochemical study. *Folia Histochemica et Cytobiologica*, *48*, 425–429. https://doi.org/10.2478/v10042-010-0058-3.

Qi, C., Wang, X., Shen, Z., et al. (2018). Anti-mitotic chemotherapeutics promote apoptosis through TL1A-activated death receptor 3 in cancer cells. *Cell Research*, *28*, 544–555. https://doi.org/10.1038/s41422-018-0018-6.

Qin, H., Srinivasula, S. M., Wu, G., et al. (1999). Structural basis of procaspase-9 recruitment by the apoptotic protease-activating factor 1. *Nature*, *399*, 549–557.

Raffo, A. J., Perlman, H., Chen, M. W., et al. (1995). Overexpression of bcl-2 protects prostate cancer cells from apoptosis in vitro and confers resistance to androgen depletion in vivo. *Cancer Research*, *55*, 4438–4445.

Rao, S., Mondragón, L., Pranjic, B., et al. (2019). AIF-regulated oxidative phosphorylation supports lung cancer development. *Cell Research*, *29*, 579–591. https://doi.org/10.1038/s41422-019-0181-4.

Reddy, R. B., Bhat, A. R., James, B. L., et al. (2016). Meta-analyses of microarray datasets identifies ANO1 and FADD as prognostic markers of head and neck cancer. *PLoS One*, *11*, 1–21. https://doi.org/10.1371/journal.pone.0147409.

Reed, J. C. (2006). Proapoptotic multidomain Bcl-2/Bax-family proteins: Mechanisms, physiological roles, and therapeutic opportunities. *Cell Death and Differentiation*, *13*, 1378–1386. https://doi.org/10.1038/sj.cdd.4401975.

Reed, J., Hakam, A., Nicosia, S. V., & Coppola, D. (2005). Significance of Fas receptor protein expression in epithelial ovarian cancer. *Human Pathology*, *36*, 971–976. https://doi.org/10.1016/j.humpath.2005.06.015.

Reimer, T., Herrnring, C., Koczan, D., et al. (2000). FasL:Fas ratio—A prognostic factor in breast carcinomas. *Cancer Research*, *60*, 822–828.

Renehan, A. R., Booth, C., & Potten, C. S. (2001). What is apoptosis, and why is it important? *British Medical Journal*, *322*, 1536–1538. https://doi.org/10.1007/978-3-030-32545-9_1.

Rinkenberger, J. L., Horning, S., Klocke, B., et al. (2000). Mcl-1 deficiency results in peri-implantation embryonic lethality. *Genes and Development*, *14*, 23–27. https://doi.org/10.1101/gad.14.1.23.

Salama, P., Phillips, M., Platell, C., & Iacopetta, B. (2011). Low expression of Granzyme B in colorectal cancer is associated with signs of early metastastic invasion. *Histopathology*, *59*, 207–215. https://doi.org/10.1111/j.1365-2559.2011.03915.x.

Sánchez-Alcázar, J. A., Khodjakov, A., & Schneider, E. (2001). Anticancer drugs induce increased mitochondrial cytochrome c expression that precedes cell death. *Cancer Research*, *61*, 1038–1044.

Schuyer, M., Van der Burg, M. E. L., Henzen-Logmans, S. C., et al. (2001). Reduced expression of BAX is associated with poor prognosis in patients with epithelial ovarian cancer: A multifactorial analysis of TP53, p21, BAX and BCL-2. *British Journal of Cancer*, *85*, 1359–1367. https://doi.org/10.1054/bjoc.2001.2101.

Schwandner, O., Schiedeck, T. H. K., Bruch, H. P., et al. (2000). p53 and Bcl-2 as significant predictors of recurrence and survival in rectal cancer. *European Journal of Cancer*, *36*, 348–356. https://doi.org/10.1016/S0959-8049(99)00271-3.

Sevrioukova, I. F. (2011). Apoptosis-inducing factor: Structure, function, and redox regulation. *Antioxidants & Redox Signaling*, *14*, 2545–2579.

Shamas-Din, A., Kale, J., Leber, B., & Andrews, D. W. (2013). Mechanisms of action of Bcl-2 family proteins. *Cold Spring Harbor Perspectives in Biology*, *5*, 1–21. https://doi.org/10.1101/cshperspect.a008714.

Sharma, A., Kashyap, D., Sak, K., et al. (2018). Therapeutic charm of quercetin and its derivatives: A review of research and patents. *Pharmaceutical Patent Analyst*, *7*, 15–32. https://doi.org/10.4155/ppa-2017-0030.

Sheen-Chen, S. M., Chen, H. S., Eng, H. L., & Chen, W. J. (2003). Circulating soluble Fas in patients with breast cancer. *World Journal of Surgery*, *27*, 10–13. https://doi.org/10.1007/s00268-002-6378-5.

Shen, X. G., Wang, C., Li, Y., et al. (2010). Downregulation of caspase-9 is a frequent event in patients with stage II colorectal cancer and correlates with poor clinical outcome. *Colorectal Disease*, *12*, 1213–1218. https://doi.org/10.1111/j.1463-1318.2009.02009.x.

Shinno, Y., Gunduz, M., Gunduz, E., et al. (2005). Identification of Caspase-6 as a candidate tumor suppressor gene at 4q25 in oral cancer. *Journal of Hard Tissue Biology*, *14*, 328–329. https://doi.org/10.2485/jhtb.14.328.

Shinoura, N., Sakurai, S., Asai, A., et al. (2001). Over-expression of APAF-1 and CASPASE-9 augments radiation-induced apoptosis in U-373MG glioma cells. *International Journal of Cancer*, *93*, 252–261.

Sierra, A., Lloveras, B., Castellsague, X., et al. (1995). Bd-2 expression is associated with lymph node metastasis. *International Journal of Cancer*, *60*, 54–60.

Silvestrini, R., Veneroni, S., Daidone, M. G., et al. (1994). The Bcl-2 protein: A prognostic indicator strongly related to p53 protein in lymph node-negative breast cancer patients. *Journal of the National Cancer Institute*, *86*, 499–504. https://doi.org/10.1093/jnci/86.7.499.

Sinha, K., Das, J., Pal, P. B., & Sil, P. C. (2013). Oxidative stress: The mitochondria-dependent and mitochondria-independent pathways of apoptosis. *Archives of Toxicology*, *87*, 1157–1180. https://doi.org/10.1007/s00204-013-1034-4.

Sirvent, J. J., Aguilar, M. C., Olona, M., et al. (2004). Prognostic value of apoptosis in breast cancer (pT1-pT2). A TUNEL, p53, bcl-2, bag-1 and Bax immunohistochemical study. *Histology and Histopathology*, *19*, 759–770. https://doi.org/10.14670/HH-19.759.

Son, J. W., Kang, H. K., Chae, M. H., et al. (2006). Polymorphisms in the caspase-8 gene and the risk of lung cancer. *Cancer Genetics and Cytogenetics*, *169*, 121–127. https://doi.org/10.1016/j.cancergencyto.2006.04.001.

Sperandio, S., De Belle, I., & Bredesen, D. E. (2000). An alternative, nonapoptotic form of programmed cell death. *Proceedings of the National Academy of Sciences of the United States of America*, *97*, 14376–14381. https://doi.org/10.1073/pnas.97.26.14376.

Spierings, D. C. J., De Vries, E. G. E., Timens, W., et al. (2003). Expression of TRAIL and TRAIL death receptors in stage III non-small cell lung cancer tumors. *Clinical Cancer Research*, *9*, 3397–3405.

Sträter, J., Herter, I., Merkel, G., et al. (2010). Expression and prognostic significance of APAF-1, caspase-8 and caspase-9 in stage II/III colon carcinoma: Caspase-8 and caspase-9 is associated with poor prognosis. *International Journal of Cancer*, *127*, 873–880. https://doi.org/10.1002/ijc.25111.

Sturm, I., Köhne, C. H., Wolff, G., et al. (1999). Analysis of the p53/BAX pathway in colorectal cancer: Low BAX is a negative prognostic factor in patients with resected liver metastases. *Journal of Clinical Oncology*, *17*, 1364–1374. https://doi.org/10.1200/jco.1999.17.5.1364.

Sun, G. G., Wang, Y. D., Cui, D. W., et al. (2014). EMP1 regulates caspase-9 and VEGFC expression and suppresses prostate cancer cell proliferation and invasion. *Tumor Biology*, *35*, 3455–3462. https://doi.org/10.1007/s13277-013-1456-x.

Sun, B. H., Zhao, X. P., Wang, B. J., et al. (2000). FADD and TRADD expression and apoptosis in primary hepatocellular carcinoma. *World Journal of Gastroenterology*, *6*, 223–227.

Tai, Y. T., Lee, S., Niloff, E., et al. (1998). BAX protein expression and clinical outcome in epithelial ovarian cancer. *Journal of Clinical Oncology*, *16*, 2583–2590. https://doi.org/10.1200/JCO.1998.16.8.2583.

Tang, D., & Kidd, V. J. (1998). Cleavage of DFF-45/ICAD by multiple caspases is essential for its function during apoptosis. *Journal of Biological Chemistry*, *273*, 28549–28552. https://doi.org/10.1074/jbc.273.44.28549.

Theodoropoulos, G. E., Gazouli, M., Vaiopoulou, A., et al. (2011). Polymorphisms of caspase 8 and caspase 9 gene and colorectal cancer susceptibility and prognosis. *International Journal of Colorectal Disease*, *26*, 1113–1118. https://doi.org/10.1007/s00384-011-1217-5.

Theodoropoulos, G. E., Michalopoulos, N. V., Panoussopoulos, S. G., et al. (2010). Effects of caspase-9 and survivin gene polymorphisms in pancreatic cancer risk and tumor characteristics. *Pancreas*, *39*, 976–980. https://doi.org/10.1097/MPA.0b013e3181d705d4.

Trapani, J. A. (2001). Granzymes: A family of lymphocyte granule serine proteases. *Genome Biology*, *2*, 1–7. https://doi.org/10.1186/gb-2001-2-12-reviews3014.

Tsujimoto, Y. (1998). Role of Bcl-2 family proteins in apoptosis: Apoptosomes or mitochondria? *Genes to Cells*, *3*, 697–707.

Tuli, H. S., Kashyap, D., Bedi, S. K., et al. (2015). Molecular aspects of metal oxide nanoparticle (MO-NPs) mediated pharmacological effects. *Life Sciences*, *143*, 71–79.

Tuli, H. S., Kashyap, D., Sharma, A. K., & Sandhu, S. S. (2015). Molecular aspects of melatonin (MLT)-mediated therapeutic effects. *Life Sciences*, *135*, 147–157.

Tuli, H. S., Sharma, A. K., Sandhu, S. S., & Kashyap, D. (2013). Cordycepin: A bioactive metabolite with therapeutic potential. *Life Sciences*, *93*, 863–869. https://doi.org/10.1016/j.lfs.2013.09.030.

Veis, D. J., Sorenson, C. M., Shutter, J. R., & Korsmeyer, S. J. (1993). Bcl-2-deficient mice demonstrate fulminant lymphoid apoptosis, polycystic kidneys, and hypopigmented hair. *Cell*, *75*, 229–240. https://doi.org/10.1016/0092-8674(93)80065-M.

Vermeulen, K., Berneman, Z. N., & Van Bockstaele, D. R. (2003). Cell cycle and apoptosis. *Cell Proliferation*, *36*, 165–175. https://doi.org/10.1046/j.1365-2184.2003.00267.x.

Wajant, H. (2019). Molecular mode of action of TRAIL receptor agonists—Common principles and their translational exploitation. *Cancers*, *11*, 954.

Wajant, H., & Siegmund, D. (2019). TNFR1 and TNFR2 in the control of the life and death balance of macrophages. *Frontiers in Cell and Development Biology*, 7, 1–14. https://doi.org/10.3389/fcell.2019.00091.

Wang, E. C. Y. (2012). On death receptor 3 and its ligands. *Immunology*, *137*, 114–116. https://doi.org/10.1096/fj.11-200618.

Wang, J., Chun, H. J., Wong, W., et al. (2001). Caspase-10 is an initiator caspase in death receptor signaling. *Proceedings of the National Academy of Sciences of the United States of America*, *98*, 13884–13888. https://doi.org/10.1073/pnas.241358198.

Wang, C., & Youle, R. J. (2009). The role of mitochondria in apoptosis. *Annual Review of Genetics*, *43*, 95–118.

Wiehe, R. S., Gole, B., Chatre, L., et al. (2018). Endonuclease G promotes mitochondrial genome cleavage and replication. *Oncotarget*, *9*, 18309–18326.

Williams, A. B., & Schumacher, B. (2016). p53 in the DNA-damage-repair process. *Cold Spring Harbor Perspectives in Medicine*, *6*, 1–15. https://doi.org/10.1101/cshperspect.a026070.

Wilson, G. D., Saunders, M. I., Dische, S., et al. (2001). bcl-2 expression in head and neck cancer: An enigmatic prognostic marker. *International Journal of Radiation Oncology Biology Physics*, *49*, 435–441. https://doi.org/10.1016/S0360-3016(00)01498-X.

Winter, R. N., Kramer, A., Borkowski, A., & Kyprianou, N. (2001). Loss of caspase-1 and caspase-3 protein expression in human prostate cancer. *Cancer Research*, *61*, 1227–1232.

Won, K. Y., Kim, G. Y., Kim, Y. W., et al. (2010). Clinicopathologic correlation of beclin-1 and bcl-2 expression in human breast cancer. *Human Pathology*, *41*, 107–112. https://doi.org/10.1016/j.humpath.2009.07.006.

Xiao, L. J., Zhao, S., Zhao, E. H., et al. (2013). Clinicopathological and prognostic significance of Ki-67, caspase-3 and p53 expression in gastric carcinomas. *Oncology Letters*, *6*, 1277–1284. https://doi.org/10.3892/ol.2013.1532.

Xu, B., Zhou, Z. G., Li, Y., et al. (2008). Clinicopathological significance of caspase-8 and caspase-10 expression in rectal cancer. *Oncology*, *74*, 229–236. https://doi.org/10.1159/000151392.

Yadav, P., Jaswal, V., Sharma, A., et al. (2018). Celastrol as a pentacyclic triterpenoid with chemopreventive properties. *Pharmaceutical Patent Analyst*, 7, 155–167.

Yamada, A., Arakaki, R., Saito, M., et al. (2017). Dual role of Fas/FasL-mediated signal in peripheral immune tolerance. *Frontiers in Immunology*, *8*, 1–10. https://doi.org/10.3389/fimmu.2017.00403.

Yamana, K., Bilim, V., Hara, N., et al. (2005). Prognostic impact of FAS/CD95/APO-1 in urothelial cancers: Decreased expression of FAS is associated with disease progression. *British Journal of Cancer*, *93*, 544–551. https://doi.org/10.1038/sj.bjc.6602732.

Yan, B., Wang, H., Xie, D., et al. (2009). Increased skin carcinogenesis in caspase-activated DNase knockout mice. *Carcinogenesis*, *30*, 1776–1780. https://doi.org/10.1093/carcin/bgp146.

Yang, S. H., Chien, C. M., Lu, M. C., et al. (2006). Up-regulation of Bax and endonuclease G, and down-modulation of Bcl-X L involved in cardiotoxin III-induced apoptosis in K562 cells. *Experimental and Molecular Medicine*, *38*, 435–444. https://doi.org/10.1038/emm.2006.51.

Yang, Y., Jiang, G., Zhang, P., & Fan, J. (2015). Programmed cell death and its role in inflammation. *Military Medical Research*, *2*, 1–12. https://doi.org/10.1186/s40779-015-0039-0.

Yang, X., Zheng, F., Xing, H., et al. (2004). Resistance to chemotherapy-induced apoptosis via decreased caspase-3 activity and overexpression of antiapoptotic proteins in ovarian cancer. *Journal of Cancer Research and Clinical Oncology*, *130*, 423–428. https://doi.org/10.1007/s00432-004-0556-9.

Yang, X., Zhong, D. N., Qin, H., et al. (2018). Caspase-3 over-expression is associated with poor overall survival and clinicopathological parameters in breast cancer: A meta-analysis of 3091 cases. *Oncotarget*, *9*, 8629–8641. https://doi.org/10.18632/oncotarget.23667.

Yao, Q., Wang, W., Jin, J., et al. (2018). Synergistic role of Caspase-8 and Caspase-3 expressions: Prognostic and predictive biomarkers in colorectal cancer. *Cancer Biomarkers*, *21*, 899–908. https://doi.org/10.3233/CBM-170967.

Yoo, N. J., Lee, J. W., Kim, Y. J., et al. (2004). Loss of caspase-2, -6 and -7 expression in gastric cancers. *APMIS*, *112*, 330–335. https://doi.org/10.1111/j.1600-0463.2004.t01-1-apm1120602.x.

Yoshino, T., Shiina, H., Urakami, S., et al. (2006). Bcl-2 expression as a predictive marker of hormone-refractory prostate cancer treated with taxane-based chemotherapy. *Clinical Cancer Research*, *12*, 6116–6124. https://doi.org/10.1158/1078-0432.CCR-06-0147.

Young, H. S., Jong, W. L., Su, Y. K., et al. (2005). CASPASE-8 gene is inactivated by somatic mutations in gastric carcinomas. *Cancer Research*, *65*, 815–821.

Yu, M., Zhou, X., Niu, L., et al. (2013). Targeting transmembrane TNF-α suppresses breast cancer growth. *Cancer Research*, *73*, 4061–4074. https://doi.org/10.1158/0008-5472.CAN-12-3946.

Zhang, L., Ren, J., Zhang, H., et al. (2016). HER2-targeted recombinant protein immuno-caspase-6 effectively induces apoptosis in HER2-overexpressing GBM cells in vitro and in vivo. *Oncology Reports*, *36*, 2689–2696. https://doi.org/10.3892/or.2016.5088.

Zhang, B., Sun, T., Xue, L., et al. (2007). Functional polymorphisms in FAS and FASL contribute to increased apoptosis of tumor infiltration lymphocytes and risk of breast cancer. *Carcinogenesis*, *28*, 1067–1073. https://doi.org/10.1093/carcin/bgl250.

Zheng, H. X., Cai, Y. D., Wang, Y. D., et al. (2013). Fas signaling promotes motility and metastasis through epithelial- mesenchymal transition in gastrointestinal cancer. *Oncogene*, *32*, 1183–1192. https://doi.org/10.1038/onc.2012.126.

Zheng, H. C., Sun, J. M., Wei, Z. L., et al. (2003). Expression of Fas ligand and Caspase-3 contributes to formation of immune escape in gastric cancer. *World Journal of Gastroenterology*, *9*, 1415–1420. https://doi.org/10.3748/wjg.v9.i7.1415.
Zhou, X., Jiang, W., Liu, Z., et al. (2017). Virus infection and death receptor-mediated apoptosis. *Viruses*, *9*, 316. https://doi.org/10.3390/v9110316.
Zhou, M., Liu, X., Li, Z., et al. (2018). Caspase-3 regulates the migration, invasion and metastasis of colon cancer cells. *International Journal of Cancer*, *143*, 921–930. https://doi.org/10.1002/ijc.31374.
Zou, H., Li, Y., Liu, X., & Wang, X. (1999). An APAF-1.cytochrome c multimeric complex is a functional apoptosome that activates procaspase-9. *The Journal of Biological Chemistry*, *274*, 11549–11556.

CHAPTER FIVE

Immunological and functional aspects of MAGEA3 cancer/testis antigen

Biswajit Das[a,b] and Shantibhusan Senapati[a,*]

[a]Tumor Microenvironment and Animal Models Lab, Department of Cancer Biology, Institute of Life Sciences, Bhubaneswar, Odisha, India

[b]Manipal Academy of Higher Education, Manipal, Karnataka, India

*Corresponding author: e-mail addresses: senapati@ils.res.in; drsantibhusan@yahoo.com

Contents

Abstract

Identification of ectopic gene activation in cancer cells serves as a basis for both gene signature-guided tumor targeting and unearthing of oncogenic mechanisms to expand the understanding of tumor biology/oncogenic process. Proteins expressed only in germ cells of testis and/or placenta (immunoprivileged organs) and in malignancies are called cancer testis antigens; they are antigenic because of the lack of antigen presentation by those specific cell types (germ cells), which limits the exposure of the proteins to the immune cells. Since the Cancer Testis Antigens (CTAs) are immunogenic and expressed in a wide variety of cancer types, CT antigens have become interesting target for immunotherapy against cancer. Among CT antigens MAGEA family is reported to have 12 members (MAGEA1 to MAGEA12). The current review highlights the studies on MAGEA3 which is a CT antigen and reported in almost all types of cancer.

Advances in Protein Chemistry and Structural Biology, Volume 125
ISSN 1876-1623
https://doi.org/10.1016/bs.apcsb.2020.08.001

MAGEA3 is well tried for cancer immunotherapy. Recent advances on its functional and immunological aspect warranted much deliberation on effective therapeutic approach, thus making it a more interesting target for cancer therapy.

List of abbreviations

CTA cancer testis antigen
CTL cytotoxic T lymphocyte
DAC decitabine
DC dendritic cell
EMT epithelial to mesenchymal transition
ESC embryonic stem cell
ESCC esophageal squamous cell carcinoma
HLA human leucocyte antigen
IFN interferon
MHC major histocompatibility complex
MSC mesenchymal stem cell
NK cell natural killer cell
NSCLC non-small cell lung cancer
rAAV recombinant adeno-associated virus
TCR T cell receptor

1. MAGEA3 and its expression in different malignancies

MAGEA3, melanoma-associated antigen A3 encoded by the gene *MAGEA3* (melanoma antigen family A, 3) located on Xq28 (151934652–151938240). MAGEA3 is a Cancer Testis antigen, as it is expressed in various cancers, testis and placenta but not in any other somatic cells. The MAGEA3 belongs to a protein family known as MAGEA-family. Like other MAGEA family members, it has a conserved MAGE Homology Domain (MHD; 116aa–286aa). MAGEA family members have significant protein sequence identity, which suggests a functional similarity among them. However, a distinct variability in the regulatory regions of *MAGEA* genes suggests a possible molecular mechanism of carrying out the same function by different members, under different transcriptional regulation (Barker & Salehi, 2002; Miranda, 2010; Sang, Lian, Zhou, & Shan, 2011).

The MAGEA3 protein is being implicated in almost all type of cancer (Table 1) since its identification.

Table 1 MAGEA3 expression in patient's tumor sample of various cancer types.

Sl no	Caner type	% Age	Reference
1	Acute lymphoblastic leukemias	37	Martinez et al. (2007)
2	Acute myeloid leukemia	31	
		57.531	Makhlouf (2011)
3	Astrocytoma	7	Sahin et al. (2000)
4	Bladder cancer	43	Dyrskjot et al. (2012)
		56.4	Ma et al. (2004)
		35	Dyrskjot et al. (2012)
5	Breast cancer	24	Fujie, Mori, Ueo, Sugimachi, and Akiyoshi (1997)
		10	Ayyoub, Scarlata, Hamai, Pignon, and Valmori (2014); Otte et al. (2001)
		50	Wascher et al. (2001)
		24	Fujie, Mori, et al. (1997)
6	Clear cell renal cell carcinoma (ccRCC)	27.6	Yin et al. (2014)
7	Colorectal cancer	28	Shantha Kumara et al. (2012)
		20	Mori et al. (1996)
		32.5	Nishimura et al. (1997)
8	Esophageal squamous-cell carcinoma	47	Quillien et al. (1997)
		57	Inoue et al. (1995)
		69	Weinert et al. (2009)
		87	Chen et al. (2017)
9	Gastric carcinoma	38	Inoue et al. (1995)
		40	Li et al. (1996)
		33	Kim et al. (2001)
		69.44	Chen et al. (2004)
10	Glioma	22	Syed, Mandigo, Killory, Canoll, and Bruce (2012)
		33	Chi et al. (1997)
		51	Guo et al. (2013)

Continued

Table 1 MAGEA3 expression in patient's tumor sample of various cancer types.—cont'd

Sl no	Caner type	% Age	Reference
11	Hepatocellular carcinoma	50	Tahara et al. (1999)
		25	Suzuki et al. (1999)
		42	Chen et al. (1999)
		42	Kariyama et al. (1999)
		39	Kobayashi et al. (2000)
12	Lung cancer	41	Tajima et al. (2003)
		45.8	Yoshimatsu et al. (1998)
13	Melanoma (primary)	40	Brasseur et al. (1995)
14	Melanoma (metastatic)	79	
15	Multiple myeloma	55	Atanackovic et al. (2007)
		70	Jungbluth et al. (2005)
16	Neuroblastoma	40	Corrias et al. (1996)
		59	Ishida et al. (1996)
		33	Soling, Schurr, and Berthold (1999)
17	Non-Hodgkin lymphoma	47	Han et al. (2010)
18	Non-small cell lung cancer	55.2	Gure et al. (2005)
		39.2	Sienel et al. (2004)
		29.8	Liu, Zhang, and Feng (2000)
		26.4	Thongprasert et al. (2016)
19	Oral squamous cell carcinoma	46	Ries et al. (2009)
		45	Lee et al. (1999)
20	Osteosarcoma	100	Zou et al. (2012)
		62.5	Sudo, Kuramoto, Komiya, Inoue, and Itoh (1997)
21	Ovarian cancer	36.6	Zhang, Zhou, Yu, and Yu (2010)
22	Pancreatic cancer	44	Kim et al. (2006)
23	Pituitary cancer	29	Zhu, Asa, and Ezzat (2008)
24	Rhabdomyosarcoma	35	Dalerba et al. (2001)
25	Testicular germ cell tumors	41.8	Cheville and Roche (1999)
26	Thyroid carcinoma	65	Milkovic, Sarcevic, and Glavan (2006)

2. Association of MAGEA3 with cancer patient's survival

The response to treatment, prognosis and patients' survival is found to be negatively correlated with MAGEA3 expression (Ayyoub et al., 2014; Chen, Wang, Liu, et al., 2017; Chen, Wang, Yue, et al., 2017; Kim et al., 2006; Lausenmeyer et al., 2019; Olarte et al., 2011; Xie et al., 2016). Tumor grade and stage are also found to be associated with MAGEA3 expression level (Ayyoub et al., 2014; Chen, Wang, Liu, et al., 2017; Chen, Wang, Yue, et al., 2017). MAGEA3 expression is correlated with disease progression in cancer patients (Nardiello et al., 2011).

MAGEA3 expression was reported to be associated with patients' survival in different cancer types (Das, Suklabaidya, Jain, Baisakh, & Senapati, 2014). We used online database analysis tool Kaplan-Meier-plotter (KM plotter: http://kmplot.com/) to analyze the association between MAGEA3 expression and the survival of different cancer patients. We observed a negative association of MAGEA3 expression with the patients' survival of different cancer types (Fig. 1). One report on clinical significance of MAGEA3 in pancreatic cancer had also showed a negative association of MAGEA3 and patients survival (Kim et al., 2006). Pancreatic cancer also reported to have immunosuppressive microenvironment (Collisson, Bailey, Chang, & Biankin, 2019; Makohon-Moore & Iacobuzio-Donahue, 2016; Semaan & Maitra, 2018) which could be the possible reason behind immune evasion of MAGEA3 expressing pancreatic cancer cells and MAGEA3 may also have a role in the immunosuppressive tumor microenvironment.

3. MAGEA3 as immunotherapeutic target

Groundbreaking finding that defined immunotherapy for cancer dates back to 1991, which identified cancer-specific unique proteins, when expressed by cancer cells could be recognized by cytotoxic T-lymphocytes and induced humoral response (De Plaen et al., 1994; Gaugler et al., 1994; van der Bruggen et al., 1991; Van Pel et al., 1995). The above report identified a family of proteins called MAGE family proteins which included MAGE-3 (later on called MAGEA3, i.e., Melanoma associated antigen family A3), named so because identified in melanoma patients. The cancer-specific nature of the MAGEA3 protein and its testis restricted expression made it a potential cancer vaccine target (Boon & Lucas, 1996; Gattoni-Celli & Cole, 1996). The immunotherapeutic advances on MAGEA3 as a candidate lead to

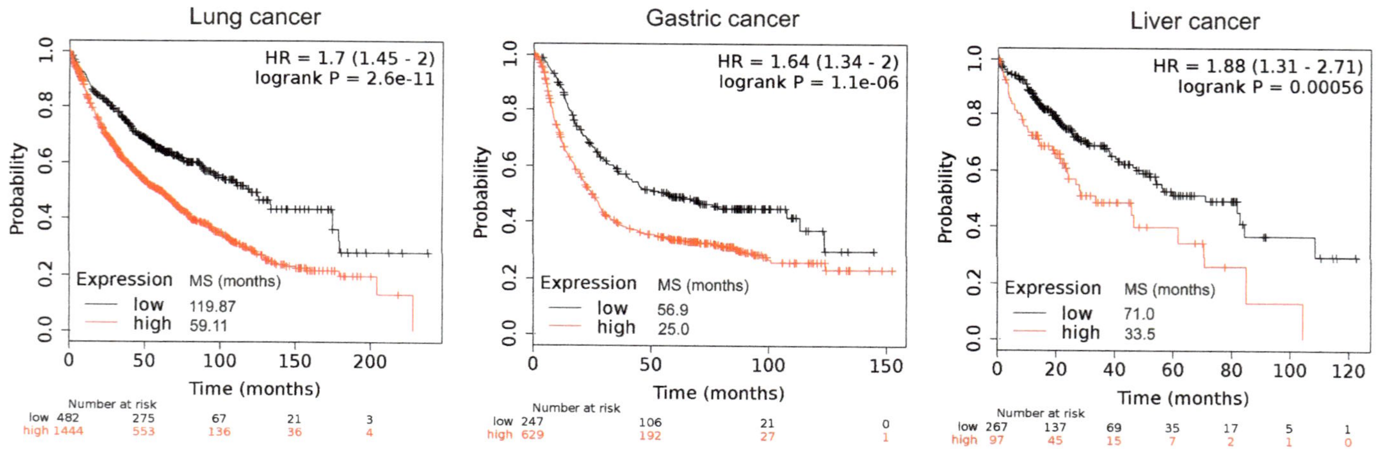

Fig. 1 KM-plot showing negative association of MAGEA3 with survival of patients of different cancer types. In silico analysis using online tool KM plotter: http://kmplot.com/. The Expression level of two-fold or more was used as cut off to distinguish between high or low expression of MAGEA3.

the identification of specific CTL epitopes/peptides of the above tumor antigen on cancer cells that recognize and effectively bind to HLA class I alleles (Celis et al., 1994; Ding, Beck, Keller, & Fenton, 1994; van der Bruggen et al., 1994). The effectiveness of MAGEA3 as an immunotherapeutic candidate was evaluated (Celis et al., 1994; Fujie et al., 1997; Tanaka et al., 1997; Valmori et al., 1997). In an esophageal cancer model, dendritic cells pulsed with MAGEA3 peptide resulted in CTL mediated cancer cell death (Kanaoka et al., 1999). Similarly, in melanoma model, a cocktail of MAGE peptide was used to improve the T cell response (Akiyama et al., 2004). Further, regression of melanoma tumors in patients treated with MAGEA3 antigenic peptide strengthened the immunotherapeutic potential of MAGEA3 (Marchand et al., 1999). Investigators successfully showed the effective response of CTLs and T-helper cells to tumor cells with unique antigens such as MAGEA3, when cross presented through dendritic cells (Berard et al., 2000; Consogno et al., 2003; Jenne, Arrighi, Jonuleit, Saurat, & Hauser, 2000; Osanto et al., 2000; Toungouz et al., 2001). Different MAGEA3 peptides presented through different MHC molecules have been evaluated in the context of eliciting T-cell response (Bilsborough et al., 2002; Coulie et al., 2002; Eifuku et al., 2001; Keiholz, Scheibenbogen, Theil, & Schadendorf, 2001; Kobayashi, Song, Hoon, Appella, & Celis, 2001; Nishiyama et al., 2001; Reynolds et al., 2000; Sadanaga et al., 2001; Schultz et al., 2000, 2001; Toungouz et al., 2001; Zhou et al., 2005).

The search for better and new immunotherapeutic target continued along with the development of the strategy to enhance the antigenicity and immunogenicity of the existing target including MAGEA3. Method to present intracellular tumor antigen is developed by expressing the MAGE-3 in DCs with a signal sequence for secretion and a cell-binding domain for receptor-mediated internalization (You et al., 2001). The antigen presentation of DCs to CTLs is enhanced by use of anti-syndecan-1 mAbs to coat the tumor cells (Dhodapkar, Krasovsky, Williamson, & Dhodapkar, 2002). Identification of epitope common to MAGEA family proteins was encouraging to use it as an immunotherapeutic target for different cancer types (Breckpot, Heirman, De Greef, van der Bruggen, & Thielemans, 2004; Graff-Dubois et al., 2002). MAGEA3 is presented to the T cells by APCs genetically modified/electroporated with MAGEA3 mRNA (Bonehill et al., 2004, 2009; Breckpot et al., 2003, 2004; Dullaers et al., 2004; Durantez et al., 2009; Lesimple et al., 2003; Minami et al., 2005; Schaft et al., 2005; Trakatelli et al., 2005). The preceding advances with promising evidence on MAGEA3 as a candidate for immunotherapy against cancer lead to experimental trials on patients using DCs pulsed with MAGEA3 peptides (Banchereau et al., 2005; Hersey et al., 2004, 2005; Lu et al., 2005; Tanaka et al., 2006; Tanzarella et al., 2004; Vantomme et al., 2004; Zhang et al., 2005). MAGEA3 turned out to be a good immunotherapeutic anticancer target and showed a better outcome in patients which was encouraging to test it in different cancer types (Bluman et al., 2007; Carrasco et al., 2008; Goodyear et al., 2008; Hiret, Senellart, & Bennouna, 2010; Tanaka, Haraguchi, Isikawa, Inoue, & Mori, 2008; Ueda et al., 2007; Zhang et al., 2007, 2008). Moving ahead, MAGEA3 was taken into clinical trial for NSCLC patients and showed some clinical benefit in terms of disease-free survival (Tyagi & Mirakhur, 2009; West, 2009). Studies also cautioned about the off-target effect of MAGEA3-immunotherapy as its other family member protein MAGEA4 is found to be expressed in mesenchymal stem cells (Saldanha-Araujo et al., 2010); further the heterogeneous expression pattern of MAGEA3 in tumor or existence of immunoregulatory mechanism to escape the immunotherapeutic strategy could be the possible reason of the limited success rate of MAGEA3 vaccine (Cecconi et al., 2010; Peled, Oton, Hirsch, & Bunn, 2009; Vourc'h-Jourdain, Volteau, Nguyen, Khammari, & Dreno, 2009). Epitopes of MAGEA3 is found to be naturally processed and presented by dendritic cells (associated with HLA-DP and -DR) to CD4 (+) T helper cells in HNSCC cancer patients (Cesson et al., 2011). Efforts have been made to enhance the potency of MAGEA3 vaccine by either modifying the immunogen (MAGEA3-HSP70 fusion DNA vaccine) (Qu et al., 2010) or inducing the expression of MAGEA3 using synthetic chromatin modifiers (Bao, Dunham, & Lucas, 2011; Cogdill et al., 2012; Moreno-Bost et al., 2011; Rao et al., 2011). Decitabine (DAC) treatment also increased the expression level of CT antigens, MHC molecules and ICAM-1 to enhance antitumor activity through CTLs (Krishnadas, Bao, Bai, Chencheri, & Lucas, 2014). Recently, a report suggested that chaetocin can enhance DCs function by upregulating HSP90 and MAGEA3, induce Th1 polarization to effectively promote antigen-specific CTLs (Vo et al., 2017). Moreover, MAGEA3 peptide-specific TCRs with high avidity were cloned and introduced into the T cells to get a better response in the context of tumor regression (Chinnasamy et al., 2011; Straetemans et al., 2012; Zhu et al., 2012). Experiments on neuroblastoma cells showed low expression of MHC-I and Antigen Processing Machinery (APM) molecules, including Endoplasmic Reticulum (ER) Aminopeptidases, thus poorly presents tumor antigens to Cytotoxic T Lymphocytes (CTL) which can be reversed by cooverexpressing NF-κB p65 and the Interferon Regulatory Factor 1

(IRF1) (Lorenzi et al., 2012). The effort to increase the uptake of MAGEA3 antigen by dendritic cells also favored the increased efficacy of CTLs (Moeller, Spagnoli, Finke, Veelken, & Houet, 2012). A phase II clinical trial on metastatic melanoma patients showed prolonged survival when vaccinated with DCs pulsed with a cocktail of antigenic peptide including MAGEA3 (gp100, tyrosinase, MAGE-A2, MAGE-A3 and MART-1 or MAGEA1) (Oshita et al., 2012; Wilgenhof et al., 2013). In NSCLC, phase II clinical trials involving MAGEA3 as antigenic peptide showed some hope (Declerck & Vansteenkiste, 2014; Reck, 2012; Thomas & Hassan, 2012). Although the TCR engineering projected immunotherapy to the next level, few reports showed nonspecific toxicity on the patients (Brichard, Louahed, & Clay, 2013; Derre et al., 2017; Hobo et al., 2013; Linette et al., 2013; Lu et al., 2017; Morgan et al., 2013; Raman et al., 2016) and some showed MAGEA3 immunotherapy within toxicity level (Destexhe, Grosdidier, et al., 2015; Destexhe, Stannard, et al., 2015; Esfandiary & Ghafouri-Fard, 2015; Pujol et al., 2015; Taborska et al., 2017). Still, further investigations found out gene signatures associated with encouraging response to MAGEA3 immunotherapy (Saiag et al., 2016; Ulloa-Montoya et al., 2013). Another strategy to present MAGEA3 efficiently through dendritic cells was pulsing with MAGEA3 fused with cell penetrating domain (Batchu, Gruzdyn, Potti, Weaver, & Gruber, 2014) and dendritic cell transduced with rAAV-6-MAGEA3 vector resulted in CTLs that secreted IFN gama and killed MAGEA3+ cancer cells (Batchu et al., 2014). The mechanistic study in an animal model with MAGEA3+ tumor showed that administration with recMAGEA3 along with immunostimulant AS15 resulted into effective anti-tumor activity through Th1 response which includes key players like CD4+ T cells, NK cells and interferon gamma (Gerard, Baudson, Ory, & Louahed, 2014). Similarly, when MAGEA3 is expressed in dendritic cells by lentiviral vector also induced CTLs to kill MAGEA3+ cancer cells (Lin et al., 2014). Combination immunotherapy including MAGEA3 peptide and immune-modulators like TLR-3 agonist and poly-ICLC showed a high frequency of T cell response in a phase II trial on myeloma patients (Rapoport et al., 2014). A phase I clinical trial on neuroblastoma and sarcoma showed chemo-immunotherapy approach using DAC/DC-CT vaccine is feasible, well tolerated and results in antitumor activity in some patients (Krishnadas et al., 2015). Again co-expression of CALR and MAGEA3 in dendritic cells resulted in better immune-clearance of cancer cells (Liu et al., 2015, 2016). The anticancer immunotherapeutic potential of MAGEA3 was also evaluated in pancreatic cancer (Batchu et al., 2016), bladder cancer (Li et al., 2016) and metastatic melanoma (Traversari & Russo, 2016). The biggest phase III clinical trial on MAGEA3 targeted immunotherapy as adjuvant therapy for NSCLC was failed (Vansteenkiste et al., 2016).

Functional studies on MAGEA3 found out the role of MAGEA3 on the modulation of the tumor microenvironment. It is shown recently, to be involved in regulating the levels of proteins related to MHC class I protein complex, interferon signaling pathways, immune response, antigen presentation and cell chemotaxis. The differentially expressed genes associated with chemokines, antigen presentation and vasculogenic mimicry formation reveal the potential impact of MAGEA3 on the immunosuppressive tumor microenvironment (Wang et al., 2018).

Recently, the generation of immortalized dendritic cells with the ability to express co-stimulatory molecules and MAGEA3 effectively primed CTLs and NK cells to hinder lung cancer pathogenesis in mice model (Wu, Zhang, Jiang, Gallo, & Cheng, 2018). A latest preclinical study involving non-human primates showed that use of adenoviral vector and the oncolytic Maraba MG1 rhabdovirus vector expressing hMAGEA3 showed CD4+ and CD8+ T cells proliferation and are specific to hMAGEA3; and circulating MAGEA3 antibodies indicated humoral immune response (Pol et al., 2019) which encouraged to go for clinical trials. One promising experimental evidence showed the use of non-pathogenic vector *Mycobacterium smegmatis* as a carrier of MAGEA3 protein to immunize mice and when blood from immunized mice transfused to nude mice bearing tumor showed significant reduction of already established tumor (Jian et al., 2018). A cartoon depicting the interaction of MAGEA3 positive cancer cells and immune cells is presented (Fig. 2).

4. Functional aspects of MAGEA3

MAGEA3 is extensively studied and implicated in immunotherapeutic aspects in different cancer types. Reports on functional role of MAGEA3 also highlighted its significance in cancer cell. The known functional implications are depicted (Figs. 3 and 4).

4.1 MAGEA3 and stem cell property

Reports recommend CT antigens to be considered as stem cell markers. Various CT antigens including MAGEA3 were reported to be expressed in human embryonic stem cells (Lifantseva et al., 2011) and mesenchymal

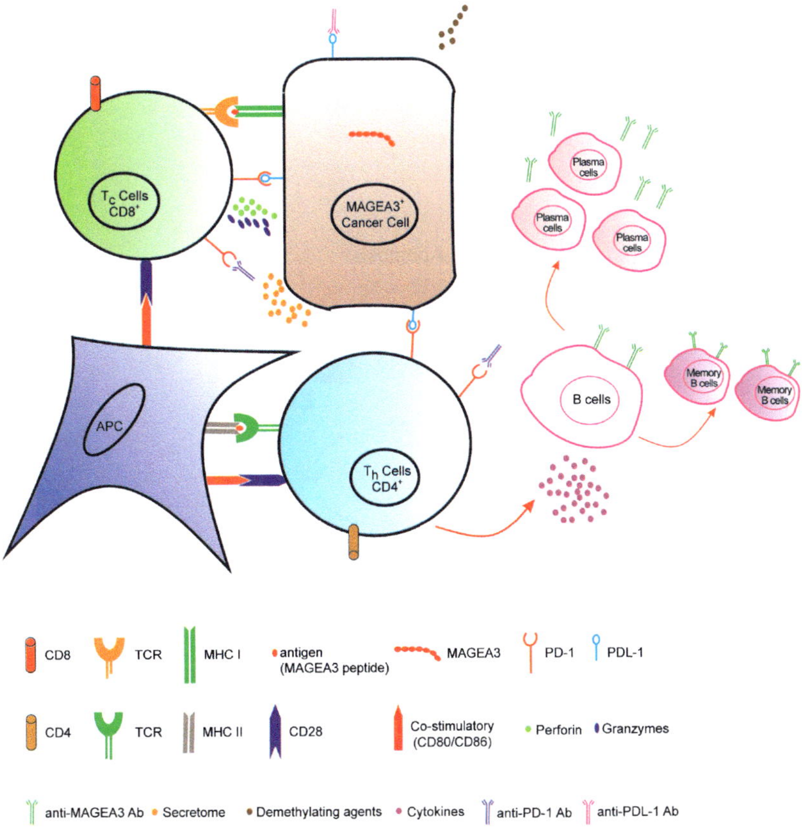

Fig. 2 Schematic diagram showing MAGEA3 positive cancer cells and interacting immune cells. MAGEA3 peptide when presented by cancer cells or antigen presenting cells, elicit cell mediated immune response or humoral response, respectively. However the clearance of the MAGEA3 positive cancer cell depends upon the immune modulating environment in the tumor microenvironment. The cancer cells through its secretory factors can regulate the tumor-microenvironment and modulate the immune composition and/or function of the tumor stroma.

stem cells (Cronwright et al., 2005; Saldanha-Araujo et al., 2010). The expression of MAGEA3 in germ cells of testis, fetal ovary and placenta suggestive of its involvement in developmental processes; further its expression in cancer stem cell-like population in bladder cancer (Yin et al., 2014), ESC (Lifantseva et al., 2011) and MSC (Saldanha-Araujo et al., 2010) makes it a conceivable molecule for maintaining stem cell population in germ cells and/or cancer cells.

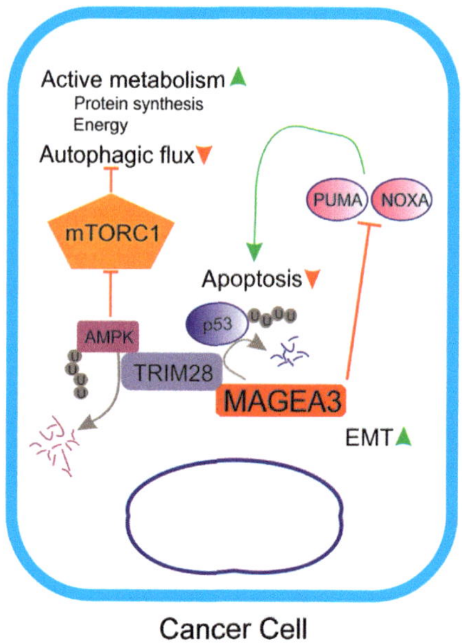

Fig. 3 Schematic illustration showing various molecular function of MAGEA3. MAGEA3 in association with TRIM28 helps degrading AMPK and provide cell survival by sustaining mTOR signaling, thus can block excess autophagy during nutrient/growth factor deprivation. MAGEA3 also degrades p53 in ubiquitin pathway and mitigates p53 mediated apoptosis. MAGEA3 blocks NOXA and PUMA and thus contributes to reduced apoptosis. MAGEA3 increases epithelial to mesenchymal transition by modulating gene expression associated with EMT. Up arrow head: upregulation/increased, Down arrow head: down regulation/decreased.

4.2 MAGEA3 and EMT

MAGEA3 is reported to be associated with lymph node metastasis in ESCC patients (Chen, Wang, Yue, et al., 2017) and lung cancer patients (Chen, Wang, Liu, et al., 2017). Supporting the above metastatic role of MAGEA3, EMT related genes like N-cadherin, Vimentin and Slug go down when MAGEA3 is depleted in ESCC cells (Chen, Wang, Yue, et al., 2017) and lung cancer cells (Chen, Wang, Liu, et al., 2017). In vivo and in vitro experiments showed involvement of MAGEA3 in increased metastasis of cancer cells (Liu, Cheng, Asa, & Ezzat, 2008).

4.3 MAGEA3 and cancer drug resistance

Drug resistance is a major obstacle in cancer therapy since cancer cells adapt alternative mechanism to survive. MAGEA3 also suggested being involved in doxorubicin resistance of colon cancer cells (Bertram, Palfner,

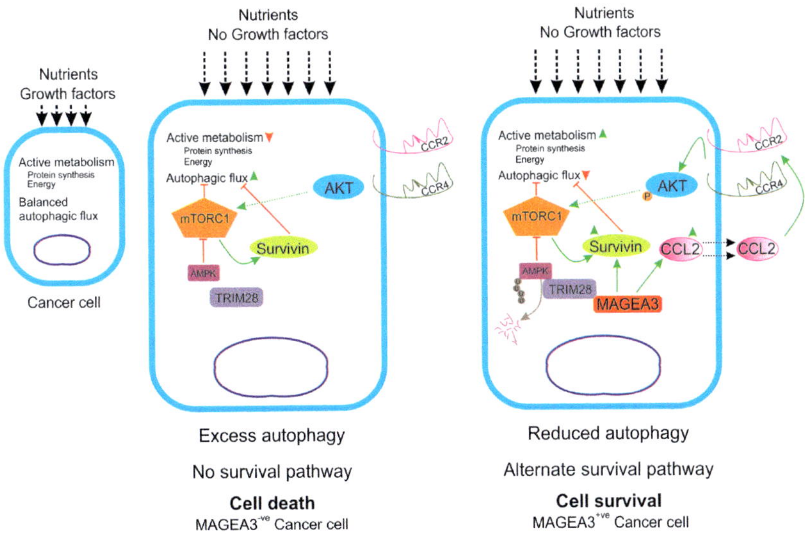

Fig. 4 Schematic illustration showing role of MAGEA3 in cancer cells survival in growth factor depleted condition. Cancer cells in presence of growth nutrients and growth factor has a balanced autophagic flux favoring active metabolism like protein synthesis and energy generation which enables cancer cells to proliferate and thrive. When the cancer cells sense limited nutrient and growth factor (AMPK is activated and pAMPK is elevated) the autophagic pathway is hyper-activated (mTOR activation is hindered) to get rid of energy stress and oxidative stress to survive in such unfavorable environment. The MAGEA3 expressing cells get survival advantage in nutrient and/or growth factor deprivation by reducing the lethal macroautophagy (AMPK gets ubiquitinated and degraded by MAGEA3-TRIM28 complex) and promoting survival signaling through increased level of survivin and CCL2. Up arrow head: upregulation/increased, Down arrow head: downregulation/decreased.

Hiddemann, & Kneba, 1998), paclitaxel/doxorubicin resistance of ovarian cancer cells (Duan et al., 2003) and tamoxifen resistance of breast cancer cells (Wong et al., 2014). Gastric cancer patient-derived cells became sensitive to docetaxel upon MAGEA3 knockdown (Xie et al., 2016). MAGEA3 provides survival advantage to multiple myeloma cells in response to chemotherapy (Atanackovic et al., 2010).

4.4 Molecular functions of MAGEA3

4.4.1 Regulating target gene expression

Regulation of gene expression by MAGEA3 was hardly studied. MAGEA3 does not have a DNA binding domain, but interacts with molecules like p53 and TRIM28/KAP1 which are transcription factor and transcriptional coregulator respectively suggesting its ability to regulate gene expression.

MAGEA3 can bind to KAP1 (TRIM28/Tif1β) and increase its ubiquitin ligase activity promoting ZNF382 (a tumor suppressor) degradation and thus increasing the ID1 (an oncogene negatively regulated by ZNF382) expression level (Xiao et al., 2011). In contrast, MAGEA3 also negatively regulates the *Ki67* (a cell proliferation marker) expression by enhancing the KAP1 binding to its target site and increased histone methylation (Xiao et al., 2011). MAGEA3 also boosts cell cycle progression (Liu et al., 2008).

4.4.2 MAGEA3 and apoptosis

Over-expression of MAGEA3, can lead to downregulation of p21 and enhanced pRb promoting cell cycle independent of p53 mutation status (Liu et al., 2008). Moreover, MAGEA3 directly interacts with and enhances the ubiquitin ligase activity of TRIM28 (a RING E3 ubiquitin ligase) and has a probable role in p53 degradation/expression, minimizing apoptosis (Marcar, Maclaine, Hupp, & Meek, 2010; Monte et al., 2006; Yang et al., 2007). MAGEA3 binds to inactive caspase-12 and blocks caspase-12 activation (Morishima, Nakanishi, Takenouchi, Shibata, & Yasuhiko, 2002). MAGEA3 also regulates NOXA and PUMA expression level, and modulates apoptosis (Xie et al., 2016).

4.4.3 MAGEA3 and autophagy

MAGEA3/KAP1 complex ubiquitinates and degrades AMPK and protects cells from autophagy (Pineda & Potts, 2015; Pineda et al., 2015; Wu et al., 2018). A recent study revealed a tumor-promoting role of MAGEA3, which degrades a tumor suppressor FBP1 in TRIM28 dependent manner and helps in reprogramming of cancer cell metabolism in hepatocellular carcinoma (Jin et al., 2017). Our recent report highlighted MAGEA3 as a prosurvival factor in pancreatic cancer cells, which provides survival advantage to the cancer cells during growth factor depleted condition and in presence of genotoxic and/or metabolic stress by minimizing macroautophagy and activating signaling pathways involving CCL2 and survivin (Das & Senapati, 2019).

5. Conclusion/future prospective

MAGEA3 is still in target for cancer treatment through the immunotherapeutic approach and achieved success to some extent. However, some challenges still exist to use MAGEA3 as an immunotherapeutic candidate. The immunomodulating tumor stroma, fibrosis at the tumor site, immune-inhibitory molecules secreted by cancer cells and limited antigen presentation by the cancer cells are some of the limiting factors for immunotherapy. Recently, functional study revealing the ability of MAGEA3 to block autophagy also supports the fact of limited antigen presentation. Apart from the immunotherapeutic approach, studies have also emphasized on its functional role in the context of cancer cell survival and tumor promotion. Thus along with immunotherapy, novel molecules that can target MAGEA3 will also be useful in combating with MAGEA3+ cancer cells, provided the targeting molecules are effectively delivered to the cancer cells.

Availability of data and materials

Data sharing is not applicable to this article as no datasets were generated during the current study.

Competing interests

The authors declare that they have no competing interests.

Funding

The project is funded by Department of Biotechnology (DBT; BT/PR13502/MED/30/1555/2015), Government of India.

Authors' contributions

B.D. and S.S. conceived and wrote the manuscript.

References

Akiyama, Y., Maruyama, K., Nara, N., Mochizuki, T., Yamamoto, A., Yamazaki, N., et al. (2004). Cytotoxic T cell induction against human malignant melanoma cells using HLA-A24-restricted melanoma peptide cocktail. *Anticancer Research*, *24*(2B), 571–577.

Atanackovic, D., Arfsten, J., Cao, Y., Gnjatic, S., Schnieders, F., Bartels, K., et al. (2007). Cancer-testis antigens are commonly expressed in multiple myeloma and induce systemic immunity following allogeneic stem cell transplantation. *Blood*, *109*(3), 1103–1112.

Atanackovic, D., Hildebrandt, Y., Jadczak, A., Cao, Y., Luetkens, T., Meyer, S., et al. (2010). Cancer-testis antigens MAGE-C1/CT7 and MAGE-A3 promote the survival of multiple myeloma cells. *Haematologica*, *95*(5), 785–793.

Ayyoub, M., Scarlata, C. M., Hamai, A., Pignon, P., & Valmori, D. (2014). Expression of MAGE-A3/6 in primary breast cancer is associated with hormone receptor negative status, high histologic grade, and poor survival. *Journal of Immunotherapy*, *37*(2), 73–76.

Banchereau, J., Ueno, H., Dhodapkar, M., Connolly, J., Finholt, J. P., Klechevsky, E., et al. (2005). Immune and clinical outcomes in patients with stage IV melanoma vaccinated with peptide-pulsed dendritic cells derived from CD34+ progenitors and activated with type I interferon. *Journal of Immunotherapy*, *28*(5), 505–516.

Bao, L., Dunham, K., & Lucas, K. (2011). MAGE-A1, MAGE-A3, and NY-ESO-1 can be upregulated on neuroblastoma cells to facilitate cytotoxic T lymphocyte-mediated tumor cell killing. *Cancer Immunology, Immunotherapy*, *60*(9), 1299–1307.

Barker, P. A., & Salehi, A. (2002). The MAGE proteins: Emerging roles in cell cycle progression, apoptosis, and neurogenetic disease. *Journal of Neuroscience Research*, *67*(6), 705–712.

Batchu, R. B., Gruzdyn, O. V., Moreno-Bost, A. M., Szmania, S., Jayandharan, G., Srivastava, A., et al. (2014). Efficient lysis of epithelial ovarian cancer cells by MAGE-A3-induced cytotoxic T lymphocytes using rAAV-6 capsid mutant vector. *Vaccine*, *32*(8), 938–943.

Batchu, R. B., Gruzdyn, O., Potti, R. B., Weaver, D. W., & Gruber, S. A. (2014). MAGE-A3 with cell-penetrating domain as an efficient therapeutic cancer vaccine. *JAMA Surgery*, *149*(5), 451–457.

Batchu, R. B., Gruzdyn, O. V., Qazi, A. M., Mahmud, E. M., Mostafa, G., Weaver, D. W., et al. (2016). Pancreatic cancer cell lysis by cell-penetrating peptide-MAGE-A3-induced cytotoxic T lymphocytes. *JAMA Surgery*, *151*(11), 1086–1088.

Berard, F., Blanco, P., Davoust, J., Neidhart-Berard, E. M., Nouri-Shirazi, M., Taquet, N., et al. (2000). Cross-priming of naive CD8 T cells against melanoma antigens using dendritic cells loaded with killed allogeneic melanoma cells. *Journal of Experimental Medicine*, *192*(11), 1535–1544.

Bertram, J., Palfner, K., Hiddemann, W., & Kneba, M. (1998). Elevated expression of S100P, CAPL and MAGE 3 in doxorubicin-resistant cell lines: Comparison of mRNA differential display reverse transcription-polymerase chain reaction and subtractive suppressive hybridization for the analysis of differential gene expression. *Anti-Cancer Drugs*, *9*(4), 311–317.

Bilsborough, J., Panichelli, C., Duffour, M. T., Warnier, G., Lurquin, C., Schultz, E. S., et al. (2002). A MAGE-3 peptide presented by HLA-B44 is also recognized by cytolytic T lymphocytes on HLA-B18. *Tissue Antigens*, *60*(1), 16–24.

Bluman, E. M., Coulie, P. G., Xiaojuan, S., Machan, J., Lin, C., Meitner, P. A., et al. (2007). Lysis of human chondrosarcoma cells by cytolytic T lymphocytes recognizing a MAGE-A3 antigen presented by HLA-A1 molecules. *Journal of Orthopaedic Research*, *25*(5), 678–684.

Bonehill, A., Heirman, C., Tuyaerts, S., Michiels, A., Breckpot, K., Brasseur, F., et al. (2004). Messenger RNA-electroporated dendritic cells presenting MAGE-A3 simultaneously in HLA class I and class II molecules. *Journal of Immunology*, *172*(11), 6649–6657.

Bonehill, A., Van Nuffel, A. M., Corthals, J., Tuyaerts, S., Heirman, C., Francois, V., et al. (2009). Single-step antigen loading and activation of dendritic cells by mRNA electroporation for the purpose of therapeutic vaccination in melanoma patients. *Clinical Cancer Research*, *15*(10), 3366–3375.

Boon, T., & Lucas, S. (1996). Rejection antigens of tumors and specific immunotherapy of cancer. *Bulletin et Mémoires de l'Académie Royale de Médecine de Belgique*, *151*(12), 469–479.

Brasseur, F., Rimoldi, D., Lienard, D., Lethe, B., Carrel, S., Arienti, F., et al. (1995). Expression of MAGE genes in primary and metastatic cutaneous melanoma. *International Journal of Cancer*, *63*(3), 375–380.

Breckpot, K., Dullaers, M., Bonehill, A., van Meirvenne, S., Heirman, C., de Greef, C., et al. (2003). Lentivirally transduced dendritic cells as a tool for cancer immunotherapy. *The Journal of Gene Medicine*, *5*(8), 654–667.

Breckpot, K., Heirman, C., De Greef, C., van der Bruggen, P., & Thielemans, K. (2004). Identification of new antigenic peptide presented by HLA-Cw7 and encoded by several MAGE genes using dendritic cells transduced with lentiviruses. *Journal of Immunology*, *172*(4), 2232–2237.

Brichard, V. G., Louahed, J., & Clay, T. M. (2013). Cancer regression and neurological toxicity cases after anti-MAGE-A3 TCR gene therapy. *Journal of Immunotherapy*, *36*(2), 79–81.

Carrasco, J., Van Pel, A., Neyns, B., Lethe, B., Brasseur, F., Renkvist, N., et al. (2008). Vaccination of a melanoma patient with mature dendritic cells pulsed with MAGE-3 peptides triggers the activity of nonvaccine anti-tumor cells. *Journal of Immunology, 180*(5), 3585–3593.

Cecconi, V., Moro, M., Del Mare, S., Sidney, J., Bachi, A., Longhi, R., et al. (2010). The CD4+ T-cell epitope-binding register is a critical parameter when generating functional HLA-DR tetramers with promiscuous peptides. *European Journal of Immunology, 40*(6), 1603–1616.

Celis, E., Fikes, J., Wentworth, P., Sidney, J., Southwood, S., Maewal, A., et al. (1994). Identification of potential CTL epitopes of tumor-associated antigen MAGE-1 for five common HLA-A alleles. *Molecular Immunology, 31*(18), 1423–1430.

Celis, E., Tsai, V., Crimi, C., DeMars, R., Wentworth, P. A., Chesnut, R. W., et al. (1994). Induction of anti-tumor cytotoxic T lymphocytes in normal humans using primary cultures and synthetic peptide epitopes. *Proceedings of the National Academy of Sciences of the United States of America, 91*(6), 2105–2109.

Cesson, V., Rivals, J. P., Escher, A., Piotet, E., Thielemans, K., Posevitz, V., et al. (2011). MAGE-A3 and MAGE-A4 specific CD4(+) T cells in head and neck cancer patients: Detection of naturally acquired responses and identification of new epitopes. *Cancer Immunology, Immunotherapy, 60*(1), 23–35.

Chen, C. H., Huang, G. T., Lee, H. S., Yang, P. M., Yan, M. D., Chen, D. S., et al. (1999). High frequency of expression of MAGE genes in human hepatocellular carcinoma. *Liver, 19*(2), 110–114.

Chen, X. H., Liu, B. Y., Zhang, D. Q., Zhang, Y., Zhang, Y., Li, J. F., et al. (2004). Expression of MAGE-1 and MAGE-3 genes in gastric cancer and gastric biopsy tissues and its clinical significance. *Xi Bao Yu Fen Zi Mian Yi Xue Za Zhi, 20*(3), 310–313.

Chen, X., Wang, L., Liu, J., Huang, L., Yang, L., Gao, Q., et al. (2017). Expression and prognostic relevance of MAGE-A3 and MAGE-C2 in non-small cell lung cancer. *Oncology Letters, 13*(3), 1609–1618.

Chen, X., Wang, L., Yue, D., Liu, J., Huang, L., Yang, L., et al. (2017). Correlation between the high expression levels of cancer-germline genes with clinical characteristics in esophageal squamous cell carcinoma. *Histology and Histopathology, 32*(8), 793–803.

Cheville, J. C., & Roche, P. C. (1999). MAGE-1 and MAGE-3 tumor rejection antigens in human germ cell tumors. *Modern Pathology, 12*(10), 974–978.

Chi, D. D., Merchant, R. E., Rand, R., Conrad, A. J., Garrison, D., Turner, R., et al. (1997). Molecular detection of tumor-associated antigens shared by human cutaneous melanomas and gliomas. *American Journal of Pathology, 150*(6), 2143–2152.

Chinnasamy, N., Wargo, J. A., Yu, Z., Rao, M., Frankel, T. L., Riley, J. P., et al. (2011). A TCR targeting the HLA-A*0201-restricted epitope of MAGE-A3 recognizes multiple epitopes of the MAGE-A antigen superfamily in several types of cancer. *Journal of Immunology, 186*(2), 685–696.

Cogdill, A. P., Frederick, D. T., Cooper, Z. A., Garber, H. R., Ferrone, C. R., Fiedler, A., et al. (2012). Targeting the MAGE A3 antigen in pancreatic cancer. *Surgery, 152*(3 Suppl. 1), S13–S18.

Collisson, E. A., Bailey, P., Chang, D. K., & Biankin, A. V. (2019). Molecular subtypes of pancreatic cancer. *Nature Review of Gastroenterology and Hepatology, 16*(4), 207–220.

Consogno, G., Manici, S., Facchinetti, V., Bachi, A., Hammer, J., Conti-Fine, B. M., et al. (2003). Identification of immunodominant regions among promiscuous HLA-DR-restricted CD4+ T-cell epitopes on the tumor antigen MAGE-3. *Blood, 101*(3), 1038–1044.

Corrias, M. V., Scaruffi, P., Occhino, M., De Bernardi, B., Tonini, G. P., & Pistoia, V. (1996). Expression of MAGE-1, MAGE-3 and MART-1 genes in neuroblastoma. *International Journal of Cancer, 69*(5), 403–407.

Coulie, P. G., Karanikas, V., Lurquin, C., Colau, D., Connerotte, T., Hanagiri, T., et al. (2002). Cytolytic T-cell responses of cancer patients vaccinated with a MAGE antigen. *Immunological Reviews*, *188*, 33–42.

Cronwright, G., Le Blanc, K., Gotherstrom, C., Darcy, P., Ehnman, M., & Brodin, B. (2005). Cancer/testis antigen expression in human mesenchymal stem cells: Down-regulation of SSX impairs cell migration and matrix metalloproteinase 2 expression. *Cancer Research*, *65*(6), 2207–2215.

Dalerba, P., Frascella, E., Macino, B., Mandruzzato, S., Zambon, A., Rosolen, A., et al. (2001). MAGE, BAGE and GAGE gene expression in human rhabdomyosarcomas. *International Journal of Cancer*, *93*(1), 85–90.

Das, B., & Senapati, S. (2019). Functional and mechanistic studies reveal MAGEA3 as a pro-survival factor in pancreatic cancer cells. *Journal of Experimental & Clinical Cancer Research*, *38*(1), 294.

Das, B., Suklabaidya, S., Jain, S., Baisakh, M. R., & Senapati, S. (2014). MAGEA3 (melanoma antigen family A, 3). *Atlas of Genetics and Cytogenetics in Oncology and Haematology*, *19*(1), 28–31.

De Plaen, E., Arden, K., Traversari, C., Gaforio, J. J., Szikora, J. P., De Smet, C., et al. (1994). Structure, chromosomal localization, and expression of 12 genes of the MAGE family. *Immunogenetics*, *40*(5), 360–369.

Declerck, S., & Vansteenkiste, J. (2014). Immunotherapy for lung cancer: Ongoing clinical trials. *Future Oncology*, *10*(1), 91–105.

Derre, L., Cesson, V., Lucca, I., Cerantola, Y., Valerio, M., Fritschi, U., et al. (2017). Intravesical Bacillus Calmette Guerin combined with a cancer vaccine increases local T-cell responses in non-muscle-invasive bladder cancer patients. *Clinical Cancer Research*, *23*(3), 717–725.

Destexhe, E., Grosdidier, E., Baudson, N., Forster, R., Gerard, C., Garcon, N., et al. (2015). Non-clinical safety evaluation of single and repeated intramuscular administrations of MAGE-A3 Cancer Immunotherapeutic in rabbits and cynomolgus monkeys. *Journal of Applied Toxicology*, *35*(7), 717–728.

Destexhe, E., Stannard, D., Wilby, O. K., Grosdidier, E., Baudson, N., Forster, R., et al. (2015). Nonclinical reproductive and developmental safety evaluation of the MAGE-A3 cancer Immunotherapeutic, a therapeutic vaccine for cancer treatment. *Reproductive Toxicology*, *51*, 90–105.

Dhodapkar, K. M., Krasovsky, J., Williamson, B., & Dhodapkar, M. V. (2002). Antitumor monoclonal antibodies enhance cross-presentation of cellular antigens and the generation of myeloma-specific killer T cells by dendritic cells. *Journal of Experimental Medicine*, *195*(1), 125–133.

Ding, M., Beck, R. J., Keller, C. J., & Fenton, R. G. (1994). Cloning and analysis of MAGE-1-related genes. *Biochemical and Biophysical Research Communications*, *202*(1), 549–555.

Duan, Z., Duan, Y., Lamendola, D. E., Yusuf, R. Z., Naeem, R., Penson, R. T., et al. (2003). Overexpression of MAGE/GAGE genes in paclitaxel/doxorubicin-resistant human cancer cell lines. *Clinical Cancer Research*, *9*(7), 2778–2785.

Dullaers, M., Breckpot, K., Van Meirvenne, S., Bonehill, A., Tuyaerts, S., Michiels, A., et al. (2004). Side-by-side comparison of lentivirally transduced and mRNA-electroporated dendritic cells: Implications for cancer immunotherapy protocols. *Molecular Therapy*, *10*(4), 768–779.

Durantez, M., Lopez-Vazquez, A. B., de Cerio, A. L., Huarte, E., Casares, N., Prieto, J., et al. (2009). Induction of multiepitopic and long-lasting immune responses against tumour antigens by immunization with peptides, DNA and recombinant adenoviruses expressing minigenes. *Scandinavian Journal of Immunology*, *69*(2), 80–89.

Dyrskjot, L., Zieger, K., Kissow Lildal, T., Reinert, T., Gruselle, O., Coche, T., et al. (2012). Expression of MAGE-A3, NY-ESO-1, LAGE-1 and PRAME in urothelial carcinoma. *British Journal of Cancer*, *107*(1), 116–122.

Eifuku, R., Takenoyama, M., Yoshino, I., Imahayashi, S., So, T., Yasuda, M., et al. (2001). Analysis of MAGE-3 derived synthetic peptide as a human lung cancer antigen recognized by cytotoxic T lymphocytes. *International Journal of Clinical Oncology*, *6*(1), 34–39.

Esfandiary, A., & Ghafouri-Fard, S. (2015). MAGE-A3: An immunogenic target used in clinical practice. *Immunotherapy*, 7(6), 683–704.

Fujie, T., Mori, M., Ueo, H., Sugimachi, K., & Akiyoshi, T. (1997). Expression of MAGE and BAGE genes in Japanese breast cancers. *Annals of Oncology*, *8*(4), 369–372.

Fujie, T., Tanaka, F., Mori, M., Takesako, K., Sugimachi, K., & Akiyoshi, T. (1997). Induction of antitumor cytotoxic T lymphocytes from the peripheral blood mononuclear cells of cancer patients using HLA-A2-restricted MAGE-3 peptide in vitro. *Clinical Cancer Research*, *3*(12 Pt. 1), 2425–2430.

Gattoni-Celli, S., & Cole, D. J. (1996). Melanoma-associated tumor antigens and their clinical relevance to immunotherapy. *Seminars in Oncology*, *23*(6), 754–758.

Gaugler, B., Van den Eynde, B., van der Bruggen, P., Romero, P., Gaforio, J. J., De Plaen, E., et al. (1994). Human gene MAGE-3 codes for an antigen recognized on a melanoma by autologous cytolytic T lymphocytes. *Journal of Experimental Medicine*, *179*(3), 921–930.

Gerard, C., Baudson, N., Ory, T., & Louahed, J. (2014). Tumor mouse model confirms MAGE-A3 cancer immunotherapeutic as an efficient inducer of long-lasting antitumoral responses. *PLoS One*, *9*(5), e94883.

Goodyear, O. C., Pratt, G., McLarnon, A., Cook, M., Piper, K., & Moss, P. (2008). Differential pattern of CD4+ and CD8+ T-cell immunity to MAGE-A1/A2/A3 in patients with monoclonal gammopathy of undetermined significance (MGUS) and multiple myeloma. *Blood*, *112*(8), 3362–3372.

Graff-Dubois, S., Faure, O., Gross, D. A., Alves, P., Scardino, A., Chouaib, S., et al. (2002). Generation of CTL recognizing an HLA-A*0201-restricted epitope shared by MAGE-A1, -A2, -A3, -A4, -A6, -A10, and -A12 tumor antigens: Implication in a broad-spectrum tumor immunotherapy. *Journal of Immunology*, *169*(1), 575–580.

Guo, L., Sang, M., Liu, Q., Fan, X., Zhang, X., & Shan, B. (2013). The expression and clinical significance of melanoma-associated antigen-A1, -A3 and -A11 in glioma. *Oncology Letters*, *6*(1), 55–62.

Gure, A. O., Chua, R., Williamson, B., Gonen, M., Ferrera, C. A., Gnjatic, S., et al. (2005). Cancer-testis genes are coordinately expressed and are markers of poor outcome in non-small cell lung cancer. *Clinical Cancer Research*, *11*(22), 8055–8062.

Han, M. H., Eom, H. S., Park, W. S., Yun, T., Park, S., Kim, H. J., et al. (2010). Detection of circulating lymphoma cells in patients with non-Hodgkin lymphoma using MAGE-A3 gene expression in peripheral blood. *Leukemia Research*, *34*(9), 1127–1131.

Hersey, P., Menzies, S. W., Coventry, B., Nguyen, T., Farrelly, M., Collins, S., et al. (2005). Phase I/II study of immunotherapy with T-cell peptide epitopes in patients with stage IV melanoma. *Cancer Immunology, Immunotherapy*, *54*(3), 208–218.

Hersey, P., Menzies, S. W., Halliday, G. M., Nguyen, T., Farrelly, M. L., DeSilva, C., et al. (2004). Phase I/II study of treatment with dendritic cell vaccines in patients with disseminated melanoma. *Cancer Immunology, Immunotherapy*, *53*(2), 125–134.

Hiret, S., Senellart, H., & Bennouna, J. (2010). Molecular biology of lung cancer series. *Revue des Maladies Respiratoires*, *27*(8), 954–958.

Hobo, W., Strobbe, L., Maas, F., Fredrix, H., Greupink-Draaisma, A., Esendam, B., et al. (2013). Immunogenicity of dendritic cells pulsed with MAGE3, Survivin and B-cell maturation antigen mRNA for vaccination of multiple myeloma patients. *Cancer Immunology and Immunotherapy*, *62*(8), 1381–1392.

Inoue, H., Mori, M., Honda, M., Li, J., Shibuta, K., Mimori, K., et al. (1995). The expression of tumor-rejection antigen "MAGE" genes in human gastric carcinoma. *Gastroenterology*, *109*(5), 1522–1525.

Inoue, H., Mori, M., Li, J., Mimori, K., Honda, M., Nakashima, H., et al. (1995). Human esophageal carcinomas frequently express the tumor-rejection antigens of MAGE genes. *International Journal of Cancer, 63*(4), 523–526.

Ishida, H., Matsumura, T., Salgaller, M. L., Ohmizono, Y., Kadono, Y., & Sawada, T. (1996). MAGE-1 and MAGE-3 or -6 expression in neuroblastoma-related pediatric solid tumors. *International Journal of Cancer, 69*(5), 375–380.

Jenne, L., Arrighi, J. F., Jonuleit, H., Saurat, J. H., & Hauser, C. (2000). Dendritic cells containing apoptotic melanoma cells prime human CD8+ T cells for efficient tumor cell lysis. *Cancer Research, 60*(16), 4446–4452.

Jian, W., Li, X., Kang, J., Lei, Y., Bai, Y., & Xue, Y. (2018). Antitumor effect of recombinant Mycobacterium smegmatis expressing MAGEA3 and SSX2 fusion proteins. *Experimenatal and Therapeutic Medicine, 16*(3), 2160–2166.

Jin, X., Pan, Y., Wang, L., Zhang, L., Ravichandran, R., Potts, P. R., et al. (2017). MAGE-TRIM28 complex promotes the Warburg effect and hepatocellular carcinoma progression by targeting FBP1 for degradation. *Oncogene, 6*(4), e312.

Jungbluth, A. A., Ely, S., DiLiberto, M., Niesvizky, R., Williamson, B., Frosina, D., et al. (2005). The cancer-testis antigens CT7 (MAGE-C1) and MAGE-A3/6 are commonly expressed in multiple myeloma and correlate with plasma-cell proliferation. *Blood, 106*(1), 167–174.

Kanaoka, S., Yamasaki, S., Okino, T., Inoue, N., Shimada, Y., Kaneko, M., et al. (1999). Induction of human leukocyte antigen (HLA)-A2-restricted and MAGE-3-gene-derived peptide-specific cytolytic T lymphocytes using cultured dendritic cells from an HLA-A2 esophageal cancer patient. *Journal of Surgical Oncology, 71*(1), 16–21.

Kariyama, K., Higashi, T., Kobayashi, Y., Nouso, K., Nakatsukasa, H., Yamano, T., et al. (1999). Expression of MAGE-1 and -3 genes and gene products in human hepatocellular carcinoma. *British Journal of Cancer, 81*(6), 1080–1087.

Keiholz, U., Scheibenbogen, C., Theil, E., & Schadendorf, D. (2001). Rational development of peptide vaccination in clinical oncology. *Onkologie, 24*(2), 174.

Kim, Y. M., Lee, Y. H., Shin, S. H., Kim, E. H., Choi, Y. W., Lee, K. M., et al. (2001). Expression of MAGE-1, -2, and -3 genes in gastric carcinomas and cancer cell lines derived from Korean patients. *Journal of Korean Medical Science, 16*(1), 62–68.

Kim, J., Reber, H. A., Hines, O. J., Kazanjian, K. K., Tran, A., Ye, X., et al. (2006). The clinical significance of MAGEA3 expression in pancreatic cancer. *International Journal of Cancer, 118*(9), 2269–2275.

Kobayashi, Y., Higashi, T., Nouso, K., Nakatsukasa, H., Ishizaki, M., Kaneyoshi, T., et al. (2000). Expression of MAGE, GAGE and BAGE genes in human liver diseases: Utility as molecular markers for hepatocellular carcinoma. *Journal of Hepatology, 32*(4), 612–617.

Kobayashi, H., Song, Y., Hoon, D. S., Appella, E., & Celis, E. (2001). Tumor-reactive T helper lymphocytes recognize a promiscuous MAGE-A3 epitope presented by various major histocompatibility complex class II alleles. *Cancer Research, 61*(12), 4773–4778.

Krishnadas, D. K., Bao, L., Bai, F., Chencheri, S. C., & Lucas, K. (2014). Decitabine facilitates immune recognition of sarcoma cells by upregulating CT antigens, MHC molecules, and ICAM-1. *Tumour Biology, 35*(6), 5753–5762.

Krishnadas, D. K., Shusterman, S., Bai, F., Diller, L., Sullivan, J. E., Cheerva, A. C., et al. (2015). A phase I trial combining decitabine/dendritic cell vaccine targeting MAGE-A1, MAGE-A3 and NY-ESO-1 for children with relapsed or therapy-refractory neuroblastoma and sarcoma. *Cancer Immunology, Immunotherapy, 64*(10), 1251–1260.

Lausenmeyer, E. M., Braun, K., Breyer, J., Gierth, M., Denzinger, S., Burger, M., et al. (2019). Strong expression of cancer testis antigens CTAG1B and MAGEA3 is correlated with unfavourable histopathological features and MAGEA3 is associated with worse progression-free survival in urothelial bladder cancer. *Urologia Internationalis, 102*(1), 77–82.

Lee, K. D., Chang, H. K., Jo, Y. K., Kim, B. S., Lee, B. H., Lee, Y. W., et al. (1999). Expression of the MAGE 3 gene product in squamous cell carcinomas of the head and neck. *Anticancer Research*, *19*(6B), 5037–5042.

Lesimple, T., Moisan, A., Carsin, A., Ollivier, I., Mousseau, M., Meunier, B., et al. (2003). Injection by various routes of melanoma antigen-associated macrophages: Biodistribution and clinical effects. *Cancer Immunology, Immunotherapy*, *52*(7), 438–444.

Li, X. Z., Han, Y., Tian, J., Ren, X., Ma, X., & Ma, M. (2016). Enhancement of dendritic cells with melanoma-associated antigen 3 for inducing cytotoxicity by cytotoxic T lymphocytes on bladder cancer BIU-87 cells. *Genetics and Molecular Research*, *15*(3), 1–11, gmr.15039001.

Li, J., Yang, Y., Fujie, T., Baba, K., Ueo, H., Mori, M., et al. (1996). Expression of BAGE, GAGE, and MAGE genes in human gastric carcinoma. *Clinical Cancer Research*, *2*(9), 1619–1625.

Lifantseva, N., Koltsova, A., Krylova, T., Yakovleva, T., Poljanskaya, G., & Gordeeva, O. (2011). Expression patterns of cancer-testis antigens in human embryonic stem cells and their cell derivatives indicate lineage tracks. *Stem Cells International*, *2011*, 795239.

Lin, L., Wei, J., Chen, Y., Huang, A., Li, K. K., & Zhang, W. (2014). Induction of antigen-specific immune responses by dendritic cells transduced with a recombinant lentiviral vector encoding MAGE-A3 gene. *Journal of Cancer Research and Clinical Oncology*, *140*(2), 281–289.

Linette, G. P., Stadtmauer, E. A., Maus, M. V., Rapoport, A. P., Levine, B. L., Emery, L., et al. (2013). Cardiovascular toxicity and titin cross-reactivity of affinity-enhanced T cells in myeloma and melanoma. *Blood*, *122*(6), 863–871.

Liu, W., Cheng, S., Asa, S. L., & Ezzat, S. (2008). The melanoma-associated antigen A3 mediates fibronectin-controlled cancer progression and metastasis. *Cancer Research*, *68*(19), 8104–8112.

Liu, X., Li, J., Liu, Y., Ding, J., Tong, Z., Liu, Y., et al. (2016). Calreticulin acts as an adjuvant to promote dendritic cell maturation and enhances antigen-specific cytotoxic T lymphocyte responses against non-small cell lung cancer cells. *Cellular Immunology*, *300*, 46–53.

Liu, X., Song, N., Liu, Y., Liu, Y., Li, J., Ding, J., et al. (2015). Efficient induction of anti-tumor immune response in esophageal squamous cell carcinoma via dendritic cells expressing MAGE-A3 and CALR antigens. *Cellular Immunology*, *295*(2), 77–82.

Liu, Q., Zhang, C., & Feng, K. (2000). Expression of MAGE-3 gene product in non-small cell lung cancer. *Zhonghua Zhong Liu Za Zhi*, *22*(2), 138–140.

Lorenzi, S., Forloni, M., Cifaldi, L., Antonucci, C., Citti, A., Boldrini, R., et al. (2012). IRF1 and NF-kB restore MHC class I-restricted tumor antigen processing and presentation to cytotoxic T cells in aggressive neuroblastoma. *PLoS One*, 7(10), e46928.

Lu, B., Liu, B. Y., Zhang, Y., Chen, X. H., Zhu, Z. G., Yin, H. R., et al. (2005). Inhibitory effect of dendritic cells pulsed with MAGE-3 peptide on transplanted murine gastric cancer in mice. *Zhonghua Yi Xue Za Zhi*, *85*(30), 2120–2123.

Lu, Y. C., Parker, L. L., Lu, T., Zheng, Z., Toomey, M. A., White, D. E., et al. (2017). Treatment of patients with metastatic cancer using a major histocompatibility complex class II-restricted T-cell receptor targeting the cancer germline antigen MAGE-A3. *Journal of Clinical Oncology*, *35*(29), 3322–3329.

Ma, M., Yu, L. Z., Na, Y. Q., Geng, L., Li, H. W., Yang, X. Y., et al. (2004). Expression of MAGE genes and MAGE gene products in human renal and urinary bladder tumor. *Beijing Da Xue Xue Bao Yi Xue Ban*, *36*(2), 159–163.

Makhlouf, M. M. (2011). Melanoma antigen encoding gene-A3 expression, its clinical significance, and its risk of developing acute myeloid leukemia. *Journal of Investigative Medicine*, *59*(7), 1131–1136.

Makohon-Moore, A., & Iacobuzio-Donahue, C. A. (2016). Pancreatic cancer biology and genetics from an evolutionary perspective. *Nature Reviews Cancer, 16*(9), 553–565.
Marcar, L., Maclaine, N. J., Hupp, T. R., & Meek, D. W. (2010). Mage-A cancer/testis antigens inhibit p53 function by blocking its interaction with chromatin. *Cancer Research, 70*(24), 10362–10370.
Marchand, M., van Baren, N., Weynants, P., Brichard, V., Dreno, B., Tessier, M. H., et al. (1999). Tumor regressions observed in patients with metastatic melanoma treated with an antigenic peptide encoded by gene MAGE-3 and presented by HLA-A1. *International Journal of Cancer, 80*(2), 219–230.
Martinez, A., Olarte, I., Mergold, M. A., Gutierrez, M., Rozen, E., Collazo, J., et al. (2007). mRNA expression of MAGE-A3 gene in leukemia cells. *Leukemia Research, 31*(1), 33–37.
Milkovic, M., Sarcevic, B., & Glavan, E. (2006). Expression of MAGE tumor-associated antigen in thyroid carcinomas. *Endocrine Pathology, 17*(1), 45–52.
Minami, K., Yamaguchi, Y., Ohshita, A., Kawabuchi, Y., Ohta, K., Hihara, J., et al. (2005). Generation of antigen-presenting cells using cultured dendritic cells and amplified autologous tumor mRNA. *Oncology, 69*(5), 399–407.
Miranda, E. I. (2010). MAGE, biological functions and potential clinical applications. *Leukemia Research, 34*(9), 1121–1122.
Moeller, I., Spagnoli, G. C., Finke, J., Veelken, H., & Houet, L. (2012). Uptake routes of tumor-antigen MAGE-A3 by dendritic cells determine priming of naive T-cell subtypes. *Cancer Immunology, Immunotherapy, 61*(11), 2079–2090.
Monte, M., Simonatto, M., Peche, L. Y., Bublik, D. R., Gobessi, S., Pierotti, M. A., et al. (2006). MAGE-A tumor antigens target p53 transactivation function through histone deacetylase recruitment and confer resistance to chemotherapeutic agents. *Proceedings of the National Academy of Sciences of the United States of America, 103*(30), 11160–11165.
Moreno-Bost, A., Szmania, S., Stone, K., Garg, T., Hoerring, A., Szymonifka, J., et al. (2011). Epigenetic modulation of MAGE-A3 antigen expression in multiple myeloma following treatment with the demethylation agent 5-azacitidine and the histone deacetylase inhibitor MGCD0103. *Cytotherapy, 13*(5), 618–628.
Morgan, R. A., Chinnasamy, N., Abate-Daga, D., Gros, A., Robbins, P. F., Zheng, Z., et al. (2013). Cancer regression and neurological toxicity following anti-MAGE-A3 TCR gene therapy. *Journal of Immunotherapy, 36*(2), 133–151.
Mori, M., Inoue, H., Mimori, K., Shibuta, K., Baba, K., Nakashima, H., et al. (1996). Expression of MAGE genes in human colorectal carcinoma. *Annals of Surgery, 224*(2), 183–188.
Morishima, N., Nakanishi, K., Takenouchi, H., Shibata, T., & Yasuhiko, Y. (2002). An endoplasmic reticulum stress-specific caspase cascade in apoptosis. Cytochrome c-independent activation of caspase-9 by caspase-12. *Journal of Biological Chemistry, 277*(37), 34287–34294.
Nardiello, T., Jungbluth, A. A., Mei, A., Diliberto, M., Huang, X., Dabrowski, A., et al. (2011). MAGE-A inhibits apoptosis in proliferating myeloma cells through repression of Bax and maintenance of survivin. *Clinical Cancer Research, 17*(13), 4309–4319.
Nishimura, S., Fujita, M., Terata, N., Tani, T., Kodama, M., & Itoh, K. (1997). Expression of MAGE genes in colorectal carcinomas. *Nihon Rinshō Men'eki Gakkai Kaishi, 20*(2), 95–101.
Nishiyama, T., Tachibana, M., Horiguchi, Y., Nakamura, K., Ikeda, Y., Takesako, K., et al. (2001). Immunotherapy of bladder cancer using autologous dendritic cells pulsed with human lymphocyte antigen-A24-specific MAGE-3 peptide. *Clinical Cancer Reseach*, 7(1), 23–31.
Olarte, I., Martinez, A., Ramos-Penafiel, C., Castellanos-Sinco, H., Zamora, J., Collazo-Jaloma, J., et al. (2011). MAGE-A3 expression is an adverse prognostic factor in diffuse large B-cell lymphoma. *Hematology, 16*(6), 368–372.

Osanto, S., Schiphorst, P. P., Weijl, N. I., Dijkstra, N., Van Wees, A., Brouwenstein, N., et al. (2000). Vaccination of melanoma patients with an allogeneic, genetically modified interleukin 2-producing melanoma cell line. *Human Gene Therapy*, *11*(5), 739–750.

Oshita, C., Takikawa, M., Kume, A., Miyata, H., Ashizawa, T., Iizuka, A., et al. (2012). Dendritic cell-based vaccination in metastatic melanoma patients: Phase II clinical trial. *Oncology Reports*, *28*(4), 1131–1138.

Otte, M., Zafrakas, M., Riethdorf, L., Pichlmeier, U., Loning, T., Janicke, F., et al. (2001). MAGE-A gene expression pattern in primary breast cancer. *Cancer Research*, *61*(18), 6682–6687.

Peled, N., Oton, A. B., Hirsch, F. R., & Bunn, P. (2009). MAGE A3 antigen-specific cancer immunotherapeutic. *Immunotherapy*, *1*(1), 19–25.

Pineda, C. T., & Potts, P. R. (2015). Oncogenic MAGEA-TRIM28 ubiquitin ligase downregulates autophagy by ubiquitinating and degrading AMPK in cancer. *Autophagy*, *11*(5), 844–846.

Pineda, C. T., Ramanathan, S., Fon Tacer, K., Weon, J. L., Potts, M. B., Ou, Y. H., et al. (2015). Degradation of AMPK by a cancer-specific ubiquitin ligase. *Cell*, *160*(4), 715–728.

Pol, J. G., Acuna, S. A., Yadollahi, B., Tang, N., Stephenson, K. B., Atherton, M. J., et al. (2019). Preclinical evaluation of a MAGE-A3 vaccination utilizing the oncolytic Maraba virus currently in first-in-human trials. *Oncoimmunology*, *8*(1), e1512329.

Pujol, J. L., Vansteenkiste, J. F., De Pas, T. M., Atanackovic, D., Reck, M., Thomeer, M., et al. (2015). Safety and immunogenicity of MAGE-A3 cancer immunotherapeutic with or without adjuvant chemotherapy in patients with resected stage IB to III MAGE-A3-positive non-small-cell lung cancer. *Journal of Thoracic Oncology*, *10*(10), 1458–1467.

Qu, P., Ma, J. H., Zhang, X. M., Huang, X. J., Yang, X. W., & Yan-Fang, S. (2010). A novel DNA vaccine constructed by heat shock protein 70 and melanoma antigen-encoding gene 3 against tumorigenesis. *Indian Journal of Experimental Biology*, *48*(5), 436–443.

Quillien, V., Raoul, J. L., Heresbach, D., Collet, B., Toujas, L., & Brasseur, F. (1997). Expression of MAGE genes in esophageal squamous-cell carcinoma. *Anticancer Research*, *17*(1A), 387–391.

Raman, M. C., Rizkallah, P. J., Simmons, R., Donnellan, Z., Dukes, J., Bossi, G., et al. (2016). Direct molecular mimicry enables off-target cardiovascular toxicity by an enhanced affinity TCR designed for cancer immunotherapy. *Scientific Reports*, *6*, 18851.

Rao, M., Chinnasamy, N., Hong, J. A., Zhang, Y., Zhang, M., Xi, S., et al. (2011). Inhibition of histone lysine methylation enhances cancer-testis antigen expression in lung cancer cells: Implications for adoptive immunotherapy of cancer. *Cancer Research*, *71*(12), 4192–4204.

Rapoport, A. P., Aqui, N. A., Stadtmauer, E. A., Vogl, D. T., Xu, Y. Y., Kalos, M., et al. (2014). Combination immunotherapy after ASCT for multiple myeloma using MAGE-A3/Poly-ICLC immunizations followed by adoptive transfer of vaccine-primed and costimulated autologous T cells. *Clinical Cancer Research*, *20*(5), 1355–1365.

Reck, M. (2012). What future opportunities may immuno-oncology provide for improving the treatment of patients with lung cancer? *Annals of Oncology*, *23* (Suppl. 8), viii28–34.

Reynolds, S. R., Celis, E., Sette, A., Oratz, R., Shapiro, R. L., Johnston, D., et al. (2000). Identification of HLA-A*03, A*11 and B*07-restricted melanoma-associated peptides that are immunogenic in vivo by vaccine-induced immune response (VIIR) analysis. *Journal of Immunological Methods*, *244*(1–2), 59–67.

Ries, J., Mollaoglu, N., Toyoshima, T., Vairaktaris, E., Neukam, F. W., Ponader, S., et al. (2009). A novel multiple-marker method for the early diagnosis of oral squamous cell carcinoma. *Disease Markers*, *27*(2), 75–84.

Sadanaga, N., Nagashima, H., Mashino, K., Tahara, K., Yamaguchi, H., Ohta, M., et al. (2001). Dendritic cell vaccination with MAGE peptide is a novel therapeutic approach for gastrointestinal carcinomas. *Clinical Cancer Research*, *7*(8), 2277–2284.

Sahin, U., Koslowski, M., Tureci, O., Eberle, T., Zwick, C., Romeike, B., et al. (2000). Expression of cancer testis genes in human brain tumors. *Clinical Cancer Research*, *6*(10), 3916–3922.

Saiag, P., Gutzmer, R., Ascierto, P. A., Maio, M., Grob, J. J., Murawa, P., et al. (2016). Prospective assessment of a gene signature potentially predictive of clinical benefit in metastatic melanoma patients following MAGE-A3 immunotherapeutic (PREDICT). *Annals of Oncology*, *27*(10), 1947–1953.

Saldanha-Araujo, F., Haddad, R., Zanette, D. L., De Araujo, A. G., Orellana, M. D., Covas, D. T., et al. (2010). Cancer/Testis antigen expression on mesenchymal stem cells isolated from different tissues. *Anticancer Research*, *30*(12), 5023–5027.

Sang, M., Lian, Y., Zhou, X., & Shan, B. (2011). MAGE-A family: Attractive targets for cancer immunotherapy. *Vaccine*, *29*(47), 8496–8500.

Schaft, N., Dorrie, J., Thumann, P., Beck, V. E., Muller, I., Schultz, E. S., et al. (2005). Generation of an optimized polyvalent monocyte-derived dendritic cell vaccine by transfecting defined RNAs after rather than before maturation. *Journal of Immunology*, *174*(5), 3087–3097.

Schultz, E. S., Lethe, B., Cambiaso, C. L., Van Snick, J., Chaux, P., Corthals, J., et al. (2000). A MAGE-A3 peptide presented by HLA-DP4 is recognized on tumor cells by CD4+ cytolytic T lymphocytes. *Cancer Research*, *60*(22), 6272–6275.

Schultz, E. S., Zhang, Y., Knowles, R., Tine, J., Traversari, C., Boon, T., et al. (2001). A MAGE-3 peptide recognized on HLA-B35 and HLA-A1 by cytolytic T lymphocytes. *Tissue Antigens*, *57*(2), 103–109.

Semaan, A., & Maitra, A. (2018). Pancreatic cancer in 2017: Rebooting pancreatic cancer knowledge and treatment options. *Nature Reviews Gastroenterology & Hepatology*, *15*(2), 76–78.

Shantha Kumara, H. M., Grieco, M. J., Caballero, O. L., Su, T., Ahmed, A., Ritter, E., et al. (2012). MAGE-A3 is highly expressed in a subset of colorectal cancer patients. *Cancer Immunity*, *12*, 16.

Sienel, W., Varwerk, C., Linder, A., Kaiser, D., Teschner, M., Delire, M., et al. (2004). Melanoma associated antigen (MAGE)-A3 expression in Stages I and II non-small cell lung cancer: Results of a multi-center study. *European Journal of Cardiothoracic Surgery*, *25*(1), 131–134.

Soling, A., Schurr, P., & Berthold, F. (1999). Expression and clinical relevance of NY-ESO-1, MAGE-1 and MAGE-3 in neuroblastoma. *Anticancer Research*, *19*(3B), 2205–2209.

Straetemans, T., van Brakel, M., van Steenbergen, S., Broertjes, M., Drexhage, J., Hegmans, J., et al. (2012). TCR gene transfer: MAGE-C2/HLA-A2 and MAGE-A3/HLA-DP4 epitopes as melanoma-specific immune targets. *Clinical & Developmental Immunology*, *2012*, 586314.

Sudo, T., Kuramoto, T., Komiya, S., Inoue, A., & Itoh, K. (1997). Expression of MAGE genes in osteosarcoma. *Journal of Orthopaedic Research*, *15*(1), 128–132.

Suzuki, K., Tsujitani, S., Konishi, I., Yamaguchi, Y., Hirooka, Y., & Kaibara, N. (1999). Expression of MAGE genes and survival in patients with hepatocellular carcinoma. *International Journal of Oncology*, *15*(6), 1227–1232.

Syed, O. N., Mandigo, C. E., Killory, B. D., Canoll, P., & Bruce, J. N. (2012). Cancer-testis and melanocyte-differentiation antigen expression in malignant glioma and meningioma. *Journal of Clinical Neuroscience*, *19*(7), 1016–1021.

Taborska, P., Stakheev, D., Strizova, Z., Vavrova, K., Podrazil, M., Bartunkova, J., et al. (2017). Personalized ex vivo multiple peptide enrichment and detection of T cells reactive to multiple tumor-associated antigens in prostate cancer patients. *Medical Oncology*, *34*(10), 173.

Tahara, K., Mori, M., Sadanaga, N., Sakamoto, Y., Kitano, S., & Makuuchi, M. (1999). Expression of the MAGE gene family in human hepatocellular carcinoma. *Cancer*, *85*(6), 1234–1240.

Tajima, K., Obata, Y., Tamaki, H., Yoshida, M., Chen, Y. T., Scanlan, M. J., et al. (2003). Expression of cancer/testis (CT) antigens in lung cancer. *Lung Cancer*, *42*(1), 23–33.

Tanaka, F., Fujie, T., Tahara, K., Mori, M., Takesako, K., Sette, A., et al. (1997). Induction of antitumor cytotoxic T lymphocytes with a MAGE-3-encoded synthetic peptide presented by human leukocytes antigen-A24. *Cancer Research*, *57*(20), 4465–4468.

Tanaka, F., Haraguchi, N., Isikawa, K., Inoue, H., & Mori, M. (2008). Potential role of dendritic cell vaccination with MAGE peptides in gastrointestinal carcinomas. *Oncology Reports*, *20*(5), 1111–1116.

Tanaka, F., Yamaguchi, H., Haraguchi, N., Mashino, K., Ohta, M., Inoue, H., et al. (2006). Efficient induction of specific cytotoxic T lymphocytes to tumor rejection peptide using functional matured 2 day-cultured dendritic cells derived from human monocytes. *International Journal of Oncology*, *29*(5), 1263–1268.

Tanzarella, S., Lionello, I., Valentinis, B., Russo, V., Lollini, P. L., & Traversari, C. (2004). Rhabdomyosarcomas are potential target of MAGE-specific immunotherapies. *Cancer Immunology, Immunotherapy*, *53*(6), 519–524.

Thomas, A., & Hassan, R. (2012). Immunotherapies for non-small-cell lung cancer and mesothelioma. *Lancet Oncology*, *13*(7), e301–e310.

Thongprasert, S., Yang, P. C., Lee, J. S., Soo, R., Gruselle, O., Myo, A., et al. (2016). The prevalence of expression of MAGE-A3 and PRAME tumor antigens in East and South East Asian non-small cell lung cancer patients. *Lung Cancer*, *101*, 137–144.

Toungouz, M., Lambermont, M., Velu, T., Buelens, C., Vanderheyde, N., Bartholome, E., et al. (2001). Anti-tumor immunotherapy based on dendritic cells. *Journal de la Societe de biologie*, *195*(1), 19–23.

Toungouz, M., Libin, M., Bulte, F., Faid, L., Lehmann, F., Duriau, D., et al. (2001). Transient expansion of peptide-specific lymphocytes producing IFN-gamma after vaccination with dendritic cells pulsed with MAGE peptides in patients with mage-A1/A3-positive tumors. *Journal of Leukocyte Biology*, *69*(6), 937–943.

Trakatelli, M., Toungouz, M., Lambermont, M., Heenen, M., Velu, T., & Bruyns, C. (2005). Immune characterization of clinical grade-dendritic cells generated from cancer patients and genetically modified by an ALVAC vector carrying MAGE minigenes. *Cancer Gene Therapy*, *12*(6), 552–559.

Traversari, C., & Russo, V. (2016). T Cells as Antigen Carriers for Anti-tumor Vaccination. *Methods in Molecular Biology*, *1393*, 97–104.

Tyagi, P., & Mirakhur, B. (2009). MAGRIT: The largest-ever phase III lung cancer trial aims to establish a novel tumor-specific approach to therapy. *Clinical Lung Cancer*, *10*(5), 371–374.

Ueda, Y., Shimizu, K., Itoh, T., Fuji, N., Naito, K., Shiozaki, A., et al. (2007). Induction of peptide-specific immune response in patients with primary malignant melanoma of the esophagus after immunotherapy using dendritic cells pulsed with MAGE peptides. *Japanese Journal of Clinical Oncology*, *37*(2), 140–145.

Ulloa-Montoya, F., Louahed, J., Dizier, B., Gruselle, O., Spiessens, B., Lehmann, F. F., et al. (2013). Predictive gene signature in MAGE-A3 antigen-specific cancer immunotherapy. *Journal of Clinical Oncology*, *31*(19), 2388–2395.

Valmori, D., Lienard, D., Waanders, G., Rimoldi, D., Cerottini, J. C., & Romero, P. (1997). Analysis of MAGE-3-specific cytolytic T lymphocytes in human leukocyte antigen-A2 melanoma patients. *Cancer Research*, *57*(4), 735–741.

van der Bruggen, P., Bastin, J., Gajewski, T., Coulie, P. G., Boel, P., De Smet, C., et al. (1994). A peptide encoded by human gene MAGE-3 and presented by HLA-A2 induces cytolytic T lymphocytes that recognize tumor cells expressing MAGE-3. *European Journal of Immunology*, *24*(12), 3038–3043.

van der Bruggen, P., Traversari, C., Chomez, P., Lurquin, C., De Plaen, E., Van den Eynde, B., et al. (1991). A gene encoding an antigen recognized by cytolytic T lymphocytes on a human melanoma. *Science*, *254*(5038), 1643–1647.

Van Pel, A., van der Bruggen, P., Coulie, P. G., Brichard, V. G., Lethe, B., van den Eynde, B., et al. (1995). Genes coding for tumor antigens recognized by cytolytic T lymphocytes. *Immunological Reviews*, *145*, 229–250.

Vansteenkiste, J. F., Cho, B. C., Vanakesa, T., De Pas, T., Zielinski, M., Kim, M. S., et al. (2016). Efficacy of the MAGE-A3 cancer immunotherapeutic as adjuvant therapy in patients with resected MAGE-A3-positive non-small-cell lung cancer (MAGRIT): A randomised, double-blind, placebo-controlled, phase 3 trial. *Lancet Oncology*, *17*(6), 822–835.

Vantomme, V., Dantinne, C., Amrani, N., Permanne, P., Gheysen, D., Bruck, C., et al. (2004). Immunologic analysis of a phase I/II study of vaccination with MAGE-3 protein combined with the AS02B adjuvant in patients with MAGE-3-positive tumors. *Journal of Immunotherapy*, *27*(2), 124–135.

Vo, M. C., Nguyen-Pham, T. N., Lee, H. J., Jung, S. H., Choi, N. R., Hoang, M. D., et al. (2017). Chaetocin enhances dendritic cell function via the induction of heat shock protein and cancer testis antigens in myeloma cells. *Oncotarget*, *8*(28), 46047–46056.

Vourc'h-Jourdain, M., Volteau, C., Nguyen, J. M., Khammari, A., & Dreno, B. (2009). Melanoma gene expression and clinical course. *Archives of Dermatological Research*, *301*(9), 673–679.

Wang, Y., Song, X., Zheng, Y., Liu, Z., Li, Y., Qian, X., et al. (2018). Cancer/testis Antigen MAGEA3 Interacts with STAT1 and remodels the tumor microenvironment. *International Journal of Medical Sciences*, *15*(14), 1702–1712.

Wascher, R. A., Bostick, P. J., Huynh, K. T., Turner, R., Qi, K., Giuliano, A. E., et al. (2001). Detection of MAGE-A3 in breast cancer patients' sentinel lymph nodes. *British Journal of Cancer*, *85*(9), 1340–1346.

Weinert, B. T., Krishnadath, K. K., Milano, F., Pedersen, A. W., Claesson, M. H., & Zocca, M. B. (2009). Real-time PCR analysis of genes encoding tumor antigens in esophageal tumors and a cancer vaccine. *Cancer Immunity*, *9*, 9.

West, H. J. (2009). Novel targeted agents for lung cancer. *Clinical Lung Cancer*, *10*(Suppl. 1), S41–S46.

Wilgenhof, S., Van Nuffel, A. M., Benteyn, D., Corthals, J., Aerts, C., Heirman, C., et al. (2013). A phase IB study on intravenous synthetic mRNA electroporated dendritic cell immunotherapy in pretreated advanced melanoma patients. *Annals of Oncology*, *24*(10), 2686–2693.

Wong, P. P., Yeoh, C. C., Ahmad, A. S., Chelala, C., Gillett, C., Speirs, V., et al. (2014). Identification of MAGEA antigens as causal players in the development of tamoxifen-resistant breast cancer. *Oncogene*, *33*(37), 4579–4588.

Wu, F., Liu, F., Dong, L., Yang, H., He, X., Li, L., et al. (2018). miR-1273g silences MAGEA3/6 to inhibit human colorectal cancer cell growth via activation of AMPK signaling. *Cancer Letters*, *435*, 1–9.

Wu, L., Zhang, H., Jiang, Y., Gallo, R. C., & Cheng, H. (2018). Induction of antitumor cytotoxic lymphocytes using engineered human primary blood dendritic cells. *Proceedings of the National Academy of Sciences of the United States of America*, *115*(19), E4453–E4462.

Xiao, T. Z., Bhatia, N., Urrutia, R., Lomberk, G. A., Simpson, A., & Longley, B. J. (2011). MAGE I transcription factors regulate KAP1 and KRAB domain zinc finger transcription factor mediated gene repression. *PLoS One*, *6*(8), e23747.

Xie, C., Subhash, V. V., Datta, A., Liem, N., Tan, S. H., Yeo, M. S., et al. (2016). Melanoma associated antigen (MAGE)-A3 promotes cell proliferation and chemotherapeutic drug resistance in gastric cancer. *Cellular Oncology (Dordrecht)*, *39*(2), 175–186.

Yang, B., O'Herrin, S. M., Wu, J., Reagan-Shaw, S., Ma, Y., Bhat, K. M., et al. (2007). MAGE-A, mMage-b, and MAGE-C proteins form complexes with KAP1 and suppress p53-dependent apoptosis in MAGE-positive cell lines. *Cancer Research*, *67*(20), 9954–9962.

Yin, B., Zeng, Y., Liu, G., Wang, X., Wang, P., & Song, Y. (2014). MAGE-A3 is highly expressed in a cancer stem cell-like side population of bladder cancer cells. *International Journal of Clinical and Experimental Pathology*, 7(6), 2934–2941.

Yin, B., Zeng, Y., Wang, X., Liu, G., Zhang, M., & Song, Y. (2014). Expression and clinical significance of cancer-testis genes in clear cell renal cell carcinoma. *International Journal of Clinical and Experimental Pathology*, 7(7), 4112–4119.

Yoshimatsu, T., Yoshino, I., Ohgami, A., Takenoyama, M., Hanagiri, T., Nomoto, K., et al. (1998). Expression of the melanoma antigen-encoding gene in human lung cancer. *Journal of Surgical Oncology*, *67*(2), 126–129.

You, Z., Hester, J., Rollins, L., Spagnoli, G. C., van der Bruggen, P., & Chen, S. Y. (2001). A retrogen strategy for presentation of an intracellular tumor antigen as an exogenous antigen by dendritic cells induces potent antitumor T helper and CTL responses. *Cancer Research*, *61*(1), 197–205.

Zhang, H. G., Chen, H. S., Peng, J. R., Shang, X. Y., Zhang, J., Xing, Q., et al. (2007). Specific CD8(+)T cell responses to HLA-A2 restricted MAGE-A3 p271-279 peptide in hepatocellular carcinoma patients without vaccination. *Cancer Immunology, Immunotherapy*, *56*(12), 1945–1954.

Zhang, Y., Renkvist, N., Sun, Z., Schuler-Thurner, B., Glaichenhaus, N., Schuler, G., et al. (2005). A polyclonal anti-vaccine CD4 T cell response detected with HLA-DP4 multimers in a melanoma patient vaccinated with MAGE-3.DP4-peptide-pulsed dendritic cells. *European Journal of Immunology*, *35*(4), 1066–1075.

Zhang, X. M., Zhang, Y. F., Huang, Y., Qu, P., Ma, B., Si, S. Y., et al. (2008). The antitumor immune response induced by a combination of MAGE-3/MAGE-n-derived peptides. *Oncology Reports*, *20*(1), 245–252.

Zhang, S., Zhou, X., Yu, H., & Yu, Y. (2010). Expression of tumor-specific antigen MAGE, GAGE and BAGE in ovarian cancer tissues and cell lines. *BMC Cancer*, *10*, 163.

Zhou, M., Peng, J. R., Zhang, H. G., Wang, H. X., Zhong, Z. H., Pan, X. Y., et al. (2005). Identification of two naturally presented MAGE antigenic peptides from a patient with hepatocellular carcinoma by mass spectrometry. *Immunology Letters*, *99*(1), 113–121.

Zhu, X., Asa, S. L., & Ezzat, S. (2008). Fibroblast growth factor 2 and estrogen control the balance of histone 3 modifications targeting MAGE-A3 in pituitary neoplasia. *Clinical Cancer Research*, *14*(7), 1984–1996.

Zhu, S., Van den Eynde, B. J., Coulie, P. G., Li, Y. F., El-Gamil, M., Rosenberg, S. A., et al. (2012). Characterization of T-cell receptors directed against HLA-A*01-restricted and C*07-restricted epitopes of MAGE-A3 and MAGE-A12. *Journal of Immunotherapy*, *35*(9), 680–688.

Zou, C., Shen, J., Tang, Q., Yang, Z., Yin, J., Li, Z., et al. (2012). Cancer-testis antigens expressed in osteosarcoma identified by gene microarray correlate with a poor patient prognosis. *Cancer*, *118*(7), 1845–1855.

CHAPTER SIX

Role of histone acetyltransferase inhibitors in cancer therapy

Muthu K. Shanmugam[a], Arunasalam Dharmarajan[b], Sudha Warrier[c], Anupam Bishayee[d], Alan Prem Kumar[a,e], Gautam Sethi[a,*], and Kwang Seok Ahn[f,*]

[a]Department of Pharmacology, Yong Loo Lin School of Medicine, National University of Singapore, Singapore, Singapore
[b]Department of Biomedical Sciences, Faculty of Biomedical Sciences Technology and Research, Sri Ramachandra Institute of Higher Education & Research, Chennai, India
[c]Division of Cancer Stem Cells and Cardiovascular Regeneration, Manipal Institute of Regenerative Medicine, Manipal University, Bangalore, India
[d]Lake Erie College of Osteopathic Medicine, Bradenton, FL, United States
[e]Cancer Science Institute of Singapore, National University of Singapore, Singapore, Singapore
[f]Department of Science in Korean Medicine, College of Korean Medicine, Kyung Hee University, Seoul, Republic of Korea
*Corresponding authors: e-mail address: phcgs@nus.edu.sg; ksahn@khu.ac.kr

Contents

Advances in Protein Chemistry and Structural Biology, Volume 125
ISSN 1876-1623
https://doi.org/10.1016/bs.apcsb.2020.08.002

Abstract

The development of cancer is a complex phenomenon driven by various extrinsic as well as intrinsic risk factors including epigenetic modifications. These post-translational modifications are encountered in diverse cancer cells and appear for a relatively short span of time. These changes can significantly affect various oncogenic genes and proteins involved in cancer initiation and progression. Histone lysine acetylation and deacetylation processes are controlled by two opposing classes of enzymes that modulate gene regulation either by adding an acetyl moiety on a histone lysine residue by histone lysine acetyltransferases (KATs) or via removing it by histone deacetylases (KDACs). Deregulated KAT activity has been implicated in the development of several diseases including cancer and can be targeted for the development of anti-neoplastic drugs. Here, we describe the predominant epigenetic changes that can affect key KAT superfamily members during carcinogenesis and briefly highlight the pharmacological potential of employing lysine acetyltransferase inhibitors (KATi) for cancer therapy.

Abbreviations

AR androgen receptor
AML acute myelogenous leukemia
ALL B-cell acute lymphocytic leukemia
BRPF1 PHD finger-containing protein 1
CDK1 cyclin-dependent kinase 1
CoA coenzyme A
CML chronic myelogenous leukemia
ER estrogen receptor
ERK1/2 extracellular signal regulated kinase
GNAT general control non-repressible 5 (GCN5)-related *N*-acetyltransferases
HIF-1 hypoxia inducible factor 1
hMOF human Males absent On First
IL-1β interleukin 1β
KATs histone lysine acetyltransferases
KDACs histone lysine deacetylases
MOZ monocytic leukemia zinc finger protein
NF-κB nuclear factor κB
NRCF nuclear receptor coactivator factor
p300/CBP CREB-binding protein and its paralog p300
p53 tumor suppressor protein
Sas2 something about silencing
TNF-α tumor necrosis factor α
Tip60 Tat-interactive protein
TAF TATA-box binding protein (TBP) associated factors

1. Introduction

It has been well established that post-translational modifications (PTMs) on histone and non-histone proteins such as phosphorylation, acetylation, methylation etc., can regulate multiple cellular processes (Baylin & Jones, 2016; Kouzarides, 2000, 2007). Epigenetic modifications in chromatin structure regulate gene expression. Euchromatin (loosely packed DNA) exhibits greater transcriptional activity whereas heterochromatin (tightly packed form) displays reduced transcriptional activity (Baylin & Jones, 2011; Jiang & Pugh, 2009; Lee, Smith, & Shilatifard, 2010). Various chromatin remodeling complexes can disrupt and remodel the nucleosome structure by increasing DNA accessibility (Martens & Winston, 2003). Two groups of chromatin-modifying complexes have been described: the ATP-dependent complex and covalent histone modifications. The homeostasis of histone modifications is established and maintained by a range of enzymes including, but not limited to, histone lysine acetyltransferases (KATs) and histone lysine deacetylases (KDACs). Histone modifications are important regulators of gene transcription and modulate dynamic processes that affect nucleosomes (Ganesan, 2018; Marmorstein & Zhou, 2014; Tessarz & Kouzarides, 2014; Venkatesh & Workman, 2015; Shanmugam et al., 2017). Several lines of evidences suggest that deregulated acetylation is implicated in cancer development (Baylin & Jones, 2016; Farria, Li, & Dent, 2015; Jones & Baylin, 2002; Jones, Issa, & Baylin, 2016; Sun, Man, Tan, Nimer, & Wang, 2015). Long noncoding RNAs have also been shown to be necessary for targeting histone-modifying activities (Wang, Yang, et al., 2011). Choudhary et al. reported the global acetylome analysis in which 3600 acetylation sites were identified on 1750 proteins and was found to be scattered across different parts of the cells and was not necessarily confined to the nucleus (Choudhary et al., 2009). Currently, public repositories such as PhosphoSitePlus database show over 35,000 acetylation sites in human cells (Hornbeck et al., 2012). Aberrant histone modifications contribute to human diseases such as cancer, neurodegenerative diseases, and infections including inflammation. Several pro-inflammatory cytokines such as tumor necrosis factor α (TNF-α), interleukin 1β (IL-1β), and other stimulants such as lipopolysaccharides can promote histone acetylation (Cantley & Haynes, 2013; Dai et al., 2015). In addition to nuclear function the levels of circulating histones as well as nucleosomes may be enhanced under pathological conditions which indicate that they can be utilized as targets for discovering novel therapeutics (Allam, Kumar, Darisipudi, & Anders, 2014; Baylin & Jones, 2016; Chen, Kang, Fan, & Tang, 2014; Jones et al., 2016).

2. Major KAT families

Based on the cellular localization KATs are classified into two types nuclear or cytoplasmic KATs (Demetriadou & Kirmizis, 2017; Kaypee et al., 2016; Roth, Denu, & Allis, 2001; Trisciuoglio, Di Martile, & Del Bufalo, 2018). Nuclear KATs families are classified based on their structural homology and the mechanism of enzyme transfer. Five distinct families have been identified with different target and functions, GCN5-related *N*-acetyltransferase (GNAT), CREB-binding protein and its paralog p300 (p300/CBP), MYST and nuclear receptor coactivator factor (NRCF) family (Borrow et al., 1996; Demetriadou & Kirmizis, 2017; Furdas, Kannan, Sippl, & Jung, 2012; Kaypee et al., 2016; Roth et al., 2001; Trisciuoglio et al., 2018). Approximately 100 protein substrates have been described for p300/CBP so far and acetylates both histone and non-histone proteins such as tumor suppressor protein p53 (Bowers et al., 2010; Dancy & Cole, 2015; Demetriadou & Kirmizis, 2017; Luo et al., 2004; Simon, Robaa, Alhalabi, Sippl, & Jung, 2016). MYST family of KATs are the largest and most diverse and are primarily involved in DNA repair and gene silencing (Voss & Thomas, 2009). They include MOZ (monocytic leukemia zinc finger protein, YBF2/Sas3, Sas2 (something about silencing), MOF, and Tip60 (Tat-interactive protein) and are characterized by the presence of a conserved 3-region histone acetyltransferase domain (containing an acetyl-CoA binding site), a C2HC zinc finger and a helix-turn-helix DNA-binding domain responsible for their catalytic histone acetyltransferase activity (Brown, Bourke, Eriksson, & Kerin, 2016; Roth et al., 2001; Trisciuoglio et al., 2018). The KAT family contains variations in their structural features such as PHD fingers, zinc fingers and chromodomains (Brown et al., 2016; Simon et al., 2016). Recently, a novel family of histone acetyltransferases known as camello has been identified that can acetylate histone H4 in zebrafish development and was found to localize in perinuclear space (Karmodiya et al., 2014). Cytoplasmic KATs are responsible for acetylation of newly synthesized histone proteins. They are HAT1, HAT2, HatB3.1, Rtt109, and HAT4 (Trisciuoglio et al., 2018; Yang & Seto, 2007).

3. p300/CBP family

The p300/CBP family consists of two members, CBP (CREB-binding protein) and its paralog p300; these proteins have similar structures and functions as well as contain a KAT domain of 500 amino acid residues with 86% sequence similarity, bromodomain and three cysteine-histidine-rich domains (TAZ, PHD, and ZZ) helping in protein-protein interactions. p300/CBP is a coactivator for several transcription factors and plays a pivotal role in the assembly and mobilization of the basal transcriptional molecules (Dancy & Cole, 2015; Demetriadou & Kirmizis, 2017; Iyer, Ozdag, & Caldas, 2004; Kaypee et al., 2016; Liu et al., 2008; Trisciuoglio et al., 2018). p300/CBP are large proteins with numerous functional domains and interact with at least 400 different cellular proteins (Bedford, Kasper, Fukuyama, & Brindle, 2010) including c-Myc, c-Myb, β-catenin, hypoxia inducible factor 1 (HIF-1), tumor suppressor protein p53, androgen receptor (AR), estrogen receptor (ER) and CREB. Interactions between CBP/p300 and these transcription factors are important events in transcribing their respective downstream genes in keeping with the functions of p300/CBP as transcriptional coactivators (Farria et al., 2015; Ying et al., 2010).

4. GNAT family

The superfamily of general control non-repressible 5 (GCN5)-related *N*-acetyltransferases (GNAT) was first reported by Neuwald and Landsman (1997). Within the GNAT superfamily, the structural information is currently available 17 distinct GNAT family members (Salah Ud-Din, Tikhomirova, & Roujeinikova, 2016). They all share a similar HAT domain of 160 amino acid residues and a conserved BRD domain at the C-terminal that can attach to lysine residues (Salah Ud-Din et al., 2016). It is of note that GCN5 (general control nonderepressible 5, KAT2A) and pCAF (p300/CBP-associated factor, KAT2B) are important proteins with functions in gene transcription and have been shown to acetylate several transcription factors and regulate their function (Nagy & Tora, 2007). Histone and non-histone protein acetylation by GNATs control several of cellular processes such as DNA replication and repair, metabolism, key signaling pathways and in the development of wide-ranging diseases including cancer (Iyer, Fairlie, & Brown, 2012; Kaypee et al., 2016; Sun et al., 2015; Wang & Dent, 2014).

5. MYST family

The MYST family of KATs are an evolutionarily conserved group of enzymes that share the conserved 370 residue MYST domain which includes zinc finger domain and an acetyl-CoA binding domain (Avvakumov & Cote, 2007; Demetriadou & Kirmizis, 2017; Sapountzi & Cote, 2011). The MYST family is comprised of five important enzymes; they are MOZ, Tip60, MOF, MORF, and HBO1. These MYST acetyltransferases are found in several protein complexes and are involved in regulating gene transcription process (Avvakumov & Cote, 2007; Demetriadou & Kirmizis, 2017). In cancers the levels of MYST acetyltransferases levels are altered. For example, Tip60 (HIV1 TAT interacting 60 kDa protein; KAT5) expression is often found to be downregulated in melanoma, cervical, colorectal and gastric cancers (Chen, Cheng, Tang, Martinka, & Li, 2012; Kaypee et al., 2016; Sakuraba et al., 2009, 2011; Subbaiah et al., 2016). hMOF (human Males absent On First; KAT8) is also an MYST family acetyltransferase and is the main enzyme implicated in acetylation of H4K16 and loss of hMOF can cause deregulation in H4K16 acetylation (Taipale et al., 2005). Loss of hMOF is also often found in gastric, renal, colorectal, ovarian, breast, medulloblastoma and hepatocellular carcinoma cells and serves as a prognostic biomarker in these cancers (Cao et al., 2014; Kaypee et al., 2016; Liu et al., 2013; Pfister et al., 2008; Wang et al., 2013; Zhang et al., 2014).

6. Nuclear receptor coactivator family and transcription factor related KAT family

The nuclear receptor coactivator family includes steroid receptor coactivators (SRC-1, -2 and -3). The SRCs are ubiquitously expressed and are the most extensively characterized. All three SRCs function in assembling multiprotein complexes to activate transcription by nuclear receptors and other transcription factor (Rohira & Lonard, 2017). The SRCs also have intrinsic KAT domain at carboxy-terminal and is specific for histone H3 and H4 (Spencer et al., 1997). The SRCs have no homology with other KATs and may represent a novel class of KAT enzymes. SRC1 and -3 predominantly acetylates histone H3 and H4 and SRC-3 can also acetylate H3 and H4 in mononucleosomes (Chen et al., 1997; Spencer et al., 1997). Transcriptional coactivator complexes possess intrinsic histone acetyltransferase activity, providing a direct link between chromatin acetylation and transcriptional activation (Grant et al., 1999; Schiltz et al., 1999). TATA-box binding protein (TBP) associated factors (TAF) TAFII250 and TFIIIC has histone lysine acetyltransferase activity (Chen, Tini, & Evans, 2001; Hsieh, Kundu, Wang, Kovelman, & Roeder, 1999; Mizzen et al., 1996). In addition, cytosolic histone acetyltransferases KAT1 acetylates histone H2A on lysine 5 and KAT4 promotes acetylation of cytosolic H4 enhancing nucleosome assembly (Parthun, 2013; Tafrova & Tafrov, 2014; Yang et al., 2011).

7. Lysine acetylation in cancer

Intrinsic and extrinsic environmental signals are the key processes that can be converted into response stimuli through the control of differential gene expression. N-ε-lysine acetylation has been recognized to play an essential role in several biological processes (Ali, Conrad, Verdin, & Ott, 2018; Benton, Fiskus, & Bhalla, 2017). Acetyl coenzyme A (acetyl-CoA) transfers an acetyl group to the ε-nitrogen on the lysine residue constitutes lysine acetylation. KATs, acetylate lysine and are known as "writers" of this modification, and "erasers" of acetylation is done by KDACs (Capell & Berger, 2013; Unnikrishnan, Gafken, & Tsukiyama, 2010; Vo & Goodman, 2001). The acetylation event was first observed 45 years ago mainly in the histones (Choudhary et al., 2009; Hirst & Marra, 2009; Kouzarides, 2000; You, Nie, Sun, Zheng, & Yang, 2012; Zhao et al., 2010). Histone acetylation is one among of the many post-translational modifications that occur in the genome of the cell and has been identified as one of the important epigenetic modifications that contribute to the development of numerous human diseases including inflammation driven cancers (Kaypee et al., 2016; Stasevich et al., 2014). A regular pattern of acetylation of histones has been observed and is associated with chromatin relaxation and DNA repair (Celic et al., 2006; Shogren-Knaak et al., 2006), protein stability, and protein to protein interactions (Jones et al., 2016; Kouzarides, 2007); and histone lysine acetylation has emerged as the most abundant PTM that is found across the entire proteome (Demetriadou & Kirmizis, 2017; Kaypee et al., 2016).

The acetylation pattern in cells has been well documented, some of them occurring in non-histone cellular signaling networks (Hornbeck et al., 2012), transcription factors, chromatin regulators, and cytoplasmic proteins, and in proteins involved in the regulation of cytoskeletal dynamics, energy metabolism, and endocytosis (Choudhary et al., 2009; Kouzarides, 2000;

You et al., 2012; Zhao et al., 2010). For example, H4K16ac is a critical acetylation mark that is involved in transcription regulation (Taylor, Eskeland, Hekimoglu-Balkan, Pradeepa, & Bickmore, 2013). H4K16ac, along with H4K20me3, has been found to be reduced in diverse malignancies (Fraga et al., 2005). In prostate cancer hypoacetylation of histone H3 at K9, K18, and H4K16 is an indicator of tumor reappearance (Seligson et al., 2005, 2009). However, in contrast increased global histone hyperacetylation is closely associated with oral cancer progression (Arif et al., 2010). Interestingly, depletion of H4K16ac may be an early indication of breast tumor development, and reduced levels of H3K9ac, H3K14ac, and H4K12ac are associated with unfavorable response to treatment (Elsheikh et al., 2009). In non-small cell lung carcinoma (NSCLC), decrease in H3K9ac is prognostic of improved survival while contrastingly, hypoacetylation at H2AK5 is associated with poor prognosis (Barlesi et al., 2007).

Furthermore, development of NSCLC was found to be correlated with abundant acetylation of H4K5 and H4K8 and reduced acetylation of H4K12 and H4K16 (Van Den Broeck et al., 2008). Moreover, dissipation of H3K9 and K18 acetylation status can be predictive of an improved outcome in glioma (Liu et al., 2010). Interestingly, the diminished levels of H3K18ac may have contrasting prognostic indications such as better survival in esophageal squamous cell carcinoma patients while in pancreatic carcinoma patients it is a potential marker for poor survival (Manuyakorn et al., 2010; Tzao et al., 2009). Thus, the overall deregulation of epigenetic mechanisms necessitates the urgency to develop epigenetic-based therapeutics. Table 1 summarizes the transcriptional and cellular role of various histone modifications in cancers.

8. Lysine acetylation in breast cancer

Access to DNA that is wrapped around histones occurs by epigenetic mechanisms such as histone lysine acetylation. 10–20 lysine residues get acetylated per histone by KATs. Acetylation of histone increases the negative charge on DNA, thus facilitating the DNA accessibility to proteins involved in replication, transcription, or DNA repair (Capell & Berger, 2013; Unnikrishnan et al., 2010; Vo & Goodman, 2001). Several reports have suggested the role of histone lysine acetylation, especially K9, K11, K18, and K56 on histone H3, and K5, K8, K13, and K16 on histone H4 is involved in constitutive transcriptional activation that is often observed in cancer cells (Berger, 2007; Grunstein, 1997). This acetylation process

Table 1 Histone acetylation by KATs in various cancer cell lines and patient derived samples.

Cancers	Cell lines	KAT family and their members involved in acetylation of histones	References
Breast	MDA-MB-231 MCF7	p300/CBP PCAF	Wang et al. (2001)
	MDA-MB-231	p300	Kim, Woo, Chong, Homenko, and Kraus (2006)
	T47D MCF7	p300	Fan et al. (2002)
	MDA-MB-231 MDA-MB-435 MCF7 ZR-75-1	p300	Liu et al. (2005)
	MDA-MB-231 SK-BR-3 T47D MCF7 MCF10A CD44+/CD24−/ESA + breast cancer stem cells	DOT1L-c-Myc-p300 complex	Cho et al. (2015)
	MDA-MB-231 MCF7 CD44+/CD24− breast cancer stem cells	CBP	Liang et al. (2015)
	Breast cancer patient's samples	p300	Xiao et al. (2011)
	Human breast cancer tissue microarray (TMA) (CBA-5-SBC) including 40 primary breast carcinomas and 8 controls of normal mammary gland	hMOF	Pfister et al. (2008)
	MCF7	HBO1	Iizuka et al. (2009)
	CD44+/CD24− breast cancer stem cells	HBO1	Duong et al. (2013)

Continued

Table 1 Histone acetylation by KATs in various cancer cell lines and patient derived samples.—cont'd

Cancers	Cell lines	KAT family and their members involved in acetylation of histones	References
Lung	Small cell lung cancer (SCC) patient's samples	p300/CBP	Gao, Geng, et al. (2014)
	H358 H69	Tip60	Van Den Broeck, Nissou, Brambilla, Eymin, and Gazzeri (2012)
	Panel of 11 lung cancer cells	Tip60	Miyamoto et al. (2008)
	NCI-H358 cells	ZEB1 associates with p300, PCAF, and Tip60	Roche et al. (2013)
	A549	p300/CBP	Ko et al. (2013)
	Fifty-nine lung cancer cell lines, consisting of 18 SCCs and 41 non-small cell lung cancers (NSCLC)	CBP	Kishimoto et al. (2005)
	43 lung tumor lines and from 8 matching normal lines	p300/CBP	Tillinghast et al. (2003)
	169 NSCLC patient's samples	p300	Hou et al. (2012)
	A549 H1299 CL (CL1-0 and CL1-5)	p300	Wang et al. (2014)
	A549	hMOF/MYST1/KAT8	Zhang, Liu, Zhang, Cheng, and Li (2013)
Hematological malignancies	Leukemia	MOZ/MORF	You et al. (2012)
	Acute myelogenous leukemia	MYST3, CBP	Diab et al. (2013)
	Acute monocytic leukemia	MOZ, p300	Chaffanet et al. (2000) and Rozman et al. (2004)
	Acute myelogenous leukemia	p300	Wang, Gural, et al. (2011)

Table 1 Histone acetylation by KATs in various cancer cell lines and patient derived samples.—cont'd

Cancers	Cell lines	KAT family and their members involved in acetylation of histones	References
	Acute myeloid leukemia	MOZ, TIF2	Deguchi et al. (2003)
	Leukemia	MOZ, CBP	Kitabayashi, Aikawa, Nguyen, Yokoyama, and Ohki (2001)
	Acute myeloid leukemia	c-Myb-p300 interaction	Pattabiraman et al. (2014)
	Hematopoietic stem and progenitor cells	c-Myb-p300 interaction	Sandberg et al. (2005)
	Human leukemia cells	c-Myb-p300 interaction	Uttarkar et al. (2015)
Gastric	AGS MKN-45 SGC-7901 Immortalized gastric mucosa epithelial GES cells	PCAF	Ying et al. (2010)
	46 gastric cancer patient samples	Tip60	Sakuraba et al. (2011)
Nasopharyngeal	209 patient's samples	p300	Liao et al. (2012)

is modulated by the lysine acetyltransferases such as p300/CBP, monocytic leukemia zinc finger protein (MOZ), histone acetyltransferase bound to ORC; MYST2/KAT7 (HBO1), and general control of amino acid synthesis 5-like 2 (GCN5) (Kaypee et al., 2016).

In addition, KATs are MOZ (a.k.a. MYST3 and KAT6A), which together with its paralog MORF (a.k.a. MYST4 and KAT6B) form tetrameric complexes with bromodomain- and PHD finger-containing protein 1 (BRPF1) and two small non-catalytic subunits (Kaypee et al., 2016). These two acetyltransferases have been found to be recurrently mutated in breast cancer (Lynch et al., 2013). Estrogen receptor alpha (ERα) is a transcription factor that gets activated upon binding to hormonal and growth factor signals. Indeed, numerous reports have shown that ERα gets acetylated post-translational by coactivator p300. Persistent activation of ERα is an associated risk factor in the development of breast cancer by increasing the abnormal growth of breast tissue. ERα gets acetylated in the hinge/ligand domain of K229, K299, K302, and K303 (Wang et al., 2001). In another study, it was reported that the acetylation of ERα by p160 and p300 coactivators at lysine K266 and K268 was upregulated in atypical breast hyperplasia (Kim et al., 2006). In BRCA1 mutated breast cancer cells, p300/CBP acetylated ERα and accelerated cell proliferation. However, the reintroduction of BRCA1 downregulated p300, but not CBP activity (Fan et al., 2002). Deacetylation of ERα in breast cancer cells by sirtuin (SIRT1) downregulated the ERα activity and proliferation of breast cancer cells, and thereby serves as an excellent alternative approach to prevent the progression of breast cancer (Popov et al., 2007). In breast cancer cells cross-talk with coactivators such as p300 or CBP is found to reduce its metastatic potential by upregulating E-cadherin expression (Liu et al., 2005).

H3 acetylation on ZEB1, ZEB2 and Snail promoters induce CSC-like features in breast cancer cells (Cho et al., 2015). Metadherin (MTDH) has been found to be involved in the development of drug resistance and metastasis of breast cancer cells. MTDH was found to interact with CBP thereby facilitating translocation to the twist family BHLH transcription factor (TWIST) promoter, thus enabling promoter proximal H3 acetylation (Liang et al., 2015). Loss of heterozygosity (LOH) at the p300 locus (22q13) has been observed in numerous cancers including breast (Iyer et al., 2004). A plethora of evidences suggest that several gene mutations can lead to an upregulation of p300/CBP in lung, colon, breast, and ovarian cancers (Kishimoto et al., 2005; Ozdag et al., 2002; Tillinghast et al., 2003) and is often found to be associated with disease recurrence and may also confer resistance to chemotherapy (Hou et al., 2012; Takeuchi et al., 2012; Xiao et al., 2011).

Loss of human males absent on the first (hMOF) is found in several cancers and the levels of hMOF may serve as a prognostic marker (Cao et al., 2014; Liu et al., 2013; Pfister et al., 2008; Wang et al., 2013; Zhang et al., 2014). In contrast, hMOF has been found to be upregulated in some cancers and was reported to be associated with HBO1 acetyltransferase, which is involved in pre-initiation complex formation and replication initiation (Iizuka, Matsui, Takisawa, & Smith, 2006). P53 negatively regulates HBO1 and inhibits replication during cellular stress (Iizuka et al., 2008). Furthermore, HBO1 is also found to be upregulated in cancer cells compared to normal cells (Iizuka et al., 2009) and phosphorylated form of

HBO1 can mediate the occurrence of cancer stem-like cells in breast cancer (Duong et al., 2013). In breast cancer cells loss of H4K16ac, as well as low levels of H3K9ac, H3K14ac, and H4K12ac are prognostic of poor outcomes (Elsheikh et al., 2009).

9. Lysine acetylation in lung cancer

In lung cancer, acetylation of lysine residues has been shown to play a major role in their initiation, progression, and metastasis. Overexpression of p300/CBP (Gao, Geng, et al., 2014) and decreased expression of histone acetyltransferase Tip60 can act as a poor prognostic indicator in lung cancer. Aberrant hyperacetylation of H4K5 and H4K8 was detected in lung cancer cells compared to normal lung epithelium, where hypoacetylation was observed (Van Den Broeck et al., 2008). In cisplatin resistant lung cancer cells Tip60 is overexpressed. Interestingly, upon Tip60 knockdown confers cisplatin-sensitivity to lung cancer cells (Miyamoto et al., 2008; Van Den Broeck et al., 2012). Barlesi et al. (2007) showed that low levels of H2AK5ac indicated poor prognosis in lung cancer patients (Barlesi et al., 2007). While in adenocarcinoma patients, reduced H3K9ac level was associated with overall survival (Barlesi et al., 2007). Moreover, upon ZEB1 stimulation, the H3K27ac mark can be modulated on the ZEB1 responsive elements thereby promoting EMT process (Roche et al., 2013). It has been shown that TGF-β can promote EMT by activating KAT3 and acetylating Smad2 and Smad3 proteins in lung cancer cells (Ko et al., 2013). In another study, it was reported that cyclin-dependent kinase 1 (CDK1) and extracellular signal regulated kinase (ERK1/2)-mediated phosphorylation of p300 on S1038 and S2039 may cause its degradation and robustly drive the process of carcinogenesis (Wang et al., 2014). Furthermore, PCAF has been reported to be reduced in solid tumors such as renal, lung, and esophageal squamous cell carcinoma (Kok, Naylor, & Buys, 1997; Qin et al., 2008; Yamakawa et al., 1991) and that hMOF can function as a key protein required for the maintenance of lung cancer cells (Zhang et al., 2013).

10. Lysine acetylation in gastric cancer

In gastric cancer either somatic mutation or loss of heterozygosity (LOH) at p300/CBP suggests a putative tumor suppressive role. Especially in gastric cancer LOH at the p300 locus (22q13) has been reported (Iyer et al., 2004). In addition, PCAF expression is found to be suppressed in gastric cancer (intestinal type) (Sunde et al., 2006; Ying et al., 2010;

Zheng et al., 2013) and loss of hMOF is a predictive marker in gastric cancer (Cao et al., 2014; Liu et al., 2013; Pfister et al., 2008).

11. Lysine acetylation in hematological malignancies

Several studies have indicated that KATs play an important role in the development of normal hematopoietic cells. The expression of KATs such as p300/CBP, MOZ, HBO1, and GCN5, can control hematopoietic stem cell (HSC) self-renewal, proliferation, as well as differentiation (Sun et al., 2015). Aberrant expression of KATs has been implicated in malignant hematopoiesis by acetylating histone and non-histone proteins. Several lines of evidence suggest that aberrant KAT activity in non-histone proteins such as oncogenes and tumor suppressor proteins is responsible for hematological malignancies. In acute myelogenous leukemia (AML), T-cell leukemia, lymphoma, B-cell acute lymphocytic leukemia (ALL), and chronic myelogenous leukemia (CML), leukemogenic proteins physically interact with the KATs p300/CBP, PCAF, MOZ, Tip60, and GCN5, and target genes such as Id1, p21, Egr1, c-Myc, p53, RARβ, PU.1, Syk, Btk, and AML1, that promote malignant transformation of cells (Sun et al., 2015). In addition, MOZ and its paralog MORF form tetrameric complexes with bromodomain and PHD finger-containing protein 1 (BRPF1) and two small non-catalytic subunits. These two acetyltransferases and BRPF1 are found to be recurrently mutated in childhood leukemia, non-hematologic malignancies, adult medulloblastoma and in multiple developmental disorders (Yang & Ullah, 2007; You et al., 2012).

In hematological malignancies such as AML and ALL, low frequency mutations may be associated with poor prognosis (Diab et al., 2013). During mitosis and meiosis, chromosomal translocation generates fusion and chimeric proteins that exhibit oncogenic potential. KAT gene translocation forms chimeric proteins that retain bromodomain and KAT catalytic activity. For example, as occurs in mixed lineage leukemia (MLL)-CBP [t(11;16) (q23;p13)] or in MLL-p300 [t(11;22)(q23;q13)] fusions (Krivtsov & Armstrong, 2007). In addition, MOZ fusions such as the MOS-CBP [t(8;16)(p11;p13)] or MOZ-p300 [t(8;22)(p11;q13)] are also reported for AML (Chaffanet et al., 2000; Iyer et al., 2004; Rozman et al., 2004). Other oncogenic fusions, in particular the most frequent seen in AMLs, AML1-ETO [t(8;21)(q22;q220)], need p300 mediated acetylation to induce leukemogenesis (Wang, Gural, et al., 2011). This clearly indicates the importance of transcriptional coactivators p300/CBP and their oncogenic role in cancer progression.

Oncogenic fusion between nuclear receptor coactivator TIF2 and MOZ is associated with AML [inv (8)(p11q13)] and recruits CBP to function as an oncoprotein (Deguchi et al., 2003; Kitabayashi et al., 2001; Yang & Ullah, 2007). Indeed, altered expression of the KATs is found in various cancers (Ozdag et al., 2006). *c-Myb* oncogene plays a critical role in hematopoiesis and a key regulator of self-renewal in hematopoietic stem cells. Deregulated *c-Myb* may cause the development of leukemia and other malignancies (Ramsay & Gonda, 2008). The interaction of c-Myb and p300 can also regulate maintenance of hematopoietic stem cells and promotion of c-Myb-mediated leukemia (Pattabiraman et al., 2014; Sandberg et al., 2005). Naphthol AS-E phosphate is pharmacological blocker that can interrupt the interaction between c-Myb and the KIX domain of p300 (Uttarkar et al., 2015). Kung et al. (2000) reported that hematological malignancies arise in mice that are heterozygous for CBP (Kung et al., 2000). Several lines of evidences suggest that double heterozygosity for CBP and p300 or CBP or p300 knock-out result in embryonic lethality in mice (Yao et al., 1998), and chimeric mice can develop neoplasms arising from the CBP −/− and p300 −/− null cells (Rebel et al., 2002).

12. Lysine acetylation in nasal cancer

Overexpression of p300/CBP is a poor prognostic indicator in nasopharyngeal cancer and is associated with severe metastasis (Liao et al., 2012).

13. KAT inhibitors (KATi)

Histone lysine acetylation has been implicated in the development and progression of several diseases thereby highlights the potential of KAT modulators as potential target for therapy intervention. However, the identification of selective and potent KATi has proven to be challenging compared to modulators of other epigenetic enzymes such as KDAC inhibitors (Merarchi et al., 2019). Several approaches such as computational methods, advanced assay technologies have been undertaken to identify small molecule inhibitors of KATs of which only a few compounds have been identified thus far with both *in vitro* and *in vivo* activity (Huang, Huang, Zheng, & Sun, 2019; Krishna et al., 2018). These KATi can be broadly classified into three distinct groups (1) bisubstrate inhibitors, (2) natural compounds and their derivatives, and (3) synthetic small molecules (Table 2).

Table 2 List of various categories of KAT inhibitors.

Types of KAT inhibitors	p300/CBP (IC_{50})	PCAF (IC_{50})	Tip60 (IC_{50})	References
Lys-CoA and its derivatives	0.5 μM	200 μM	–	Lau et al. (2000)
	0.98 μM	108.3 μM	29.8 μM	Wu, Xie, Wu, Zhang, and Zheng (2009)
	0.05 μM	–	–	Zheng et al. (2005)
	30 μM	–	–	Costi et al. (2007)
	0.0032 μM	–	–	Tohyama et al. (2012)
H3-CoA-20	200 μM	0.3 μM	–	Lau et al. (2000)
Boc-C5-CoA	0.07 μM	–	–	Kwie et al. (2011)
H4K16-CoA	6.62 μM	58.4 μM	17.5 μM	Wu et al. (2009)
3′-dephospho-Lys-CoA	1.6 μM	–	–	Cebrat, Kim, Thompson, Daugherty, and Cole (2003)
Spd(N1)-CoA	30–50 μM	–	–	Bandyopadhyay et al. (2009) and Cullis, Wolfenden, Cousens, and Alberts (1982)
Curcumin and its derivatives	25 μM	–	–	Balasubramanyam, Varier, et al. (2004)
	6.5 μM	–	–	Castellano et al. (2015)
	>40 μM	–	>200 μM	Wu et al. (2009)
	>400 μM	–	–	Costi et al. (2007)
	6.5 μM	–	–	Milite et al. (2015)
Anacardic acid and its derivatives	8.5 μM	5 μM	–	Balasubramanyam, Swaminathan, Ranganathan, and Kundu (2003)
	>1000 μM	667.1 μM	347.6 μM	Wu et al. (2009)
	>200 μM	>200 μM	64 μM	Ghizzoni et al. (2012)
	–	1–2 mM	–	Wapenaar et al. (2015)
	–	33.9 μM	–	Castellano et al. (2015)
	–	4.15 μM	–	Gao et al. (2013)
	–	662 μM	–	Wapenaar et al. (2015)

Table 2 List of various categories of KAT inhibitors.—cont'd

Types of KAT inhibitors	p300/CBP (IC_{50})	PCAF (IC_{50})	Tip60 (IC_{50})	References
Garcinol	7 μM	5 μM		Balasubramanyam, Altaf, et al. (2004)
LTK14	5–7 μM	–	–	Mantelingu et al. (2007)
Embelin	25 μM	7.2 μM	25 μM	Modak et al. (2013)
Plumbagin	20–25 μM	50 μM	–	Ravindra et al. (2009)
PU139	5.35 μM [p300]/2.49 [CBP]	9.74 μM	–	Furdas et al. (2014) and Furdas et al. (2011)
PU141	5.92 μM [p300]/2.85 [CBP]	130 μM	–	Furdas et al. (2014) and Furdas et al. (2011)
MB-3	500 μM	–	–	Biel, Kretsovali, Karatzali, Papamatheakis, and Giannis (2004)
C646	1.6 μM	10 μM	–	Bowers et al. (2010)
C646 derivative 1r	0.16 μM	–	–	Liu et al. (2019)
L002	150 μM	>100 μM	149 μM	Yang et al. (2013)
NU9056	58 μM	35 μM	2 μM	Coffey et al. (2012)
TH1834	–	–	500 μM	Gao, Bourke, et al. (2014)
I-CBP112	170 nmol/L	–	–	Picaud et al. (2015)
A-485	9.8 nM	–	–	Lasko et al. (2017)
	0.59 μmol/L 0.93 μmol/L 0.96 μmol/L	–	–	Wang et al. (2018)
Thiobarbituric acid derivative DCH36_06	0.6 μM 3.2 μM	–	–	Lu et al. (2018)
Spiro Oxazolidinediones derivatives	0.24–27.8 μM	–	–	Michaelides et al. (2018)

14. Bisubstrate inhibitors

The very first KAT inhibitor was a bisubstrate analog that mimic the ternary complex of cofactor acetyl-Co and lysine substrate. This concept was adopted by several other groups to identify specific KATi. Lys-CoA is one such compound that was synthesized by linking coenzyme A (CoA) and a single lysine residue via a methylene linker (Lau et al., 2000). Lys-CoA was reported to have potent p300 inhibitory activity compared to PCAF. Compound H3-CoA-20 was found to be specific to PCAF (Lau et al., 2000). In addition, another compound Boc-C5-CoA was found to inhibit p300 by occupying two binding pockets within the enzyme active site (Kwie et al., 2011). Similarly, another bisubstrate analog H4K16-CoA was found to be a potent inhibitor of MYST family enzyme Tip60 and its yeast homolog Esa1 with IC50 values in micromolar range (Wu et al., 2009). However, bisubstrate inhibitors have poor cell permeability and metabolic instability due to the polar phosphate group and the partial peptidic structure. These limitations can be overcome by adopting cell membrane penetration techniques such as cell micro-injection or lipid permeabilization (Simon et al., 2016). Several potential prodrugs of coenzyme A analogs targeting p300 have been developed (Cebrat et al., 2003). Modifications such as coupling of inhibitors amino acid backbone to Tat protein transduction domain or to arginine rich peptides enhanced transmembrane delivery and potent p300 inhibitory activity (Wadia & Dowdy, 2005; Zheng et al., 2005). Using the approach described by Cullis et al. (1982) polyamine spermidine (Spd) was linked covalently to the S atom of coenzyme A through a thioglycolic acid linkage forming Spd(N1)-CoA, which can function as non-toxic agent and internalize into the cells via polyamine transporter. Spd(N1)-CoA can alter DNA damage repair pathways and thereby sensitize cells to chemotherapeutic agents as well as radiation (Bandyopadhyay et al., 2009; Cullis et al., 1982). In a recent study it was further reported that a novel peptide-CoA conjugate bisubstrate inhibitor exhibited submicromolar potency against histone acetyltransferase 1 (Ngo, Brown, & Zheng, 2019).

15. Natural product based KATi

Several lines of evidences suggest that compound isolated from natural product have proven to be important in modulating KAT activity/expression. These agents have been found to be potent KATi at nanomolar concentrations in different biological assays. However, due to unfavorable physicochemical properties limit their further development. In addition, natural products with novel structures have been used as a template to develop semi-synthetic derivatives and synthetic molecules. We briefly discuss about the KATi derived from natural agents below:

16. Curcumin (PubChem CID: 969516)

Numerous studies have reported that KATs can be inhibited by several natural agents (Delage & Dashwood, 2008; Kaypee et al., 2016; Seidel, Florean, Schnekenburger, Dicato, & Diederich, 2012; Seidel, Schnekenburger, Dicato, & Diederich, 2012; Selvi, Mohankrishna, Ostwal, & Kundu, 2010). Curcumin was found to be a natural KATi and has also shown significant efficacy against diverse malignancies (Bishayee & Sethi, 2016; Deng et al., 2019; Kamat, Sethi, & Aggarwal, 2007; Kumar et al., 2014; Malhotra, Nair, & Dhawan, 2014; Mishra et al., 2019; Shanmugam et al., 2016; Shanmugam, Kannaiyan, & Sethi, 2011; Shanmugam, Warrier, Kumar, Sethi, & Arfuso, 2017; Sung et al., 2009; Tewari et al., 2018; Yarla et al., 2016). Novel curcumin formulations are also being investigated, such as nanoparticles and sustained release capsules for therapy of cancer and/or inflammatory conditions (Di Costanzo, Del Gaudio, Migliaccio, & Altucci, 2014; Gupta, Patchva, & Aggarwal, 2013; Shanmugam et al., 2015, 2019). Interestingly, it was found that curcumin inhibits p300/CREB-binding protein (CBP) but not of p300/CBP-associated factors *in vitro* and *in vivo* (Balasubramanyam, Varier, et al., 2004). Curcumin was shown to abrogate acetyltransferase activity of pure p300 by using histone H3 or p53 as substrate. In addition, curcumin also blocked MS-275 (HDAC inhibitor) induced hyperacetylation in PC3-M prostate cancer cells (Marcu et al., 2006). Curcumin was shown to inhibit proliferation, HDAC activity, and the expression of tumor suppressor protein p53 thus leading to apoptosis in promyelocytic leukemia NB4 cells (Li, Chen, Wu, & Liu, 2005). Similarly, in human lymphoma Raji cells curcumin was noted to acetylate histone H3 at the site of p21 (WAF1/CIP1), and induce cell cycle arrest (Li, Chen, Wu, & Sun, 2006). Curcumin treatment also upregulated global levels of acetylated H3K18 and H4K16 and was less effective in inducing DNA damage markers (Collins et al., 2013).

Interestingly, curcumin acetylated histone H3 and H4 in a dose-dependent manner, upregulated p53, and induced apoptosis in LNCaP prostate cancer cells (Shankar & Srivastava, 2007). In brain glioma cells, curcumin caused both histone hypoacetylation and apoptosis (Kang, Cha, & Jeon, 2006). In tumor associated macrophages, triggering receptor expressed on myeloid cells 1 (TREM-1) can induce toll-like receptor initiated inflammatory response and may be overexpressed in these cells. TREM-1 can also serve as a poor prognostic indicator in patients with NSCLC (Nguyen, Berim, & Agrawal, 2015; Yuan et al., 2012). Curcumin was reported to modulate TREM-1 expression by suppressing p300 acetylation of H3 and H4 in the TREM-1 promoter region (Yuan et al., 2012). In another study, curcumin could abolish KAT activity in THP-1 (human monocytic cell line) and blocked acetylation of nuclear factor κB (NF-κB) at Lys310 as well as subsequent nuclear translocation and transcription activation of its downstream target genes (Yun, Jialal, & Devaraj, 2011). Currently curcumin, one of the least specific KATi, is the only compound undergoing clinical trials for various ailments (Gupta et al., 2013; Manzo, Tambaro, Mai, & Altucci, 2009).

17. Anacardic acid (PubChem CID: 167551)

Anacardic acid (6-pentadecylsalicylic acid) was initially isolated from *Anacardium occidentale* (cashewnut) shell liquid (Balasubramanyam et al., 2003). Anacardic acid was found to be a non-selective and non-competitive p300/CBP and PCAF inhibitor. Several studies indicated that anacardic acid also inhibited Tip60 under similar assay conditions (Ghizzoni, Boltjes, Graaf, Haisma, & Dekker, 2010; Ghizzoni et al., 2012). Anacardic acid was also demonstrated to target NF-κB signaling pathway by affecting both the acetylation and nuclear localization of RelA subunit resulting in abrogation of tumorigenesis (Hemshekhar, Sebastin Santhosh, Kemparaju, & Girish, 2012). Due to anacardic acid high lipophilicity and difficult physiochemical properties, several novel 6-alkyl salicylic acid and phenoxyacetic acid analogs have been tested for improved solubility, cell permeability and KAT inhibitory activity.

It was reported that phenoxyacetic acid analogs showed potent KAT inhibitory activity depending on the position and length of the alkyl chain (Eliseeva, Valkov, Jung, & Jung, 2007). Anacardic acid was also reported to mitigate Tip60 mediated DNA damage thereby augmenting the response of tumor cells to radiation therapy (Sun, Jiang, Chen, & Price, 2006). In addition, variations in alkyl chain or the salicylic acid residue shifted specificity toward MYST family KATs MOF inhibition (Wapenaar et al., 2015; Zhang et al., 2018). Also, another derivative of anacardic acid,

4-cyano-3-trifluoromethylphenylbenzamides can display potent KAT3 inhibitory activity (Souto et al., 2008). A simplified analog of anacardic acid, pentadecylidenemalonate was originally discovered as a KAT activator/inhibitor. It was shown to activate PCAF while inhibit p300/CBP and recombinant CBP. It effectively induced apoptosis in U937 human leukemia cells (Sbardella et al., 2008). In a recent study, Castellano et al. (2015) showed that modification of the carbonyl function of diethyl pentadecylidenemalonate (SPV106) enhanced KAT inhibitory activity, induced cell cycle arrest and apoptosis in U937 leukemia cells (Castellano et al., 2015).

18. Garcinol (PubChem CID: 5281560)

Garcinol (polyisoprenylated benzophenone) is isolated from *Garcinia indica* (Kokum fruit) (Li et al., 2013, 2015; Liu et al., 2015; Manu et al., 2011; Ravindra et al., 2009; Sandur, Ichikawa, Sethi, Ahn, & Aggarwal, 2006; Sethi et al., 2014) and is a potent non-specific KATi (IC50 $\cong$ 5 μM for p300 and 7 μm for PCAF) (Balasubramanyam, Altaf, et al., 2004). Garcinol mediated cancer cell death was associated with suppression of histone acetylation and cellular apoptosis (Arif et al., 2009). To improve the pharmacokinetic profile of garcinol, several derivatives were synthesized with reduced toxicity, improved potency and specificity. Isogarcinol was obtained by intramolecular cyclization which served as a template to design several novel KATi. Monosubstitution at C14 yielded 14-isopropoxy isogarcinol (LTK-13), C14-methoxy isogarcinol (LTK-14), and di-substitution at C13 and C14 yielded C13, C14 disulfoxy isogarcinol (LTK-19). All these compounds showed potent p300 inhibitory activity but did not have any activity against PCAF (Mantelingu et al., 2007). Using molecular pruning approach with isosteric replacement of the benzophenone core Milite et al. (2015) recently reported benzylidene barbituric acid derivative (EML425) as a selective blocker of p300/CBP with an inhibitory potency in the low micromolar range (IC50 2.9 μM for p300 and 1.1 μM for CBP) (Milite et al., 2015). In human leukemia U937 cells, EML425 caused cell cycle arrest and promoted H3K9 and H4K5 hypoacetylation (Milite et al., 2015).

19. Plumbagin (PubChem CID: 10205)

Plumbagin (hydroxynaphthoquinone) is another natural compound isolated from the roots of medicinal herb *Plumbago rosea* is a specific KATi (Manu et al., 2011; Ravindra et al., 2009; Sandur et al., 2006).

Plumbagin can specifically inhibit p300-dependent acetylation of histone H3, H4, and p53 in hepatocellular carcinoma cells without any effect on PCAF (Ravindra et al., 2009). Plumbagin is also shown to possess anti-cancer activity (Manu et al., 2011). A novel 1,4-naphthoquinone derivative of plumbagin with an addition of methyl group at C3 position abrogated its thiol reactivity and cytotoxicity while retaining the KAT inhibitory potential (Vasudevarao et al., 2014).

20. Epigallocatechin-3-gallate (EGCG) (PubChem CID: 65064)

EGCG is a naturally occurring polyphenol present in green tea has been found to have KATi activity. It is non-specific and inhibits NF-κB signaling pathway by suppressing p300-mediated RelA acetylation (Choi, Jung, et al., 2009).

21. Other natural product compounds

Among other agents, 3-alkyl dihydroxybenzoquinone (Embelin) derived from *Embelia ribes* has been shown to specifically inhibit PCAF-mediated MyoD acetylation in HEK239T cells and to promote H3K9 hypoacetylation (Huang et al., 2014; Modak et al., 2013; Poojari, 2014). Procyanidin B3 is a proanthocyanidin which can be found in red wine, beer, barley and peach. It is a p300 specific KATi. It is also effective in inhibiting prostate cancer cell proliferation and induce apoptosis (Choi et al., 2011). Gallic acid is a trihydroxybenzoic acid isolated from various natural products such as rose flowers (*Rosa rugosa*), tea leaves, gallnuts, apple peels, grapes, strawberries, pineapples and lemons (Choi, Lee, et al., 2009). Recently gallic acid has been shown to be a p300 specific KATi and inhibits RelA acetylation and subsequent NF-κB activation in lung cancer cells (Choi, Lee, et al., 2009). Sanguinarine isolated from the roots of *Sanguinaria canadensis* and *Argemone mexicana* can also affect epigenetic enzymes such as KAT3, CARM1, G9a thereby modulating epigenetic mediated gene transcription (Selvi et al., 2009; Slaninová, Pěnčíková, Urbanová, Slanina, & Táborská, 2014). Delphinidin is an anthocyanidin that is found abundantly in pigmented vegetables and fruits such as blueberry and pomegranate. Delphinidin has been shown to be a p300 specific KATi without any effect on KDAC or methyltransferases. Delphinidin is reported to induce hypoacetylation of RelA resulting in retention of NF-κB in the cytoplasm thus suppressing gene transcription in fibroblast like synoviocyte MH7A cells (Seong et al., 2011). In addition to compounds isolated from natural product sources, fungal metabolite isolated from penicillium strain containing a peptide structure with citric acid moiety was shown to be a potent KATi with specificity toward p300, and Tip60. Furthermore, it was also reported to possess anti-cancer activity against a variety of cancer cells that overexpress androgen and estrogen receptors (Furumai et al., 2001; Tohyama et al., 2012).

22. Synthetic inhibitors

Although natural compounds have shown great promise as KATi, several of these compounds have limited physiochemical properties, cell permeability and poor bioavailability. Several synthetic molecules with improved drug-like properties containing novel scaffold and enzyme inhibitory activity has been synthesized as effective KATi. High throughput screening of chemical libraries has led to identification of several KATi molecules. One such is pyrazolone-containing compound C646 a potent p300 inhibitor (Bowers et al., 2010). C646 can sensitize NSCLC cells to radiotherapy (Oike et al., 2014) and can enhance chemo-responsiveness in melanoma cells (Yan et al., 2013). In a recent study by Shrimp et al. (2016) showed that C646 may form covalent conjugates with a several cysteine containing proteins (Shrimp et al., 2016). Compound C646 has intrinsic fluorescence hence hinders utility in fluorescence-based assays. Dancy et al. (2012) resolved this issue by substituting the furan group with a phenyl ring without any loss in KATi activity (Dancy et al., 2012). In AML1-ETO positive leukemia cells C646 induced hypoacetylation of histone H3 and H4 and inhibited cell growth, cell cycle arrest and caused apoptosis (Gao et al., 2013). Recently, a novel series of C646 derivatives were tested for their p300 inhibitory activity. It was reported that compound 1r of the series showed potent p300 inhibitory activity with improved drug-like properties compared to C646 (Liu et al., 2019). C646 was also shown to inhibit the proliferation of gastric cancer SGC-7901, MKN45, MGC-803, BGC-823 and KATO III cells compared with human normal gastric epithelial cell line (GES-1 cells) and induced apoptosis (Wang et al., 2017). L002 is another compound with specific p300 KATi identified by high throughput screening of approximately 622,079 compound chemical library (Baell et al., 2018; Baell & Miao, 2016; Yang et al., 2013). A selective and highly potent compound I-CBP112 was reported to target p300/CBP bromodomain and could suppress the leukemia initiating potential of MLL-AF9+ acute myeloid leukemia cells in a dose-dependent manner *in vitro* and *in vivo* (Picaud et al., 2015).

Similarly, isothiazolone-based compound KATi, CCT077791 and CCT077792, were developed as an antagonist of both p300 and PCAF (Stimson et al., 2005). In another study, isothiazolone based inhibitors, 5-chloroisothiazolone was identified as a PCAF specific KATi (Ghizzoni, Haisma, & Dekker, 2009). Biel et al. showed that MB-3 (α-methylene-γ-butyrolactone) as a GCN5-selective KATi with improved cell permeability (Biel et al., 2004). In neuroblastoma and glioblastoma cells thiazole-based compound BF1 (1-(4-(4-chlorophenyl) thiazol-2-yl)-2-(propan-2-ylidene) hydrazine) was shown to inhibit p300 and GCN5 KATs *in vitro* (Secci et al., 2014). Zeng et al. (2013) demonstrated that 4-acetyl-2-methyl-N-morpholino-3,4-dihydro-2H-benzo(b)(1,4)thiazine-7-sulfonamide was shown to have specific p300 inhibitory activity in the micromolar range (Zeng et al., 2013). Another pan-inhibitor of p300, CBP, PCAF and GCN5 compound, PU139 and PU141 a selective KAT3 inhibitor was found to abrogate the growth of SK-N-SH neuroblastoma xenograft in nude mice. In combination with doxorubicin, the compound PU139 also synergistically inhibited tumor growth (Gajer et al., 2015).

Remarkably, Wu et al. (2011) reported identification of novel Tip60 inhibitors by virtual ligand screening by utilizing the knowledge of the crystal structure of Esa1(the yeast homolog of Tip60) (Wu et al., 2011). In prostate cancer cells, a novel isothiazole-based compound NU9056 mitigated cellular growth and induced apoptosis by selectively suppressing Tip60 activity (Coffey et al., 2012). In another study, compound TH1834 was identified using computational tools that can target Tip60 in breast cancer cell and cause sensitization to the ionizing radiation (Gao, Bourke, et al., 2014). Therefore, Tip60 might be a likely therapeutic target for cancer treatment (Judes et al., 2015; Zhang, Wu, & Luan, 2017). High throughput screening and in-silico computational approaches have provided with novel KATi with greater specificity, bioavailability and low toxicity. However, these compounds need to be tested in human clinical trials to determine their potential as novel and efficacious KATi.

Furdas et al. (2011) showed that pyridoisothiazolones as novel KATi with IC_{50} values at micromolar range (Furdas et al., 2011). Furthermore, N-aryl substituted compounds were found to inhibit KATs without any specificity such as PCAF, Gcn5, p300, CBP and MOF while N-benzyl or N-alkyl substituents led to distinct KAT selectivity (Furdas et al.,

2014). A series of novel quinoline compound were tested for KAT inhibitory activity, of which compound 22 was found to be a KATi. Compound 22 was found to potently inhibit histone H3K9 and acetylation in U937 leukemia cells compared to anacardic acid, curcumin and MB-3 (Fiorentino, Mai, & Rotili, 2018; Mai et al., 2006). Thiazole derivative CPTH6 was shown to inhibit Gcn5 and pCAF KAT activity. Moreover, it inhibits H3/H4 histones and non-histone α-tubulin acetylation in a panel of leukemia cell lines (Chimenti et al., 2009; Trisciuoglio et al., 2012). In another study, compound A-485 was reported to selectively inhibit several hematological malignancies and androgen-sensitive prostate cancer by competing with acetyl-CoA. Remarkably, compound A-485 inhibited the growth of castration-resistant prostate tumor in a xenograft mouse model (Lasko et al., 2017). In addition, compound A-485 was shown to inhibit p300/CBP and induced apoptosis in human multiple myeloma cells (Wang et al., 2018). High throughput screening for KAT6A inhibitors revealed that an aryl sulfonohydrazide derivative (CTX-0124143) inhibited KAT6A (Huang, 2018; Leaver et al., 2019). In another study by Baell et al. (2018) they reported that inhibitors of KAT6A/B induced senescence and induced apoptosis in several cancer cells *in vitro* and *in vivo* increased the median survival oflymphoma bearing mice with concomitant suppression of tumor growth (Baell et al., 2018). Through virtual screening and iterative optimization, a new thiobarbituric acid derivative DCH36_06 was found to be a potent p300/CBP inhibitor in several leukemic cells and induced apoptosis *in vitro* and *in vivo* (Lu et al., 2018). High throughput screening produced a novel GCN5 inhibitor DC_G16 with 1,8-acridinedione scaffold. It was shown to inhibit proliferation and induce cell cycle arrest and apoptosis in cancer cells while it presented minimal effects on normal cells (Xiong et al., 2018). Furthermore, a novel and highly selective, orally bioavailable p300/CBP KATi spiro oxazolidinediones has been identified through virtual ligand (Michaelides et al., 2018). Therefore, these new and novel molecules provide new opportunities to target KATs as a potential therapy for cancer.

23. Conclusion and perspectives

Interestingly, reversible lysine acetylation has become apparent as one of the most important epigenetic modifiers and has developed into a promising class of new target for cancer therapy. Epigenetic acetylation has unquestionably emerged as the main modification in maintaining cellular homeostasis. Any deregulation in "writer" KATs and "eraser" KDACs have always led to development and progression of chronic inflammation driven diseases to malignant phenotype including cancer. Thus, the development of KATi is the need of the hour. Several natural compound inhibitors and synthetic molecules have been described in this book chapter. However not many KATi have entered clinical trials. This is primarily due to the physiochemical properties of these compounds, non-specificity, toxicity and bioavailability are major roadblocks in the development of KATi.

Of note curcumin is the only natural product based KATi that has been tested in clinical trials. Several synthetic compounds currently in pre-clinical studies have exhibited potent KAT inhibitory activities against malignant cells. Moreover, they can effectively augment the anti-cancer activities of standard chemotherapeutic agents such as paclitaxel, doxorubicin and cisplatin and sensitizing drug resistant cells to radiation therapy. Drug-like KATi with fewer adverse effects have the potential for clinical use. Small molecules targeting conserved BRD and KIX domains may have vast promising roles in clinical use. PRI-724 (ICG-001 derivative), a small molecule antagonist of CBP/β-catenin interaction for the treatment of advanced cancers including myeloid malignancies (NCT01606579) and pancreatic adenocarcinomas (NCT01764477) has been completed (https://clinicaltrials.gov/). Therefore, negatively regulating epigenetic enzymes has been identified as important strategy in anti-neoplastic therapeutics. Regulation of lysine acetylation by KAT inhibition either alone or in combination has opened enormous opportunities for cancer treatment.

Acknowledgments

K.S.A. was supported by a National Research Foundation of Korea (NRF) grant funded by the Korean government (MSIP) (NRF-2018R1D1A1B07042969). A.P.K. was supported by a grant from the National Medical Research Council of Singapore. A.P.K. was also supported by the National Medical Research Council of Singapore and the Singapore Ministry of Education under its Research Centres of Excellence initiative to Cancer Science Institute of Singapore, National University of Singapore.

References

Ali, I., Conrad, R. J., Verdin, E., & Ott, M. (2018). Lysine acetylation goes global: From epigenetics to metabolism and therapeutics. *Chemical Reviews, 118*(3), 1216–1252. https://doi.org/10.1021/acs.chemrev.7b00181.

Allam, R., Kumar, S. V., Darisipudi, M. N., & Anders, H. J. (2014). Extracellular histones in tissue injury and inflammation. *Journal of Molecular Medicine (Berlin, Germany), 92*(5), 465–472. https://doi.org/10.1007/s00109-014-1148-z.

Arif, M., Pradhan, S. K., Thanuja, G. R., Vedamurthy, B. M., Agrawal, S., Dasgupta, D., et al. (2009). Mechanism of p300 specific histone acetyltransferase inhibition by small molecules. *Journal of Medicinal Chemistry*, *52*(2), 267–277. https://doi.org/10.1021/jm800657z.

Arif, M., Vedamurthy, B. M., Choudhari, R., Ostwal, Y. B., Mantelingu, K., Kodaganur, G. S., et al. (2010). Nitric oxide-mediated histone hyperacetylation in oral cancer: Target for a water-soluble HAT inhibitor, CTK7A. *Chemistry & Biology*, *17*(8), 903–913. https://doi.org/10.1016/j.chembiol.2010.06.014.

Avvakumov, N., & Cote, J. (2007). The MYST family of histone acetyltransferases and their intimate links to cancer. *Oncogene*, *26*(37), 5395–5407. https://doi.org/10.1038/sj.onc.1210608.

Baell, J. B., Leaver, D. J., Hermans, S. J., Kelly, G. L., Brennan, M. S., Downer, N. L., et al. (2018). Inhibitors of histone acetyltransferases KAT6A/B induce senescence and arrest tumour growth. *Nature*, *560*(7717), 253–257. https://doi.org/10.1038/s41586-018-0387-5.

Baell, J. B., & Miao, W. (2016). Histone acetyltransferase inhibitors: Where art thou? *Future Medicinal Chemistry*, *8*(13), 1525–1528. https://doi.org/10.4155/fmc-2016-0151.

Balasubramanyam, K., Altaf, M., Varier, R. A., Swaminathan, V., Ravindran, A., Sadhale, P. P., et al. (2004). Polyisoprenylated benzophenone, garcinol, a natural histone acetyltransferase inhibitor, represses chromatin transcription and alters global gene expression. *The Journal of Biological Chemistry*, *279*(32), 33716–33726. https://doi.org/10.1074/jbc.M402839200.

Balasubramanyam, K., Swaminathan, V., Ranganathan, A., & Kundu, T. K. (2003). Small molecule modulators of histone acetyltransferase p300. *The Journal of Biological Chemistry*, *278*(21), 19134–19140. https://doi.org/10.1074/jbc.M301580200.

Balasubramanyam, K., Varier, R. A., Altaf, M., Swaminathan, V., Siddappa, N. B., Ranga, U., et al. (2004). Curcumin, a novel p300/CREB-binding protein-specific inhibitor of acetyltransferase, represses the acetylation of histone/nonhistone proteins and histone acetyltransferase-dependent chromatin transcription. *The Journal of Biological Chemistry*, *279*(49), 51163–51171. https://doi.org/10.1074/jbc.M409024200.

Bandyopadhyay, K., Baneres, J. L., Martin, A., Blonski, C., Parello, J., & Gjerset, R. A. (2009). Spermidinyl-CoA-based HAT inhibitors block DNA repair and provide cancer-specific chemo- and radiosensitization. *Cell Cycle*, *8*(17), 2779–2788. https://doi.org/10.4161/cc.8.17.9416.

Barlesi, F., Giaccone, G., Gallegos-Ruiz, M. I., Loundou, A., Span, S. W., Lefesvre, P., et al. (2007). Global histone modifications predict prognosis of resected non small-cell lung cancer. *Journal of Clinical Oncology*, *25*(28), 4358–4364. https://doi.org/10.1200/JCO.2007.11.2599.

Baylin, S. B., & Jones, P. A. (2011). A decade of exploring the cancer epigenome—Biological and translational implications. *Nature Reviews Cancer*, *11*(10), 726–734. https://doi.org/10.1038/nrc3130.

Baylin, S. B., & Jones, P. A. (2016). Epigenetic determinants of cancer. *Cold Spring Harbor Perspectives in Biology*, *8*(9), a019505. https://doi.org/10.1101/cshperspect.a019505.

Bedford, D. C., Kasper, L. H., Fukuyama, T., & Brindle, P. K. (2010). Target gene context influences the transcriptional requirement for the KAT3 family of CBP and p300 histone acetyltransferases. *Epigenetics*, *5*(1), 9–15. https://doi.org/10.4161/epi.5.1.10449.

Benton, C. B., Fiskus, W., & Bhalla, K. N. (2017). Targeting histone acetylation: Readers and writers in leukemia and cancer. *Cancer Journal*, *23*(5), 286–291. https://doi.org/10.1097/PPO.0000000000000284.

Berger, S. L. (2007). The complex language of chromatin regulation during transcription. *Nature*, *447*(7143), 407–412. https://doi.org/10.1038/nature05915.

Biel, M., Kretsovali, A., Karatzali, E., Papamatheakis, J., & Giannis, A. (2004). Design, synthesis, and biological evaluation of a small-molecule inhibitor of the histone acetyltransferase Gcn5. *Angewandte Chemie (International Ed. in English)*, *43*(30), 3974–3976. https://doi.org/10.1002/anie.200453879.

Bishayee, A., & Sethi, G. (2016). Bioactive natural products in cancer prevention and therapy: Progress and promise. *Seminars in Cancer Biology*, *40–41*, 1–3. https://doi.org/10.1016/j.semcancer.2016.08.006.

Borrow, J., Stanton, V. P., Jr., Andresen, J. M., Becher, R., Behm, F. G., Chaganti, R. S., et al. (1996). The translocation t(8,16)(p11;p13) of acute myeloid leukaemia fuses a putative acetyltransferase to the CREB-binding protein. *Nature Genetics*, *14*(1), 33–41. https://doi.org/10.1038/ng0996-33.

Bowers, E. M., Yan, G., Mukherjee, C., Orry, A., Wang, L., Holbert, M. A., et al. (2010). Virtual ligand screening of the p300/CBP histone acetyltransferase: Identification of a selective small molecule inhibitor. *Chemistry & Biology*, *17*(5), 471–482. https://doi.org/10.1016/j.chembiol.2010.03.006.

Brown, J. A., Bourke, E., Eriksson, L. A., & Kerin, M. J. (2016). Targeting cancer using KAT inhibitors to mimic lethal knockouts. *Biochemical Society Transactions*, *44*(4), 979–986. https://doi.org/10.1042/BST20160081.

Cantley, M. D., & Haynes, D. R. (2013). Epigenetic regulation of inflammation: Progressing from broad acting histone deacetylase (HDAC) inhibitors to targeting specific HDACs. *Inflammopharmacology*, *21*(4), 301–307. https://doi.org/10.1007/s10787-012-0166-0.

Cao, L., Zhu, L., Yang, J., Su, J., Ni, J., Du, Y., et al. (2014). Correlation of low expression of hMOF with clinicopathological features of colorectal carcinoma, gastric cancer and renal cell carcinoma. *International Journal of Oncology*, *44*(4), 1207–1214. https://doi.org/10.3892/ijo.2014.2266.

Capell, B. C., & Berger, S. L. (2013). Genome-wide epigenetics. *The Journal of Investigative Dermatology*, *133*(6), e9. https://doi.org/10.1038/jid.2013.173.

Castellano, S., Milite, C., Feoli, A., Viviano, M., Mai, A., Novellino, E., et al. (2015). Identification of structural features of 2-alkylidene-1,3-dicarbonyl derivatives that induce inhibition and/or activation of histone acetyltransferases KAT3B/p300 and KAT2B/PCAF. *ChemMedChem*, *10*(1), 144–157. https://doi.org/10.1002/cmdc.201402371.

Cebrat, M., Kim, C. M., Thompson, P. R., Daugherty, M., & Cole, P. A. (2003). Synthesis and analysis of potential prodrugs of coenzyme A analogues for the inhibition of the histone acetyltransferase p300. *Bioorganic & Medicinal Chemistry*, *11*(15), 3307–3313.

Celic, I., Masumoto, H., Griffith, W. P., Meluh, P., Cotter, R. J., Boeke, J. D., et al. (2006). The sirtuins hst3 and Hst4p preserve genome integrity by controlling histone h3 lysine 56 deacetylation. *Current Biology*, *16*(13), 1280–1289. https://doi.org/10.1016/j.cub.2006.06.023.

Chaffanet, M., Gressin, L., Preudhomme, C., Soenen-Cornu, V., Birnbaum, D., & Pebusque, M. J. (2000). MOZ is fused to p300 in an acute monocytic leukemia with t(8,22). *Genes, Chromosomes & Cancer*, *28*(2), 138–144.

Chen, G., Cheng, Y., Tang, Y., Martinka, M., & Li, G. (2012). Role of Tip60 in human melanoma cell migration, metastasis, and patient survival. *The Journal of Investigative Dermatology*, *132*(11), 2632–2641. https://doi.org/10.1038/jid.2012.193.

Chen, R., Kang, R., Fan, X. G., & Tang, D. (2014). Release and activity of histone in diseases. *Cell Death & Disease*, *5*, e1370. https://doi.org/10.1038/cddis.2014.337.

Chen, H., Lin, R. J., Schiltz, R. L., Chakravarti, D., Nash, A., Nagy, L., et al. (1997). Nuclear receptor coactivator ACTR is a novel histone acetyltransferase and forms a multimeric activation complex with P/CAF and CBP/p300. *Cell*, *90*(3), 569–580.

Chen, H., Tini, M., & Evans, R. M. (2001). HATs on and beyond chromatin. *Current Opinion in Cell Biology*, *13*(2), 218–224.

Chimenti, F., Bizzarri, B., Maccioni, E., Secci, D., Bolasco, A., Chimenti, P., et al. (2009). A novel histone acetyltransferase inhibitor modulating Gcn5 network: Cyclopentylidene-[4-(4′-chlorophenyl)thiazol-2-yl]hydrazone. *Journal of Medicinal Chemistry*, *52*(2), 530–536. https://doi.org/10.1021/jm800885d.

Cho, M. H., Park, J. H., Choi, H. J., Park, M. K., Won, H. Y., Park, Y. J., et al. (2015). DOT1L cooperates with the c-Myc-p300 complex to epigenetically derepress CDH1 transcription factors in breast cancer progression. *Nature Communications*, *6*, 7821. https://doi.org/10.1038/ncomms8821.

Choi, K. C., Jung, M. G., Lee, Y. H., Yoon, J. C., Kwon, S. H., Kang, H. B., et al. (2009). Epigallocatechin-3-gallate, a histone acetyltransferase inhibitor, inhibits EBV-induced B lymphocyte transformation via suppression of RelA acetylation. *Cancer Research*, *69*(2), 583–592. https://doi.org/10.1158/0008-5472.CAN-08-2442.

Choi, K. C., Lee, Y. H., Jung, M. G., Kwon, S. H., Kim, M. J., Jun, W. J., et al. (2009). Gallic acid suppresses lipopolysaccharide-induced nuclear factor-kappaB signaling by preventing RelA acetylation in A549 lung cancer cells. *Molecular Cancer Research*, 7(12), 2011–2021. https://doi.org/10.1158/1541-7786.MCR-09-0239.

Choi, K. C., Park, S., Lim, B. J., Seong, A. R., Lee, Y. H., Shiota, M., et al. (2011). Procyanidin B3, an inhibitor of histone acetyltransferase, enhances the action of antagonist for prostate cancer cells via inhibition of p300-dependent acetylation of androgen receptor. *The Biochemical Journal*, *433*(1), 235–244. https://doi.org/10.1042/BJ20100980.

Choudhary, C., Kumar, C., Gnad, F., Nielsen, M. L., Rehman, M., Walther, T. C., et al. (2009). Lysine acetylation targets protein complexes and co-regulates major cellular functions. *Science*, *325*(5942), 834–840. https://doi.org/10.1126/science.1175371.

Coffey, K., Blackburn, T. J., Cook, S., Golding, B. T., Griffin, R. J., Hardcastle, I. R., et al. (2012). Characterisation of a Tip60 specific inhibitor, NU9056, in prostate cancer. *PLoS One*, 7(10), e45539. https://doi.org/10.1371/journal.pone.0045539.

Collins, H. M., Abdelghany, M. K., Messmer, M., Yue, B., Deeves, S. E., Kindle, K. B., et al. (2013). Differential effects of garcinol and curcumin on histone and p53 modifications in tumour cells. *BMC Cancer*, *13*, 37. https://doi.org/10.1186/1471-2407-13-37.

Costi, R., Di Santo, R., Artico, M., Miele, G., Valentini, P., Novellino, E., et al. (2007). Cinnamoyl compounds as simple molecules that inhibit p300 histone acetyltransferase. *Journal of Medicinal Chemistry*, *50*(8), 1973–1977. https://doi.org/10.1021/jm060943s.

Cullis, P. M., Wolfenden, R., Cousens, L. S., & Alberts, B. M. (1982). Inhibition of histone acetylation by N-[2-(S-coenzyme A)acetyl] spermidine amide, a multisubstrate analog. *The Journal of Biological Chemistry*, *257*(20), 12165–12169.

Dai, X., Zhang, J., Arfuso, F., Chinnathambi, A., Zayed, M. E., Alharbi, S. A., et al. (2015). Targeting TNF-related apoptosis-inducing ligand (TRAIL) receptor by natural products as a potential therapeutic approach for cancer therapy. *Experimental Biology and Medicine (Maywood, N.J.)*, *240*(6), 760–773. https://doi.org/10.1177/1535370215579167.

Dancy, B. M., & Cole, P. A. (2015). Protein lysine acetylation by p300/CBP. *Chemical Reviews*, *115*(6), 2419–2452. https://doi.org/10.1021/cr500452k.

Dancy, B. M., Crump, N. T., Peterson, D. J., Mukherjee, C., Bowers, E. M., Ahn, Y. H., et al. (2012). Live-cell studies of p300/CBP histone acetyltransferase activity and inhibition. *Chembiochem*, *13*(14), 2113–2121. https://doi.org/10.1002/cbic.201200381.

Deguchi, K., Ayton, P. M., Carapeti, M., Kutok, J. L., Snyder, C. S., Williams, I. R., et al. (2003). MOZ-TIF2-induced acute myeloid leukemia requires the MOZ nucleosome binding motif and TIF2-mediated recruitment of CBP. *Cancer Cell*, *3*(3), 259–271.

Delage, B., & Dashwood, R. H. (2008). Dietary manipulation of histone structure and function. *Annual Review of Nutrition*, *28*, 347–366. https://doi.org/10.1146/annurev.nutr.28.061807.155354.

Demetriadou, C., & Kirmizis, A. (2017). Histone acetyltransferases in cancer: Guardians or hazards? *Critical Reviews in Oncogenesis*, *22*(3–4), 195–218. https://doi.org/10.1615/CritRevOncog.2017024506.

Deng, S., Shanmugam, M. K., Kumar, A. P., Yap, C. T., Sethi, G., & Bishayee, A. (2019). Targeting autophagy using natural compounds for cancer prevention and therapy. *Cancer*, *125*(8), 1228–1246. https://doi.org/10.1002/cncr.31978.

Di Costanzo, A., Del Gaudio, N., Migliaccio, A., & Altucci, L. (2014). Epigenetic drugs against cancer: An evolving landscape. *Archives of Toxicology*, *88*(9), 1651–1668. https://doi.org/10.1007/s00204-014-1315-6.

Diab, A., Zickl, L., Abdel-Wahab, O., Jhanwar, S., Gulam, M. A., Panageas, K. S., et al. (2013). Acute myeloid leukemia with translocation t(8,16) presents with features which mimic acute promyelocytic leukemia and is associated with poor prognosis. *Leukemia Research*, *37*(1), 32–36. https://doi.org/10.1016/j.leukres.2012.08.025.

Duong, M. T., Akli, S., Macalou, S., Biernacka, A., Debeb, B. G., Yi, M., et al. (2013). Hbo1 is a cyclin E/CDK2 substrate that enriches breast cancer stem-like cells. *Cancer Research*, *73*(17), 5556–5568. https://doi.org/10.1158/0008-5472.CAN-13-0013.

Eliseeva, E. D., Valkov, V., Jung, M., & Jung, M. O. (2007). Characterization of novel inhibitors of histone acetyltransferases. *Molecular Cancer Therapeutics*, *6*(9), 2391–2398. https://doi.org/10.1158/1535-7163.MCT-07-0159.

Elsheikh, S. E., Green, A. R., Rakha, E. A., Powe, D. G., Ahmed, R. A., Collins, H. M., et al. (2009). Global histone modifications in breast cancer correlate with tumor phenotypes, prognostic factors, and patient outcome. *Cancer Research*, *69*(9), 3802–3809. https://doi.org/10.1158/0008-5472.CAN-08-3907.

Fan, S., Ma, Y. X., Wang, C., Yuan, R. Q., Meng, Q., Wang, J. A., et al. (2002). p300 Modulates the BRCA1 inhibition of estrogen receptor activity. *Cancer Research*, *62*(1), 141–151.

Farria, A., Li, W., & Dent, S. Y. (2015). KATs in cancer: Functions and therapies. *Oncogene*, *34*(38), 4901–4913. https://doi.org/10.1038/onc.2014.453.

Fiorentino, F., Mai, A., & Rotili, D. (2018). Lysine acetyltransferase inhibitors: Structure-activity relationships and potential therapeutic implications. *Future Medicinal Chemistry*, *10*(9), 1067–1091. https://doi.org/10.4155/fmc-2017-0244.

Fraga, M. F., Ballestar, E., Villar-Garea, A., Boix-Chornet, M., Espada, J., Schotta, G., et al. (2005). Loss of acetylation at Lys16 and trimethylation at Lys20 of histone H4 is a common hallmark of human cancer. *Nature Genetics*, *37*(4), 391–400. https://doi.org/10.1038/ng1531.

Furdas, S. D., Hoffmann, I., Robaa, D., Herquel, B., Malinka, W., Świątek, P., et al. (2014). Pyrido- and benzisothiazolones as inhibitors of histone acetyltransferases (HATs). *Medicinal Chemistry Communications*, *5*, 1856–1862.

Furdas, S. D., Kannan, S., Sippl, W., & Jung, M. (2012). Small molecule inhibitors of histone acetyltransferases as epigenetic tools and drug candidates. *Archiv der Pharmazie*, *345*(1), 7–21. https://doi.org/10.1002/ardp.201100209.

Furdas, S. D., Shekfeh, S., Bissinger, E. M., Wagner, J. M., Schlimme, S., Valkov, V., et al. (2011). Synthesis and biological testing of novel pyridoisothiazolones as histone acetyltransferase inhibitors. *Bioorganic & Medicinal Chemistry*, *19*(12), 3678–3689. https://doi.org/10.1016/j.bmc.2011.01.063.

Furumai, R., Komatsu, Y., Nishino, N., Khochbin, S., Yoshida, M., & Horinouchi, S. (2001). Potent histone deacetylase inhibitors built from trichostatin A and cyclic tetrapeptide antibiotics including trapoxin. *Proceedings of the National Academy of Sciences of the United States of America*, *98*(1), 87–92. https://doi.org/10.1073/pnas.011405598.

Gajer, J. M., Furdas, S. D., Grunder, A., Gothwal, M., Heinicke, U., Keller, K., et al. (2015). Histone acetyltransferase inhibitors block neuroblastoma cell growth in vivo. *Oncogene*, *4*, e137. https://doi.org/10.1038/oncsis.2014.51.

Ganesan, A. (2018). Epigenetic drug discovery: A success story for cofactor interference. *Philosophical Transactions of the Royal Society of London. Series B, Biological Sciences, 373*(1748), 20170069. https://doi.org/10.1098/rstb.2017.0069.

Gao, C., Bourke, E., Scobie, M., Famme, M. A., Koolmeister, T., Helleday, T., et al. (2014). Rational design and validation of a Tip60 histone acetyltransferase inhibitor. *Scientific Reports, 4*, 5372. https://doi.org/10.1038/srep05372.

Gao, Y., Geng, J., Hong, X., Qi, J., Teng, Y., Yang, Y., et al. (2014). Expression of p300 and CBP is associated with poor prognosis in small cell lung cancer. *International Journal of Clinical and Experimental Pathology*, 7(2), 760–767.

Gao, X. N., Lin, J., Ning, Q. Y., Gao, L., Yao, Y. S., Zhou, J. H., et al. (2013). A histone acetyltransferase p300 inhibitor C646 induces cell cycle arrest and apoptosis selectively in AML1-ETO-positive AML cells. *PLoS One, 8*(2), e55481. https://doi.org/10.1371/journal.pone.0055481.

Ghizzoni, M., Boltjes, A., Graaf, C., Haisma, H. J., & Dekker, F. J. (2010). Improved inhibition of the histone acetyltransferase PCAF by an anacardic acid derivative. *Bioorganic & Medicinal Chemistry, 18*(16), 5826–5834. https://doi.org/10.1016/j.bmc.2010.06.089.

Ghizzoni, M., Haisma, H. J., & Dekker, F. J. (2009). Reactivity of isothiazolones and isothiazolone-1-oxides in the inhibition of the PCAF histone acetyltransferase. *European Journal of Medicinal Chemistry, 44*(12), 4855–4861. https://doi.org/10.1016/j.ejmech.2009.07.025.

Ghizzoni, M., Wu, J., Gao, T., Haisma, H. J., Dekker, F. J., & George Zheng, Y. (2012). 6-alkylsalicylates are selective Tip60 inhibitors and target the acetyl-CoA binding site. *European Journal of Medicinal Chemistry, 47*(1), 337–344. https://doi.org/10.1016/j.ejmech.2011.11.001.

Grant, P. A., Eberharter, A., John, S., Cook, R. G., Turner, B. M., & Workman, J. L. (1999). Expanded lysine acetylation specificity of Gcn5 in native complexes. *The Journal of Biological Chemistry, 274*(9), 5895–5900. https://doi.org/10.1074/jbc.274.9.5895.

Grunstein, M. (1997). Histone acetylation in chromatin structure and transcription. *Nature, 389*(6649), 349–352. https://doi.org/10.1038/38664.

Gupta, S. C., Patchva, S., & Aggarwal, B. B. (2013). Therapeutic roles of curcumin: Lessons learned from clinical trials. *The AAPS Journal, 15*(1), 195–218. https://doi.org/10.1208/s12248-012-9432-8.

Hemshekhar, M., Sebastin Santhosh, M., Kemparaju, K., & Girish, K. S. (2012). Emerging roles of anacardic acid and its derivatives: A pharmacological overview. *Basic & Clinical Pharmacology & Toxicology, 110*(2), 122–132. https://doi.org/10.1111/j.1742-7843.2011.00833.x.

Hirst, M., & Marra, M. A. (2009). Epigenetics and human disease. *The International Journal of Biochemistry & Cell Biology, 41*(1), 136–146. https://doi.org/10.1016/j.biocel.2008.09.011.

Hornbeck, P. V., Kornhauser, J. M., Tkachev, S., Zhang, B., Skrzypek, E., Murray, B., et al. (2012). PhosphoSitePlus: A comprehensive resource for investigating the structure and function of experimentally determined post-translational modifications in man and mouse. *Nucleic Acids Research, 40*(Database issue), D261–D270. https://doi.org/10.1093/nar/gkr1122.

Hou, X., Li, Y., Luo, R. Z., Fu, J. H., He, J. H., Zhang, L. J., et al. (2012). High expression of the transcriptional co-activator p300 predicts poor survival in resectable non-small cell lung cancers. *European Journal of Surgical Oncology, 38*(6), 523–530. https://doi.org/10.1016/j.ejso.2012.02.180.

Hsieh, Y. J., Kundu, T. K., Wang, Z., Kovelman, R., & Roeder, R. G. (1999). The TFIIIC90 subunit of TFIIIC interacts with multiple components of the RNA polymerase III machinery and contains a histone-specific acetyltransferase activity. *Molecular and Cellular Biology, 19*(11), 7697–7704. https://doi.org/10.1128/mcb.19.11.7697.

Huang, F. (2018). New KAT6 inhibitors induce senescence and arrest cancer growth. *Synthetic and Systems Biotechnology*, *3*(4), 244–245. https://doi.org/10.1016/j.synbio.2018.10.006.

Huang, M., Huang, J., Zheng, Y., & Sun, Q. (2019). Histone acetyltransferase inhibitors: An overview in synthesis, structure-activity relationship and molecular mechanism. *European Journal of Medicinal Chemistry*, *178*, 259–286. https://doi.org/10.1016/j.ejmech.2019.05.078.

Huang, M., Tang, S. N., Upadhyay, G., Marsh, J. L., Jackman, C. P., Shankar, S., et al. (2014). Embelin suppresses growth of human pancreatic cancer xenografts, and pancreatic cancer cells isolated from KrasG12D mice by inhibiting Akt and Sonic hedgehog pathways. *PLoS One*, *9*(4), e92161. https://doi.org/10.1371/journal.pone.0092161.

Iizuka, M., Matsui, T., Takisawa, H., & Smith, M. M. (2006). Regulation of replication licensing by acetyltransferase Hbo1. *Molecular and Cellular Biology*, *26*(3), 1098–1108. https://doi.org/10.1128/MCB.26.3.1098-1108.2006.

Iizuka, M., Sarmento, O. F., Sekiya, T., Scrable, H., Allis, C. D., & Smith, M. M. (2008). Hbo1 Links p53-dependent stress signaling to DNA replication licensing. *Molecular and Cellular Biology*, *28*(1), 140–153. https://doi.org/10.1128/MCB.00662-07.

Iizuka, M., Takahashi, Y., Mizzen, C. A., Cook, R. G., Fujita, M., Allis, C. D., et al. (2009). Histone acetyltransferase Hbo1: Catalytic activity, cellular abundance, and links to primary cancers. *Gene*, *436*(1–2), 108–114. https://doi.org/10.1016/j.gene.2009.01.020.

Iyer, A., Fairlie, D. P., & Brown, L. (2012). Lysine acetylation in obesity, diabetes and metabolic disease. *Immunology and Cell Biology*, *90*(1), 39–46. https://doi.org/10.1038/icb.2011.99.

Iyer, N. G., Ozdag, H., & Caldas, C. (2004). p300/CBP and cancer. *Oncogene*, *23*(24), 4225–4231. https://doi.org/10.1038/sj.onc.1207118.

Jiang, C., & Pugh, B. F. (2009). Nucleosome positioning and gene regulation: Advances through genomics. *Nature Reviews. Genetics*, *10*(3), 161–172. https://doi.org/10.1038/nrg2522.

Jones, P. A., & Baylin, S. B. (2002). The fundamental role of epigenetic events in cancer. *Nature Reviews. Genetics*, *3*(6), 415–428. https://doi.org/10.1038/nrg816.

Jones, P. A., Issa, J. P., & Baylin, S. (2016). Targeting the cancer epigenome for therapy. *Nature Reviews. Genetics*, *17*(10), 630–641. https://doi.org/10.1038/nrg.2016.93.

Judes, G., Rifai, K., Ngollo, M., Daures, M., Bignon, Y. J., Penault-Llorca, F., et al. (2015). A bivalent role of TIP60 histone acetyl transferase in human cancer. *Epigenomics*, 7(8), 1351–1363. https://doi.org/10.2217/epi.15.76.

Kamat, A. M., Sethi, G., & Aggarwal, B. B. (2007). Curcumin potentiates the apoptotic effects of chemotherapeutic agents and cytokines through down-regulation of nuclear factor-kappaB and nuclear factor-kappaB-regulated gene products in IFN-alpha-sensitive and IFN-alpha-resistant human bladder cancer cells. *Molecular Cancer Therapeutics*, *6*(3), 1022–1030. https://doi.org/10.1158/1535-7163.MCT-06-0545.

Kang, S. K., Cha, S. H., & Jeon, H. G. (2006). Curcumin-induced histone hypoacetylation enhances caspase-3-dependent glioma cell death and neurogenesis of neural progenitor cells. *Stem Cells and Development*, *15*(2), 165–174. https://doi.org/10.1089/scd.2006.15.165.

Karmodiya, K., Anamika, K., Muley, V., Pradhan, S. J., Bhide, Y., & Galande, S. (2014). Camello, a novel family of Histone acetyltransferases that acetylate histone H4 and is essential for zebrafish development. *Scientific Reports*, *4*, 6076. https://doi.org/10.1038/srep06076.

Kaypee, S., Sudarshan, D., Shanmugam, M. K., Mukherjee, D., Sethi, G., & Kundu, T. K. (2016). Aberrant lysine acetylation in tumorigenesis: Implications in the development of therapeutics. *Pharmacology & Therapeutics*, *162*, 98–119. https://doi.org/10.1016/j.pharmthera.2016.01.011.

Kim, M. Y., Woo, E. M., Chong, Y. T., Homenko, D. R., & Kraus, W. L. (2006). Acetylation of estrogen receptor alpha by p300 at lysines 266 and 268 enhances the deoxyribonucleic acid binding and transactivation activities of the receptor. *Molecular Endocrinology*, *20*(7), 1479–1493. https://doi.org/10.1210/me.2005-0531.

Kishimoto, M., Kohno, T., Okudela, K., Otsuka, A., Sasaki, H., Tanabe, C., et al. (2005). Mutations and deletions of the CBP gene in human lung cancer. *Clinical Cancer Research*, *11*(2 Pt. 1), 512–519.

Kitabayashi, I., Aikawa, Y., Nguyen, L. A., Yokoyama, A., & Ohki, M. (2001). Activation of AML1-mediated transcription by MOZ and inhibition by the MOZ-CBP fusion protein. *The EMBO Journal*, *20*(24), 7184–7196. https://doi.org/10.1093/emboj/20.24.7184.

Ko, H., So, Y., Jeon, H., Jeong, M. H., Choi, H. K., Ryu, S. H., et al. (2013). TGF-β1-induced epithelial-mesenchymal transition and acetylation of Smad2 and Smad3 are negatively regulated by EGCG in human A549 lung cancer cells. *Cancer Letters*, *335*(1), 205–213. https://doi.org/10.1016/j.canlet.2013.02.018.

Kok, K., Naylor, S. L., & Buys, C. H. (1997). Deletions of the short arm of chromosome 3 in solid tumors and the search for suppressor genes. *Advances in Cancer Research*, *71*, 27–92.

Kouzarides, T. (2000). Acetylation: A regulatory modification to rival phosphorylation? *The EMBO Journal*, *19*(6), 1176–1179. https://doi.org/10.1093/emboj/19.6.1176.

Kouzarides, T. (2007). Chromatin modifications and their function. *Cell*, *128*(4), 693–705. https://doi.org/10.1016/j.cell.2007.02.005.

Krishna, S., Kumar, S., Singh, D. K., Lakra, A. D., Banerjee, D., & Siddiqi, M. I. (2018). Multiple machine learning based-chemoinformatics models for identification of histone acetyl transferase inhibitors. *Molecular Informatics*, *37*(8), e1700150. https://doi.org/10.1002/minf.201700150.

Krivtsov, A. V., & Armstrong, S. A. (2007). MLL translocations, histone modifications and leukaemia stem-cell development. *Nature Reviews. Cancer*, 7(11), 823–833. https://doi.org/10.1038/nrc2253.

Kumar, B., Yadav, A., Hideg, K., Kuppusamy, P., Teknos, T. N., & Kumar, P. (2014). A novel curcumin analog (H-4073) enhances the therapeutic efficacy of cisplatin treatment in head and neck cancer. *PLoS One*, *9*(3), e93208. https://doi.org/10.1371/journal.pone.0093208.

Kung, A. L., Rebel, V. I., Bronson, R. T., Ch'ng, L. E., Sieff, C. A., Livingston, D. M., et al. (2000). Gene dose-dependent control of hematopoiesis and hematologic tumor suppression by CBP. *Genes & Development*, *14*(3), 272–277.

Kwie, F. H., Briet, M., Soupaya, D., Hoffmann, P., Maturano, M., Rodriguez, F., et al. (2011). New potent bisubstrate inhibitors of histone acetyltransferase p300: Design, synthesis and biological evaluation. *Chemical Biology & Drug Design*, 77(1), 86–92. https://doi.org/10.1111/j.1747-0285.2010.01056.x.

Lasko, L. M., Jakob, C. G., Edalji, R. P., Qiu, W., Montgomery, D., Digiammarino, E. L., et al. (2017). Discovery of a selective catalytic p300/CBP inhibitor that targets lineage-specific tumours. *Nature*, *550*(7674), 128–132. https://doi.org/10.1038/nature24028.

Lau, O. D., Kundu, T. K., Soccio, R. E., Ait-Si-Ali, S., Khalil, E. M., Vassilev, A., et al. (2000). HATs off: Selective synthetic inhibitors of the histone acetyltransferases p300 and PCAF. *Molecular Cell*, *5*(3), 589–595.

Leaver, D. J., Cleary, B., Nguyen, N., Priebbenow, D. L., Lagiakos, H. R., Sanchez, J., et al. (2019). Discovery of benzoylsulfonohydrazides as potent inhibitors of the histone acetyltransferase KAT6A. *Journal of Medicinal Chemistry*, *62*(15), 7146–7159. https://doi.org/10.1021/acs.jmedchem.9b00665.

Lee, J. S., Smith, E., & Shilatifard, A. (2010). The language of histone crosstalk. *Cell*, *142*(5), 682–685. https://doi.org/10.1016/j.cell.2010.08.011.

Li, X. G., Chen, Y., Wu, Q., & Liu, H. L. (2005). Effects of curcumin on the acetylation of histone H3, P53 and the proliferation of NB4 cells. *Zhonghua Xue Ye Xue Za Zhi*, *26*(9), 551–553.

Li, X. G., Chen, Y., Wu, Q., & Sun, C. Y. (2006). Effect of curcumin on acetylation of histone H3 in human lymphoma cell line Raji. *Ai Zheng*, *25*(5), 582–586.

Li, F., Shanmugam, M. K., Chen, L., Chatterjee, S., Basha, J., Kumar, A. P., et al. (2013). Garcinol, a polyisoprenylated benzophenone modulates multiple proinflammatory signaling cascades leading to the suppression of growth and survival of head and neck carcinoma. *Cancer Prevention Research (Philadelphia, Pa.)*, *6*(8), 843–854. https://doi.org/10.1158/1940-6207.CAPR-13-0070.

Li, F., Shanmugam, M. K., Siveen, K. S., Wang, F., Ong, T. H., Loo, S. Y., et al. (2015). Garcinol sensitizes human head and neck carcinoma to cisplatin in a xenograft mouse model despite downregulation of proliferative biomarkers. *Oncotarget*, *6*(7), 5147–5163. https://dx.doi.org/10.18632/oncotarget.2881.

Liang, Y., Hu, J., Li, J., Liu, Y., Yu, J., Zhuang, X., et al. (2015). Epigenetic activation of TWIST1 by MTDH promotes cancer stem-like cell traits in breast cancer. *Cancer Research*, *75*(17), 3672–3680. https://doi.org/10.1158/0008-5472.CAN-15-0930.

Liao, Z. W., Zhou, T. C., Tan, X. J., Song, X. L., Liu, Y., Shi, X. Y., et al. (2012). High expression of p300 is linked to aggressive features and poor prognosis of nasopharyngeal carcinoma. *Journal of Translational Medicine*, *10*, 110. https://doi.org/10.1186/1479-5876-10-110.

Liu, B. L., Cheng, J. X., Zhang, X., Wang, R., Zhang, W., Lin, H., et al. (2010). Global histone modification patterns as prognostic markers to classify glioma patients. *Cancer Epidemiology, Biomarkers & Prevention*, *19*(11), 2888–2896. https://doi.org/10.1158/1055-9965.EPI-10-0454.

Liu, C., Ho, P. C., Wong, F. C., Sethi, G., Wang, L. Z., & Goh, B. C. (2015). Garcinol: Current status of its anti-oxidative, anti-inflammatory and anti-cancer effects. *Cancer Letters*, *362*(1), 8–14. https://doi.org/10.1016/j.canlet.2015.03.019.

Liu, Y. N., Lee, W. W., Wang, C. Y., Chao, T. H., Chen, Y., & Chen, J. H. (2005). Regulatory mechanisms controlling human E-cadherin gene expression. *Oncogene*, *24*(56), 8277–8290. https://doi.org/10.1038/sj.onc.1208991.

Liu, X., Wang, L., Zhao, K., Thompson, P. R., Hwang, Y., Marmorstein, R., et al. (2008). The structural basis of protein acetylation by the p300/CBP transcriptional coactivator. *Nature*, *451*(7180), 846–850. https://doi.org/10.1038/nature06546.

Liu, R., Zhang, Z., Yang, H., Zhou, K., Geng, M., Zhou, W., et al. (2019). Design, synthesis, and biological evaluation of a new class of histone acetyltransferase p300 inhibitors. *European Journal of Medicinal Chemistry*, *180*, 171–190. https://doi.org/10.1016/j.ejmech.2019.07.026.

Liu, N., Zhang, R., Zhao, X., Su, J., Bian, X., Ni, J., et al. (2013). A potential diagnostic marker for ovarian cancer: Involvement of the histone acetyltransferase, human males absent on the first. *Oncology Letters*, *6*(2), 393–400. https://doi.org/10.3892/ol.2013.1380.

Lu, W., Xiong, H., Chen, Y., Wang, C., Zhang, H., Xu, P., et al. (2018). Discovery and biological evaluation of thiobarbituric derivatives as potent p300/CBP inhibitors. *Bioorganic & Medicinal Chemistry*, *26*(20), 5397–5407. https://doi.org/10.1016/j.bmc.2018.07.048.

Luo, J., Li, M., Tang, Y., Laszkowska, M., Roeder, R. G., & Gu, W. (2004). Acetylation of p53 augments its site-specific DNA binding both in vitro and in vivo. *Proceedings of the National Academy of Sciences of the United States of America*, *101*(8), 2259–2264. https://doi.org/10.1073/pnas.0308762101.

Lynch, H., Wen, H., Kim, Y. C., Snyder, C., Kinarsky, Y., Chen, P. X., et al. (2013). Can unknown predisposition in familial breast cancer be family-specific? *The Breast Journal*, *19*(5), 520–528. https://doi.org/10.1111/tbj.12145.

Mai, A., Rotili, D., Tarantino, D., Ornaghi, P., Tosi, F., Vicidomini, C., et al. (2006). Small-molecule inhibitors of histone acetyltransferase activity: Identification and biological properties. *Journal of Medicinal Chemistry*, *49*(23), 6897–6907. https://doi.org/10.1021/jm060601m.

Malhotra, A., Nair, P., & Dhawan, D. K. (2014). Study to evaluate molecular mechanics behind synergistic chemo-preventive effects of curcumin and resveratrol during lung carcinogenesis. *PLoS One*, *9*(4), e93820. https://doi.org/10.1371/journal.pone.0093820.

Mantelingu, K., Reddy, B. A., Swaminathan, V., Kishore, A. H., Siddappa, N. B., Kumar, G. V., et al. (2007). Specific inhibition of p300-HAT alters global gene expression and represses HIV replication. *Chemistry & Biology*, *14*(6), 645–657. https://doi.org/10.1016/j.chembiol.2007.04.011.

Manu, K. A., Shanmugam, M. K., Rajendran, P., Li, F., Ramachandran, L., Hay, H. S., et al. (2011). Plumbagin inhibits invasion and migration of breast and gastric cancer cells by downregulating the expression of chemokine receptor CXCR4. *Molecular Cancer*, *10*, 107. https://doi.org/10.1186/1476-4598-10-107.

Manuyakorn, A., Paulus, R., Farrell, J., Dawson, N. A., Tze, S., Cheung-Lau, G., et al. (2010). Cellular histone modification patterns predict prognosis and treatment response in resectable pancreatic adenocarcinoma: Results from RTOG 9704. *Journal of Clinical Oncology*, *28*(8), 1358–1365. https://doi.org/10.1200/JCO.2009.24.5639.

Manzo, F., Tambaro, F. P., Mai, A., & Altucci, L. (2009). Histone acetyltransferase inhibitors and preclinical studies. *Expert Opinion on Therapeutic Patents*, *19*(6), 761–774. https://doi.org/10.1517/13543770902895727.

Marcu, M. G., Jung, Y. J., Lee, S., Chung, E. J., Lee, M. J., Trepel, J., et al. (2006). Curcumin is an inhibitor of p300 histone acetyltransferase. *Medicinal Chemistry*, *2*(2), 169–174.

Marmorstein, R., & Zhou, M. M. (2014). Writers and readers of histone acetylation: Structure, mechanism, and inhibition. *Cold Spring Harbor Perspectives in Biology*, *6*(7), a018762. https://doi.org/10.1101/cshperspect.a018762.

Martens, J. A., & Winston, F. (2003). Recent advances in understanding chromatin remodeling by Swi/Snf complexes. *Current Opinion in Genetics & Development*, *13*(2), 136–142.

Merarchi, M., Sethi, G., Shanmugam, M. K., Lu, F., Arfuso, F., & Ahn, K. S. (2019). Role of natural products in modulating histone deacetylases in cancer. *Molecules*, *24*(6), 1047–1061. https://doi.org/10.3390/molecules24061047.

Michaelides, M. R., Kluge, A., Patane, M., Van Drie, J. H., Wang, C., Hansen, T. M., et al. (2018). Discovery of spiro oxazolidinediones as selective, orally bioavailable inhibitors of p300/CBP histone acetyltransferases. *ACS Medicinal Chemistry Letters*, *9*(1), 28–33. https://doi.org/10.1021/acsmedchemlett.7b00395.

Milite, C., Feoli, A., Sasaki, K., La Pietra, V., Balzano, A. L., Marinelli, L., et al. (2015). A novel cell-permeable, selective, and noncompetitive inhibitor of KAT3 histone acetyltransferases from a combined molecular pruning/classical isosterism approach. *Journal of Medicinal Chemistry*, *58*(6), 2779–2798. https://doi.org/10.1021/jm5019687.

Mishra, S., Verma, S. S., Rai, V., Awasthee, N., Chava, S., Hui, K. M., et al. (2019). Long non-coding RNAs are emerging targets of phytochemicals for cancer and other chronic diseases. *Cellular and Molecular Life Sciences*, *76*(10), 1947–1966. https://doi.org/10.1007/s00018-019-03053-0.

Miyamoto, N., Izumi, H., Noguchi, T., Nakajima, Y., Ohmiya, Y., Shiota, M., et al. (2008). Tip60 is regulated by circadian transcription factor clock and is involved in cisplatin resistance. *The Journal of Biological Chemistry*, *283*(26), 18218–18226. https://doi.org/10.1074/jbc.M802332200.

Mizzen, C. A., Yang, X. J., Kokubo, T., Brownell, J. E., Bannister, A. J., Owen-Hughes, T., et al. (1996). The TAF(II)250 subunit of TFIID has histone acetyltransferase activity. *Cell*, *87*(7), 1261–1270.

Modak, R., Basha, J., Bharathy, N., Maity, K., Mizar, P., Bhat, A. V., et al. (2013). Probing p300/CBP associated factor (PCAF)-dependent pathways with a small molecule inhibitor. *ACS Chemical Biology*, *8*(6), 1311–1323. https://doi.org/10.1021/cb4000597.

Nagy, Z., & Tora, L. (2007). Distinct GCN5/PCAF-containing complexes function as co-activators and are involved in transcription factor and global histone acetylation. *Oncogene*, *26*(37), 5341–5357. https://doi.org/10.1038/sj.onc.1210604.

Neuwald, A. F., & Landsman, D. (1997). GCN5-related histone N-acetyltransferases belong to a diverse superfamily that includes the yeast SPT10 protein. *Trends in Biochemical Sciences*, *22*(5), 154–155.

Ngo, L., Brown, T., & Zheng, Y. G. (2019). Bisubstrate inhibitors to target histone acetyltransferase 1. *Chemical Biology & Drug Design*, *93*(5), 865–873. https://doi.org/10.1111/cbdd.13476.

Nguyen, A. H., Berim, I. G., & Agrawal, D. K. (2015). Chronic inflammation and cancer: Emerging roles of triggering receptors expressed on myeloid cells. *Expert Review of Clinical Immunology*, *11*(7), 849–857. https://doi.org/10.1586/1744666X.2015.1043893.

Oike, T., Komachi, M., Ogiwara, H., Amornwichet, N., Saitoh, Y., Torikai, K., et al. (2014). C646, a selective small molecule inhibitor of histone acetyltransferase p300, radiosensitizes lung cancer cells by enhancing mitotic catastrophe. *Radiotherapy and Oncology*, *111*(2), 222–227. https://doi.org/10.1016/j.radonc.2014.03.015.

Ozdag, H., Batley, S. J., Forsti, A., Iyer, N. G., Daigo, Y., Boutell, J., et al. (2002). Mutation analysis of CBP and PCAF reveals rare inactivating mutations in cancer cell lines but not in primary tumours. *British Journal of Cancer*, *87*(10), 1162–1165. https://doi.org/10.1038/sj.bjc.6600554.

Ozdag, H., Teschendorff, A. E., Ahmed, A. A., Hyland, S. J., Blenkiron, C., Bobrow, L., et al. (2006). Differential expression of selected histone modifier genes in human solid cancers. *BMC Genomics*, 7, 90. https://doi.org/10.1186/1471-2164-7-90.

Parthun, M. R. (2013). Histone acetyltransferase 1: More than just an enzyme? *Biochimica et Biophysica Acta*, *1819*(3–4), 256–263.

Pattabiraman, D. R., McGirr, C., Shakhbazov, K., Barbier, V., Krishnan, K., Mukhopadhyay, P., et al. (2014). Interaction of c-Myb with p300 is required for the induction of acute myeloid leukemia (AML) by human AML oncogenes. *Blood*, *123*(17), 2682–2690. https://doi.org/10.1182/blood-2012-02-413187.

Pfister, S., Rea, S., Taipale, M., Mendrzyk, F., Straub, B., Ittrich, C., et al. (2008). The histone acetyltransferase hMOF is frequently downregulated in primary breast carcinoma and medulloblastoma and constitutes a biomarker for clinical outcome in medulloblastoma. *International Journal of Cancer*, *122*(6), 1207–1213. https://doi.org/10.1002/ijc.23283.

Picaud, S., Fedorov, O., Thanasopoulou, A., Leonards, K., Jones, K., Meier, J., et al. (2015). Generation of a selective small molecule inhibitor of the CBP/p300 bromodomain for leukemia therapy. *Cancer Research*, *75*(23), 5106–5119. https://doi.org/10.1158/0008-5472.CAN-15-0236.

Poojari, R. (2014). Embelin—A drug of antiquity: Shifting the paradigm towards modern medicine. *Expert Opinion on Investigational Drugs*, *23*(3), 427–444. https://doi.org/10.1517/13543784.2014.867016.

Popov, V. M., Wang, C., Shirley, L. A., Rosenberg, A., Li, S., Nevalainen, M., et al. (2007). The functional significance of nuclear receptor acetylation. *Steroids*, *72*(2), 221–230. https://doi.org/10.1016/j.steroids.2006.12.001.

Qin, Y. R., Fu, L., Sham, P. C., Kwong, D. L., Zhu, C. L., Chu, K. K., et al. (2008). Single-nucleotide polymorphism-mass array reveals commonly deleted regions at 3p22 and 3p14.2 associate with poor clinical outcome in esophageal squamous cell carcinoma. *International Journal of Cancer*, *123*(4), 826–830. https://doi.org/10.1002/ijc.23577.

Ramsay, R. G., & Gonda, T. J. (2008). MYB function in normal and cancer cells. *Nature Reviews. Cancer*, *8*(7), 523–534. https://doi.org/10.1038/nrc2439.

Ravindra, K. C., Selvi, B. R., Arif, M., Reddy, B. A., Thanuja, G. R., Agrawal, S., et al. (2009). Inhibition of lysine acetyltransferase KAT3B/p300 activity by a naturally occurring hydroxynaphthoquinone, plumbagin. *The Journal of Biological Chemistry*, *284*(36), 24453–24464. https://doi.org/10.1074/jbc.M109.023861.

Rebel, V. I., Kung, A. L., Tanner, E. A., Yang, H., Bronson, R. T., & Livingston, D. M. (2002). Distinct roles for CREB-binding protein and p300 in hematopoietic stem cell self-renewal. *Proceedings of the National Academy of Sciences of the United States of America*, *99*(23), 14789–14794. https://doi.org/10.1073/pnas.232568499.

Roche, J., Nasarre, P., Gemmill, R., Baldys, A., Pontis, J., Korch, C., et al. (2013). Global decrease of histone H3K27 acetylation in ZEB1-induced epithelial to mesenchymal transition in lung cancer cells. *Cancers*, *5*(2), 334–356. https://doi.org/10.3390/cancers5020334.

Rohira, A. D., & Lonard, D. M. (2017). Steroid receptor coactivators present a unique opportunity for drug development in hormone-dependent cancers. *Biochemical Pharmacology*, *140*, 1–7. https://doi.org/10.1016/j.bcp.2017.04.005.

Roth, S. Y., Denu, J. M., & Allis, C. D. (2001). Histone acetyltransferases. *Annual Review of Biochemistry*, *70*, 81–120. https://doi.org/10.1146/annurev.biochem.70.1.81.

Rozman, M., Camos, M., Colomer, D., Villamor, N., Esteve, J., Costa, D., et al. (2004). Type I MOZ/CBP (MYST3/CREBBP) is the most common chimeric transcript in acute myeloid leukemia with t(8,16)(p11;p13) translocation. *Genes, Chromosomes & Cancer*, *40*(2), 140–145. https://doi.org/10.1002/gcc.20022.

Sakuraba, K., Yasuda, T., Sakata, M., Kitamura, Y. H., Shirahata, A., Goto, T., et al. (2009). Down-regulation of Tip60 gene as a potential marker for the malignancy of colorectal cancer. *Anticancer Research*, *29*(10), 3953–3955.

Sakuraba, K., Yokomizo, K., Shirahata, A., Goto, T., Saito, M., Ishibashi, K., et al. (2011). TIP60 as a potential marker for the malignancy of gastric cancer. *Anticancer Research*, *31*(1), 77–79.

Salah Ud-Din, A. I., Tikhomirova, A., & Roujeinikova, A. (2016). Structure and functional diversity of GCN5-related N-acetyltransferases (GNAT). *International Journal of Molecular Sciences*, *17*(7), 1018. https://doi.org/10.3390/ijms17071018.

Sandberg, M. L., Sutton, S. E., Pletcher, M. T., Wiltshire, T., Tarantino, L. M., Hogenesch, J. B., et al. (2005). c-Myb and p300 regulate hematopoietic stem cell proliferation and differentiation. *Developmental Cell*, *8*(2), 153–166. https://doi.org/10.1016/j.devcel.2004.12.015.

Sandur, S. K., Ichikawa, H., Sethi, G., Ahn, K. S., & Aggarwal, B. B. (2006). Plumbagin (5-hydroxy-2-methyl-1,4-naphthoquinone) suppresses NF-kappaB activation and NF-kappaB-regulated gene products through modulation of p65 and IkappaBalpha kinase activation, leading to potentiation of apoptosis induced by cytokine and chemotherapeutic agents. *The Journal of Biological Chemistry*, *281*(25), 17023–17033. https://doi.org/10.1074/jbc.M601595200.

Sapountzi, V., & Cote, J. (2011). MYST-family histone acetyltransferases: Beyond chromatin. *Cellular and Molecular Life Sciences*, *68*(7), 1147–1156. https://doi.org/10.1007/s00018-010-0599-9.

Sbardella, G., Castellano, S., Vicidomini, C., Rotili, D., Nebbioso, A., Miceli, M., et al. (2008). Identification of long chain alkylidenemalonates as novel small molecule modulators of histone acetyltransferases. *Bioorganic & Medicinal Chemistry Letters*, *18*(9), 2788–2792. https://doi.org/10.1016/j.bmcl.2008.04.017.

Schiltz, R. L., Mizzen, C. A., Vassilev, A., Cook, R. G., Allis, C. D., & Nakatani, Y. (1999). Overlapping but distinct patterns of histone acetylation by the human coactivators p300 and PCAF within nucleosomal substrates. *The Journal of Biological Chemistry*, *274*(3), 1189–1192. https://doi.org/10.1074/jbc.274.3.1189.

Secci, D., Carradori, S., Bizzarri, B., Bolasco, A., Ballario, P., Patramani, Z., et al. (2014). Synthesis of a novel series of thiazole-based histone acetyltransferase inhibitors. *Bioorganic & Medicinal Chemistry*, *22*(5), 1680–1689. https://doi.org/10.1016/j.bmc.2014.01.022.

Seidel, C., Florean, C., Schnekenburger, M., Dicato, M., & Diederich, M. (2012). Chromatin-modifying agents in anti-cancer therapy. *Biochimie*, *94*(11), 2264–2279. https://doi.org/10.1016/j.biochi.2012.05.012.

Seidel, C., Schnekenburger, M., Dicato, M., & Diederich, M. (2012). Histone deacetylase modulators provided by Mother Nature. *Genes & Nutrition*, 7(3), 357–367. https://doi.org/10.1007/s12263-012-0283-9.

Seligson, D. B., Horvath, S., McBrian, M. A., Mah, V., Yu, H., Tze, S., et al. (2009). Global levels of histone modifications predict prognosis in different cancers. *The American Journal of Pathology*, *174*(5), 1619–1628. https://doi.org/10.2353/ajpath.2009.080874.

Seligson, D. B., Horvath, S., Shi, T., Yu, H., Tze, S., Grunstein, M., et al. (2005). Global histone modification patterns predict risk of prostate cancer recurrence. *Nature*, *435*(7046), 1262–1266. https://doi.org/10.1038/nature03672.

Selvi, B. R., Mohankrishna, D. V., Ostwal, Y. B., & Kundu, T. K. (2010). Small molecule modulators of histone acetylation and methylation: A disease perspective. *Biochimica et Biophysica Acta*, *1799*(10 − 12), 810–828. https://doi.org/10.1016/j.bbagrm.2010.09.005.

Selvi, B. R., Pradhan, S. K., Shandilya, J., Das, C., Sailaja, B. S., Shankar, G. N., et al. (2009). Sanguinarine interacts with chromatin, modulates epigenetic modifications, and transcription in the context of chromatin. *Chemistry & Biology*, *16*(2), 203–216. https://doi.org/10.1016/j.chembiol.2008.12.006.

Seong, A. R., Yoo, J. Y., Choi, K., Lee, M. H., Lee, Y. H., Lee, J., et al. (2011). Delphinidin, a specific inhibitor of histone acetyltransferase, suppresses inflammatory signaling via prevention of NF-kappaB acetylation in fibroblast-like synoviocyte MH7A cells. *Biochemical and Biophysical Research Communications*, *410*(3), 581–586. https://doi.org/10.1016/j.bbrc.2011.06.029.

Sethi, G., Chatterjee, S., Rajendran, P., Li, F., Shanmugam, M. K., Wong, K. F., et al. (2014). Inhibition of STAT3 dimerization and acetylation by garcinol suppresses the growth of human hepatocellular carcinoma in vitro and in vivo. *Molecular Cancer*, *13*, 66. https://doi.org/10.1186/1476-4598-13-66.

Shankar, S., & Srivastava, R. K. (2007). Involvement of Bcl-2 family members, phosphatidylinositol 3′-kinase/AKT and mitochondrial p53 in curcumin (diferulolylmethane)-induced apoptosis in prostate cancer. *International Journal of Oncology*, *30*(4), 905–918.

Shanmugam, M. K., Arfuso, F., Arumugam, S., Chinnathambi, A., Jinsong, B., Warrier, S., et al. (2017). Role of novel histone modifications in cancer. *Oncotarget*, *9*(13), 11414–11426.

Shanmugam, M. K., Arfuso, F., Sng, J. C., Bishayee, A., Kumar, A. P., & Sethi, G. (2019). Epigenetic effects of curcumin in cancer prevention. In P. Anupam Bishayee, & P. Deepak Bhatia (Eds.), *Epigenetics of cancer prevention* (pp. 107–128). United Kingdom: Elsevier.

Shanmugam, M. K., Kannaiyan, R., & Sethi, G. (2011). Targeting cell signaling and apoptotic pathways by dietary agents: Role in the prevention and treatment of cancer. *Nutrition and Cancer*, *63*(2), 161–173. https://doi.org/10.1080/01635581.2011.523502.

Shanmugam, M. K., Lee, J. H., Chai, E. Z., Kanchi, M. M., Kar, S., Arfuso, F., et al. (2016). Cancer prevention and therapy through the modulation of transcription factors by bioactive natural compounds. *Seminars in Cancer Biology*, *40–41*, 35–47. https://doi.org/10.1016/j.semcancer.2016.03.005.

Shanmugam, M. K., Rane, G., Kanchi, M. M., Arfuso, F., Chinnathambi, A., Zayed, M. E., et al. (2015). The multifaceted role of curcumin in cancer prevention and treatment. *Molecules*, *20*(2), 2728–2769. https://doi.org/10.3390/molecules20022728.

Shanmugam, M. K., Warrier, S., Kumar, A. P., Sethi, G., & Arfuso, F. (2017). Potential role of natural compounds as anti-angiogenic agents in cancer. *Current Vascular Pharmacology*, *15*(6), 503–519. https://doi.org/10.2174/1570161115666170713094319.

Shogren-Knaak, M., Ishii, H., Sun, J. M., Pazin, M. J., Davie, J. R., & Peterson, C. L. (2006). Histone H4-K16 acetylation controls chromatin structure and protein interactions. *Science*, *311*(5762), 844–847. https://doi.org/10.1126/science.1124000.

Shrimp, J. H., Sorum, A. W., Garlick, J. M., Guasch, L., Nicklaus, M. C., & Meier, J. L. (2016). Characterizing the covalent targets of a small molecule inhibitor of the lysine acetyltransferase P300. *ACS Medicinal Chemistry Letters*, 7(2), 151–155. https://doi.org/10.1021/acsmedchemlett.5b00385.

Simon, R. P., Robaa, D., Alhalabi, Z., Sippl, W., & Jung, M. (2016). KATching-up on small molecule modulators of lysine acetyltransferases. *Journal of Medicinal Chemistry*, *59*(4), 1249–1270. https://doi.org/10.1021/acs.jmedchem.5b01502.

Slaninová, I., Pěnčíková, K., Urbanová, J., Slanina, J., & Táborská, E. (2014). Antitumour activities of sanguinarine and related alkaloids. *Phytochemistry Reviews*, *13*, 51–68.

Souto, J. A., Conte, M., Alvarez, R., Nebbioso, A., Carafa, V., Altucci, L., et al. (2008). Synthesis of benzamides related to anacardic acid and their histone acetyltransferase (HAT) inhibitory activities. *ChemMedChem*, *3*(9), 1435–1442. https://doi.org/10.1002/cmdc.200800096.

Spencer, T. E., Jenster, G., Burcin, M. M., Allis, C. D., Zhou, J., Mizzen, C. A., et al. (1997). Steroid receptor coactivator-1 is a histone acetyltransferase. *Nature*, *389*(6647), 194–198. https://doi.org/10.1038/38304.

Stasevich, T. J., Hayashi-Takanaka, Y., Sato, Y., Maehara, K., Ohkawa, Y., Sakata-Sogawa, K., et al. (2014). Regulation of RNA polymerase II activation by histone acetylation in single living cells. *Nature*, *516*(7530), 272–275. https://doi.org/10.1038/nature13714.

Stimson, L., Rowlands, M. G., Newbatt, Y. M., Smith, N. F., Raynaud, F. I., Rogers, P., et al. (2005). Isothiazolones as inhibitors of PCAF and p300 histone acetyltransferase activity. *Molecular Cancer Therapeutics*, *4*(10), 1521–1532. https://doi.org/10.1158/1535-7163.MCT-05-0135.

Subbaiah, V. K., Zhang, Y., Rajagopalan, D., Abdullah, L. N., Yeo-Teh, N. S., Tomaic, V., et al. (2016). E3 ligase EDD1/UBR5 is utilized by the HPV E6 oncogene to destabilize tumor suppressor TIP60. *Oncogene*, *35*(16), 2062–2074. https://doi.org/10.1038/onc.2015.268.

Sun, Y., Jiang, X., Chen, S., & Price, B. D. (2006). Inhibition of histone acetyltransferase activity by anacardic acid sensitizes tumor cells to ionizing radiation. *FEBS Letters*, *580*(18), 4353–4356. https://doi.org/10.1016/j.febslet.2006.06.092.

Sun, X. J., Man, N., Tan, Y., Nimer, S. D., & Wang, L. (2015). The role of histone acetyltransferases in normal and malignant hematopoiesis. *Frontiers in Oncology*, *5*, 108. https://doi.org/10.3389/fonc.2015.00108.

Sunde, J. S., Donninger, H., Wu, K., Johnson, M. E., Pestell, R. G., Rose, G. S., et al. (2006). Expression profiling identifies altered expression of genes that contribute to the inhibition of transforming growth factor-beta signaling in ovarian cancer. *Cancer Research*, *66*(17), 8404–8412. https://doi.org/10.1158/0008-5472.CAN-06-0683.

Sung, B., Kunnumakkara, A. B., Sethi, G., Anand, P., Guha, S., & Aggarwal, B. B. (2009). Curcumin circumvents chemoresistance in vitro and potentiates the effect of thalidomide and bortezomib against human multiple myeloma in nude mice model. *Molecular Cancer Therapeutics*, *8*(4), 959–970. https://doi.org/10.1158/1535-7163.MCT-08-0905.

Tafrova, J. I., & Tafrov, S. T. (2014). Human histone acetyltransferase 1 (Hat1) acetylates lysine 5 of histone H2A in vivo. *Molecular and Cellular Biochemistry*, *392*(1–2), 259–272. https://doi.org/10.1007/s11010-014-2036-0.

Taipale, M., Rea, S., Richter, K., Vilar, A., Lichter, P., Imhof, A., et al. (2005). hMOF histone acetyltransferase is required for histone H4 lysine 16 acetylation in mammalian cells. *Molecular and Cellular Biology*, *25*(15), 6798–6810. https://doi.org/10.1128/MCB.25.15.6798-6810.2005.

Takeuchi, A., Shiota, M., Tatsugami, K., Yokomizo, A., Tanaka, S., Kuroiwa, K., et al. (2012). p300 mediates cellular resistance to doxorubicin in bladder cancer. *Molecular Medicine Reports*, *5*(1), 173–176. https://doi.org/10.3892/mmr.2011.593.

Taylor, G. C., Eskeland, R., Hekimoglu-Balkan, B., Pradeepa, M. M., & Bickmore, W. A. (2013). H4K16 acetylation marks active genes and enhancers of embryonic stem cells, but does not alter chromatin compaction. *Genome Research*, *23*(12), 2053–2065. https://doi.org/10.1101/gr.155028.113.

Tessarz, P., & Kouzarides, T. (2014). Histone core modifications regulating nucleosome structure and dynamics. *Nature Reviews. Molecular Cell Biology*, *15*(11), 703–708. https://doi.org/10.1038/nrm3890.

Tewari, D., Nabavi, S. F., Nabavi, S. M., Sureda, A., Farooqi, A. A., Atanasov, A. G., et al. (2018). Targeting activator protein 1 signaling pathway by bioactive natural agents: Possible therapeutic strategy for cancer prevention and intervention. *Pharmacological Research*, *128*, 366–375. https://doi.org/10.1016/j.phrs.2017.09.014.

Tillinghast, G. W., Partee, J., Albert, P., Kelley, J. M., Burtow, K. H., & Kelly, K. (2003). Analysis of genetic stability at the EP300 and CREBBP loci in a panel of cancer cell lines. *Genes, Chromosomes & Cancer*, *37*(2), 121–131. https://doi.org/10.1002/gcc.10195.

Tohyama, S., Tomura, A., Ikeda, N., Hatano, M., Odanaka, J., Kubota, Y., et al. (2012). Discovery and characterization of NK13650s, naturally occurring p300-selective histone acetyltransferase inhibitors. *The Journal of Organic Chemistry*, 77(20), 9044–9052. https://doi.org/10.1021/jo301534b.

Trisciuoglio, D., Di Martile, M., & Del Bufalo, D. (2018). Emerging role of histone acetyltransferase in stem cells and cancer. *Stem Cells International*, *2018*, 8908751. https://doi.org/10.1155/2018/8908751.

Trisciuoglio, D., Ragazzoni, Y., Pelosi, A., Desideri, M., Carradori, S., Gabellini, C., et al. (2012). CPTH6, a thiazole derivative, induces histone hypoacetylation and apoptosis in human leukemia cells. *Clinical Cancer Research*, *18*(2), 475–486. https://doi.org/10.1158/1078-0432.CCR-11-0579.

Tzao, C., Tung, H. J., Jin, J. S., Sun, G. H., Hsu, H. S., Chen, B. H., et al. (2009). Prognostic significance of global histone modifications in resected squamous cell carcinoma of the esophagus. *Modern Pathology*, *22*(2), 252–260. https://doi.org/10.1038/modpathol.2008.172.

Unnikrishnan, A., Gafken, P. R., & Tsukiyama, T. (2010). Dynamic changes in histone acetylation regulate origins of DNA replication. *Nature Structural & Molecular Biology*, *17*(4), 430–437. https://doi.org/10.1038/nsmb.1780.

Uttarkar, S., Dukare, S., Bopp, B., Goblirsch, M., Jose, J., & Klempnauer, K. H. (2015). Naphthol AS-E phosphate inhibits the activity of the transcription factor Myb by blocking the interaction with the KIX domain of the coactivator p300. *Molecular Cancer Therapeutics*, *14*(6), 1276–1285. https://doi.org/10.1158/1535-7163.MCT-14-0662.

Van Den Broeck, A., Brambilla, E., Moro-Sibilot, D., Lantuejoul, S., Brambilla, C., Eymin, B., et al. (2008). Loss of histone H4K20 trimethylation occurs in preneoplasia and influences prognosis of non-small cell lung cancer. *Clinical Cancer Research*, *14*(22), 7237–7245. https://doi.org/10.1158/1078-0432.CCR-08-0869.

Van Den Broeck, A., Nissou, D., Brambilla, E., Eymin, B., & Gazzeri, S. (2012). Activation of a Tip60/E2F1/ERCC1 network in human lung adenocarcinoma cells exposed to cisplatin. *Carcinogenesis*, *33*(2), 320–325. https://doi.org/10.1093/carcin/bgr292.

Vasudevarao, M. D., Mizar, P., Kumari, S., Mandal, S., Siddhanta, S., Swamy, M. M., et al. (2014). Naphthoquinone-mediated inhibition of lysine acetyltransferase KAT3B/p300, basis for non-toxic inhibitor synthesis. *The Journal of Biological Chemistry*, *289*(11), 7702–7717. https://doi.org/10.1074/jbc.M113.486522.

Venkatesh, S., & Workman, J. L. (2015). Histone exchange, chromatin structure and the regulation of transcription. *Nature Reviews. Molecular Cell Biology*, *16*(3), 178–189. https://doi.org/10.1038/nrm3941.

Vo, N., & Goodman, R. H. (2001). CREB-binding protein and p300 in transcriptional regulation. *The Journal of Biological Chemistry*, *276*(17), 13505–13508. https://doi.org/10.1074/jbc.R000025200.

Voss, A. K., & Thomas, T. (2009). MYST family histone acetyltransferases take center stage in stem cells and development. *Bioessays*, *31*(10), 1050–1061. https://doi.org/10.1002/bies.200900051.

Wadia, J. S., & Dowdy, S. F. (2005). Transmembrane delivery of protein and peptide drugs by TAT-mediated transduction in the treatment of cancer. *Advanced Drug Delivery Reviews*, *57*(4), 579–596. https://doi.org/10.1016/j.addr.2004.10.005.

Wang, L., & Dent, S. Y. (2014). Functions of SAGA in development and disease. *Epigenomics*, *6*(3), 329–339. https://doi.org/10.2217/epi.14.22.

Wang, C., Fu, M., Angeletti, R. H., Siconolfi-Baez, L., Reutens, A. T., Albanese, C., et al. (2001). Direct acetylation of the estrogen receptor alpha hinge region by p300 regulates transactivation and hormone sensitivity. *The Journal of Biological Chemistry*, *276*(21), 18375–18383. https://doi.org/10.1074/jbc.M100800200.

Wang, Y. M., Gu, M. L., Meng, F. S., Jiao, W. R., Zhou, X. X., Yao, H. P., et al. (2017). Histone acetyltransferase p300/CBP inhibitor C646 blocks the survival and invasion pathways of gastric cancer cell lines. *International Journal of Oncology*, *51*(6), 1860–1868. https://doi.org/10.3892/ijo.2017.4176.

Wang, L., Gural, A., Sun, X. J., Zhao, X., Perna, F., Huang, G., et al. (2011). The leukemogenicity of AML1-ETO is dependent on site-specific lysine acetylation. *Science*, *333*(6043), 765–769. https://doi.org/10.1126/science.1201662.

Wang, R., He, Y., Robinson, V., Yang, Z., Hessler, P., Lasko, L. M., et al. (2018). Targeting lineage-specific MITF pathway in human melanoma cell lines by A-485, the selective small-molecule inhibitor of p300/CBP. *Molecular Cancer Therapeutics*, *17*(12), 2543–2550. https://doi.org/10.1158/1535-7163.MCT-18-0511.

Wang, S. A., Hung, C. Y., Chuang, J. Y., Chang, W. C., Hsu, T. I., & Hung, J. J. (2014). Phosphorylation of p300 increases its protein degradation to enhance the lung cancer progression. *Biochimica et Biophysica Acta*, *1843*(6), 1135–1149. https://doi.org/10.1016/j.bbamcr.2014.02.001.

Wang, K. C., Yang, Y. W., Liu, B., Sanyal, A., Corces-Zimmerman, R., Chen, Y., et al. (2011). A long noncoding RNA maintains active chromatin to coordinate homeotic gene expression. *Nature*, *472*(7341), 120–124. https://doi.org/10.1038/nature09819.

Wang, Y., Zhang, R., Wu, D., Lu, Z., Sun, W., Cai, Y., et al. (2013). Epigenetic change in kidney tumor: Downregulation of histone acetyltransferase MYST1 in human renal cell carcinoma. *Journal of Experimental & Clinical Cancer Research*, *32*, 8. https://doi.org/10.1186/1756-9966-32-8.

Wapenaar, H., van der Wouden, P. E., Groves, M. R., Rotili, D., Mai, A., & Dekker, F. J. (2015). Enzyme kinetics and inhibition of histone acetyltransferase KAT8. *European Journal of Medicinal Chemistry*, *105*, 289–296. https://doi.org/10.1016/j.ejmech.2015.10.016.

Wu, J., Wang, J., Li, M., Yang, Y., Wang, B., & Zheng, Y. G. (2011). Small molecule inhibitors of histone acetyltransferase Tip60. *Bioorganic Chemistry*, *39*(1), 53–58. https://doi.org/10.1016/j.bioorg.2010.11.003.

Wu, J., Xie, N., Wu, Z., Zhang, Y., & Zheng, Y. G. (2009). Bisubstrate inhibitors of the MYST HATs Esa1 and Tip60. *Bioorganic & Medicinal Chemistry*, *17*(3), 1381–1386. https://doi.org/10.1016/j.bmc.2008.12.014.

Xiao, X. S., Cai, M. Y., Chen, J. W., Guan, X. Y., Kung, H. F., Zeng, Y. X., et al. (2011). High expression of p300 in human breast cancer correlates with tumor recurrence and predicts adverse prognosis. *Chinese Journal of Cancer Research*, *23*(3), 201–207. https://doi.org/10.1007/s11670-011-0201-5.

Xiong, H., Han, J., Wang, J., Lu, W., Wang, C., Chen, Y., et al. (2018). Discovery of 1,8-acridinedione derivatives as novel GCN5 inhibitors via high throughput screening. *European Journal of Medicinal Chemistry*, *151*, 740–751. https://doi.org/10.1016/j.ejmech.2018.02.005.

Yamakawa, K., Morita, R., Takahashi, E., Hori, T., Ishikawa, J., & Nakamura, Y. (1991). A detailed deletion mapping of the short arm of chromosome 3 in sporadic renal cell carcinoma. *Cancer Research*, *51*(17), 4707–4711.

Yan, G., Eller, M. S., Elm, C., Larocca, C. A., Ryu, B., Panova, I. P., et al. (2013). Selective inhibition of p300 HAT blocks cell cycle progression, induces cellular senescence, and inhibits the DNA damage response in melanoma cells. *The Journal of Investigative Dermatology*, *133*(10), 2444–2452. https://doi.org/10.1038/jid.2013.187.

Yang, H., Pinello, C. E., Luo, J., Li, D., Wang, Y., Zhao, L. Y., et al. (2013). Small-molecule inhibitors of acetyltransferase p300 identified by high-throughput screening are potent anticancer agents. *Molecular Cancer Therapeutics*, *12*(5), 610–620. https://doi.org/10.1158/1535-7163.MCT-12-0930.

Yang, X. J., & Seto, E. (2007). HATs and HDACs: From structure, function and regulation to novel strategies for therapy and prevention. *Oncogene*, *26*(37), 5310–5318. https://doi.org/10.1038/sj.onc.1210599.

Yang, X. J., & Ullah, M. (2007). MOZ and MORF, two large MYSTic HATs in normal and cancer stem cells. *Oncogene*, *26*(37), 5408–5419. https://doi.org/10.1038/sj.onc.1210609.

Yang, X., Yu, W., Shi, L., Sun, L., Liang, J., Yi, X., et al. (2011). HAT4, a Golgi apparatus-anchored B-type histone acetyltransferase, acetylates free histone H4 and facilitates chromatin assembly. *Molecular Cell*, *44*(1), 39–50. https://doi.org/10.1016/j.molcel.2011.07.032.

Yao, T. P., Oh, S. P., Fuchs, M., Zhou, N. D., Ch'ng, L. E., Newsome, D., et al. (1998). Gene dosage-dependent embryonic development and proliferation defects in mice lacking the transcriptional integrator p300. *Cell*, *93*(3), 361–372.

Yarla, N. S., Bishayee, A., Sethi, G., Reddanna, P., Kalle, A. M., Dhananjaya, B. L., et al. (2016). Targeting arachidonic acid pathway by natural products for cancer prevention and therapy. *Seminars in Cancer Biology*, *40–41*, 48–81. https://doi.org/10.1016/j.semcancer.2016.02.001.

Ying, M. Z., Wang, J. J., Li, D. W., Yu, G., Wang, X., Pan, J., et al. (2010). The p300/CBP associated factor is frequently downregulated in intestinal-type gastric carcinoma and constitutes a biomarker for clinical outcome. *Cancer Biology & Therapy*, *9*(4), 312–320. https://doi.org/10.4161/cbt.9.4.10748.

You, L., Nie, J., Sun, W. J., Zheng, Z. Q., & Yang, X. J. (2012). Lysine acetylation: Enzymes, bromodomains and links to different diseases. *Essays in Biochemistry*, *52*, 1–12. https://doi.org/10.1042/bse0520001.

Yuan, Z., Syed, M. A., Panchal, D., Rogers, D., Joo, M., & Sadikot, R. T. (2012). Curcumin mediated epigenetic modulation inhibits TREM-1 expression in response to lipopolysaccharide. *The International Journal of Biochemistry & Cell Biology*, *44*(11), 2032–2043. https://doi.org/10.1016/j.biocel.2012.08.001.

Yun, J. M., Jialal, I., & Devaraj, S. (2011). Epigenetic regulation of high glucose-induced proinflammatory cytokine production in monocytes by curcumin. *The Journal of Nutritional Biochemistry*, *22*(5), 450–458. https://doi.org/10.1016/j.jnutbio.2010.03.014.

Zeng, F. Q., Peng, S. M., Li, L., Mu, L. B., Zhang, Z. H., Zhang, Z. Y., et al. (2013). Structure-based identification of drug-like inhibitors of p300 histone acetyltransferase. *Yao Xue Xue Bao*, *48*(5), 700–708.

Zhang, J., Liu, H., Pan, H., Yang, Y., Huang, G., Yang, Y., et al. (2014). The histone acetyltransferase hMOF suppresses hepatocellular carcinoma growth. *Biochemical and Biophysical Research Communications*, *452*(3), 575–580. https://doi.org/10.1016/j.bbrc.2014.08.122.

Zhang, S., Liu, X., Zhang, Y., Cheng, Y., & Li, Y. (2013). RNAi screening identifies KAT8 as a key molecule important for cancer cell survival. *International Journal of Clinical and Experimental Pathology*, *6*(5), 870–877.

Zhang, R., Wang, J., Zhao, L., Liu, S., Du, D., Ding, H., et al. (2018). Identification of novel inhibitors of histone acetyltransferase hMOF through high throughput screening. *European Journal of Medicinal Chemistry*, *157*, 867–876. https://doi.org/10.1016/j.ejmech.2018.08.026.

Zhang, X., Wu, J., & Luan, Y. (2017). Tip60: Main functions and its inhibitors. *Mini Reviews in Medicinal Chemistry*, *17*(8), 675–682. https://doi.org/10.2174/1389557516666160923125031.

Zhao, S., Xu, W., Jiang, W., Yu, W., Lin, Y., Zhang, T., et al. (2010). Regulation of cellular metabolism by protein lysine acetylation. *Science*, *327*(5968), 1000–1004. https://doi.org/10.1126/science.1179689.

Zheng, Y., Balasubramanyam, K., Cebrat, M., Buck, D., Guidez, F., Zelent, A., et al. (2005). Synthesis and evaluation of a potent and selective cell-permeable p300 histone acetyltransferase inhibitor. *Journal of the American Chemical Society*, *127*(49), 17182–17183. https://doi.org/10.1021/ja0558544.

Zheng, X., Gai, X., Ding, F., Lu, Z., Tu, K., Yao, Y., et al. (2013). Histone acetyltransferase PCAF up-regulated cell apoptosis in hepatocellular carcinoma via acetylating histone H4 and inactivating AKT signaling. *Molecular Cancer*, *12*(1), 96. https://doi.org/10.1186/1476-4598-12-96.

CHAPTER SEVEN

Triggering the downstream apoptotic signal inside human parasitic organisms demonstrates a promising approach for anti-parasitic drug development: A mechanistic perspective

Nikhilesh Joardar[a,†], Niladri Mukherjee[a,†], Satyajit Halder[b], Kuladip Jana[b], and Santi P. Sinha Babu[a,*]

[a]Parasitology Laboratory, Department of Zoology, Siksha-Bhavana, Visva-Bharati, Santiniketan, West Bengal, India

[b]Division of Molecular Medicine, Bose Institute, Kolkata, India

*Corresponding author: e-mail address: spsinhababu@gmail.com

Contents

Abstract

Parasitic organisms of various genera have threatened humankind. Although they are not always fatal but can damage the well-being of an individual in terms of both economic and societal crisis. Marked progress has been made toward eliminating those pathogenic organisms, however, complete removal is still not possible. Several

† Equal contribution.

ISSN 1876-1623
https://doi.org/10.1016/bs.apcsb.2020.12.002

antiparasitic drug moieties have been largely commercialized and are routinely used at the same time novel drug candidates are still required. Programmed cell death (PCD) is a vital biological phenomenon inside every organism. Particularly, induction of the death signaling inside the parasitic species through selective targeting of effective drug candidates is one of the major strategies to combat these infectious organisms.

In this chapter significance of apoptosis induction to eliminate the parasitic disease has been illustrated with suitable references. Moreover, we have shared our own experiences of apoptosis induction in eliminating a World Health Organization enlisted Neglected Tropical Disease, lymphatic filariasis. On the other hand, we have also tried to put some light on the mechanism of apoptosis in different parasites.

1. Introduction

Since the mid-19th century, many interpretations signified that cell death is indeed a very important physiological process of multicellular organisms, predominantly during embryogenesis and metamorphosis provided that the cell deaths are controlled and sequential (Lockshin & Zakeri, 2001). The term programmed cell death (PCD) was first introduced in 1964 with the hypothesis that cell death during development is not accidental but a sequence of controlled steps toward self-destruction that bears spatial as well as temporal dependencies (Lockshin & Zakeri, 2001). In due course, the term apoptosis had been first introduced by Kerr, Wyllie, and Currie (1972) to describe these morphological processes leading to controlled cellular self-destruction (Kerr et al., 1972). Apoptosis is often termed as a suicide program used by animals to get rid of cells that are either superfluous or potentially dangerous (Hengartner, 2006). On the other hand, development in living organisms is characterized by rapid cell growth and cell division along with high rates of cell death (Glücksmann, 1951). Controlled cell death is also important for homeostatic functions, or to eliminate aberrant, damaged, or potentially harmful cells (Baehrecke, 2002; Meier, Finch, & Evan, 2000). Apoptotic cells are usually associated with characteristic morphological changes, including chromatin condensation and DNA laddering, loss of mitochondrial membrane potential and plasma-membrane phospholipids asymmetry, and detachment from the cellular matrix (Kerr et al., 1972; Wyllie, 1980). During apoptosis, the cell activates suicide machinery that executes systematic cell disassembly, including condensation and fragmentation of the chromosomal DNA. Many key components that initiate and execute apoptosis are conserved across species (Elmore, 2007). PCD can occur through different biochemical programs.

This process occurs through a coordinated dismantling of a cell in a matter of hours. Cells shrink, detach and end up forming small pieces or "apoptotic bodies," which are immediately cleared by phagocytes and are thus removed from the body in a silent way (Fink & Cookson, 2005). During apoptosis, cells maintain the integrity of their plasma membrane, which helps avoid inflammation. In contrast, cell death by "necrosis," which occurs in situations of uncontrolled tissue damage (for instance after traumatic injury due to heat shock or radiation) is a "passive" form of cell death that triggers inflammation (Rock & Kono, 2008).

In addition to its extensive plasma membrane blabbing followed by karyorrhexis and ultimately separation of cell fragments resulting in the formation of apoptotic bodies during a process called "budding." Apoptotic bodies consist of cytoplasm with tightly packed organelles with or without a nuclear fragment (Ziegler & Groscurth, 2004). However, the integrity of the organelle is still maintained, and all of this is enclosed within an intact plasma membrane. These bodies are subsequently phagocytosed by macrophages, parenchymal cells, or neoplastic cells and degraded within phagolysosomes. The cell junctions in the plasma membrane get disrupted whereby the plasma membrane becomes convoluted (Elmore, 2007; Lawen, 2003).

One should remember that apoptosis is an extremely well-defined and possibly the most frequent form of PCD, however, other non-apoptotic types of cell death are also of biological significance (Leist & Jaattela, 2001).

Intriguingly, apoptosis also plays a defensive role. There are a wide variety of stimuli such as physiological, environmental, and pathological conditions that can trigger apoptosis, but this does not imply that all cells will necessarily die in response to the same stimulus. For instance, irradiation or drugs that are used for cancer chemotherapy at times result in DNA damage in some cells, which ultimately can lead to apoptosis through the p53-dependent pathway (Borges, Linden, & Wang, 2008). Moreover, some hormones, such as corticosteroids, may lead to apoptotic death in some cells (e.g., thymocytes) although other cells are unaffected or even stimulated in the process (Elmore, 2007).

Fascinating insights into the origin and evolution of programmed cell death might be given by the fact, that PCD is also an integral part of the life cycle of unicellular eukaryotes (such as the kinetoplastid parasite *Trypanosoma brucei brucei*, the ciliate *Tetrahymena thermophila*, and the slime mold *Dictyostelium discoideum*) and even in prokaryotes (such as *Bacillus subtilis*, *Streptomyces* and Myxobacteria) (Ameisen, 2002). Apoptosis requires several genes that are well conserved across animal evolution. Multiple genetic and cell biological experiments have shown that the apoptotic event is executed through similar biochemical pathways in vertebrates, *Drosophila*, and *C. elegans*. These biochemical pathways lead to the activation of "caspases" proteins. What distinguishes apoptosis from other forms of cell death in many morphological and biochemical features which are the consequence of the activity of caspases (Saraste & Pulkki, 2000).

The therapeutic potential by the controlled manipulation of the apoptotic process has elevated great potential in the biomedical area (Barcinski & DosReis, 1999).

In this chapter, we will discuss the recent findings of apoptosis among pathogenic protozoan and metazoan parasites and the clinical possibilities of apoptosis manipulation. The proposition is that pharmacological and immunological exploitation of apoptosis can lead us to a new therapeutic advance against chronic parasitic complications. As a final point, recognition of PCD in parasites will allow us a more comprehensive view of the mechanisms of anti-parasite drugs and will suggest to us new therapeutic possibilities.

2. Biochemical and structural basis of apoptosis in parasitic organisms

Cells following an apoptotic pathway not only have exhibited the characteristic morphology but also display distinguishing biochemical changes that are exclusive to the process. For example, an endonuclease cleaves the chromosomal DNA of the apoptotic cell into fragments of distinctive sizes (Ishikawa, Handa, & Kobayashi, 2009). A very important change occurs in the plasma membrane of apoptotic cells. The negatively charged phospholipid phosphatidylserine, which is normally exclusively located in the inner leaflet of the lipid bilayer of the plasma membrane, flips to the outer leaflet during apoptosis, whereby it can serve as a marker of these cells (Kay & Grinstein, 2011). Besides, cells undergoing apoptosis generally lose their electrical potential that normally exists across the inner membrane of their mitochondria. Proteins, for example, cytochrome-c usually are released from the inter-membrane space of mitochondria during apoptosis (Wang & Youle, 2009). At the same time, extensive protein cross-linking occurs within the apoptotic cells and this is achieved through the expression and activation of tissue transglutaminase (Elmore, 2007).

Apoptosis can be triggered by a wide variety of different signals, many of which induce cell differentiation and proliferation in other situations. Both apoptosis and cell proliferation may be regulated by the same genes, and many genes involved in the induction or suppression of apoptosis act as tumor suppressor genes or proto-oncogenes, respectively. Some of the genes that regulate mammalian apoptosis have homologs in nematodes and insects, suggesting considerable evolutionary conservation within the animal kingdom (Fuchs & Steller, 2011). However, the typical mammalian cytological features of apoptosis may not be present during PCD in invertebrates, in which autophagy and autolysis may be prominent (Fuchs & Steller, 2011). The primary role of mammalian apoptosis is to kill unwanted, damaged, or pathogen-invaded cells in such a manner that they are phagocytosed and do not cause any inflammatory response. Apoptosis also has been suggested to be the default surveillance strategy that promptly eliminates the unhealthy cells that do not receive survival signals (Wang & Youle, 2009).

The mechanisms of apoptosis are highly complex and involve an energy-dependent cascade of molecular events. So far, two major pathways have been described, each of which possesses distinct features. These two pathways are the extrinsic or death receptor pathway and the intrinsic or mitochondrial pathway. These pathways can be triggered by various stimuli from outside or inside the cell such as ligation of cell surface receptors, DNA damage as a cause of defects in DNA repair mechanisms, treatment with cytotoxic drugs or irradiation, a lack of survival signals, improper cell cycle, or induction of developmental death signals (Green & Llambi, 2015). Cells following an apoptotic pathway not only have exhibited the characteristic morphology but also display distinguishing biochemical changes that are exclusive to the process. For example, an endonuclease cleaves the chromosomal DNA of the apoptotic cell into fragments of distinctive sizes (Ishikawa et al., 2009). A very important change occurs in the plasma membrane of apoptotic cells. The negatively charged phospholipid phosphatidylserine, which is normally exclusively located in the inner leaflet of the lipid bilayer of the plasma membrane, flips to the outer leaflet during apoptosis, whereby it can serve as a marker of these cells (Kay & Grinstein, 2011). Besides, cells undergoing apoptosis generally lose their electrical potential that normally exists across the inner membrane of their mitochondria. Proteins, for example, cytochrome-c usually are released from the inter-membrane space of mitochondria during apoptosis (Wang & Youle, 2009). At the same time, extensive protein cross-linking occurs within the apoptotic cells and this is achieved through the expression and activation of tissue transglutaminase (Elmore, 2007).

Apoptosis can be triggered by a wide variety of different signals, many of which induce cell differentiation and proliferation in other situations. Both apoptosis and cell proliferation may be regulated by the same genes, and many genes involved in the induction or suppression of apoptosis act as tumor suppressor genes or proto-oncogenes, respectively. Some of the genes that regulate mammalian apoptosis have homologs in nematodes and insects, suggesting considerable evolutionary conservation within the animal kingdom (Fuchs & Steller, 2011). However, the typical mammalian cytological features of apoptosis may not be present during PCD in invertebrates, in which autophagy and autolysis may be prominent (Fuchs & Steller, 2011). The primary role of mammalian apoptosis is to kill unwanted, damaged, or pathogen-invaded cells in such a manner that they are phagocytosed and do not cause any inflammatory response. Apoptosis also has been suggested to be the default surveillance strategy that promptly eliminates the unhealthy cells that do not receive survival signals (Wang & Youle, 2009).

The mechanisms of apoptosis are highly complex and involve an energy-dependent cascade of molecular events. So far, two major pathways have been described, each of which possesses distinct features. These two pathways are the extrinsic or death receptor pathway and the intrinsic or mitochondrial pathway. These pathways can be triggered by various stimuli from outside or inside the cell such as ligation of cell surface receptors, DNA damage as a cause of defects in DNA repair mechanisms, treatment with cytotoxic drugs or irradiation, a lack of survival signals, improper cell cycle, or induction of developmental death signals (Green & Llambi, 2015).

2.1 Basic concept of the extrinsic and intrinsic pathways of apoptosis

In the extrinsic apoptotic pathway, the procaspase-8 is recruited by its death effector domains (DEDs) to the death-inducing signaling complex (DISC). A membrane receptor complex is thereafter formed following the ligation of a member of the tumor necrosis factor receptor (TNFR) family. After binding to the DISC, several procaspase-8 molecules come close to each other. This arrangement of the procaspase-8 molecules is assumed to activate each other by autoproteolysis (Chang, Xing, Capacio, Peter, & Yang, 2003; Tummers & Green, 2017).

In the extrinsic pathway, a death ligand such as TNF (Tumor necrosis factor), TRAIL (Tumor necrosis factor-related apoptosis-inducing ligand), or Fas Ligand activates a death receptor, which recruits FADD

(FAS-associated death domain protein) through its death domain (DD). FADD recruits caspase-8 through death effector domain (DED) interactions. Caspase-8 can directly cleave and activate effectors caspases such as caspase-3 and caspase-7, or cleave Bid to induce Bax/Bak activation. In the mitochondrial pathway, a BH3-only protein such as Bid, Puma, Noxa, etc. activates Bax and/or Bak which permeabilize mitochondria triggering cytochrome-c release (depicted as small spheres). Cytochrome-c triggers the oligomerization of Apaf-1 (Apoptotic protease activating factor 1) and formation of the apoptosome, where caspase-9 is activated. Caspase-9 cleaves and activates effector caspases (Bratton & Salvesen, 2010).

The intrinsic apoptotic pathway involves procaspase-9. This is activated downstream of mitochondrial proapoptotic events within a structure called apoptosome. This apoptosome is a cytosolic death signaling protein complex that is formed upon the release of cytochrome-c from the mitochondria (D'Amelio, Tino, & Cecconi, 2008). In this pathway, the dimerization of procaspase-9 molecules at the Apaf-1 scaffold helps in the activation of caspase-9. When the initiator caspases get activated, they can in turn proteolytically activate the effector procaspases-3, procaspases-6, and procaspases-7 (Shi, 2004). This subsequently cleaves a specific set of protein substrates, including the procaspases themselves, which results in the generation and amplification of the death signal and eventually in the execution of cell death which exhibit all the usual morphological and biochemical features (Wang & Youle, 2009).

Besides the above two pathways, there is another pathway of apoptosis called the Perforin/granzyme pathway. However, this pathway is mainly concerned with the immune response of the body and essentially involves T-cells (Elmore, 2007).

The final piece of the apoptotic machinery is the engulfment and degradation of the apoptotic corpse which is induced by certain signals expressed on the membrane of the dying cells (Birge, Boeltz, Kumar, et al., 2016). The above discussed three pathways have been depicted in Fig. 1.

2.2 General characteristic features of caspases

Initially, the history of caspases started with the discovery of the ced-3 gene in the nematode *C. elegans* (Malin & Shaham, 2015). This gene was proposed to encode a homolog of the interleukin-1βprocessing enzyme (ICE) that was independently identified in humans (Liang, Li, Guo, et al., 2016).

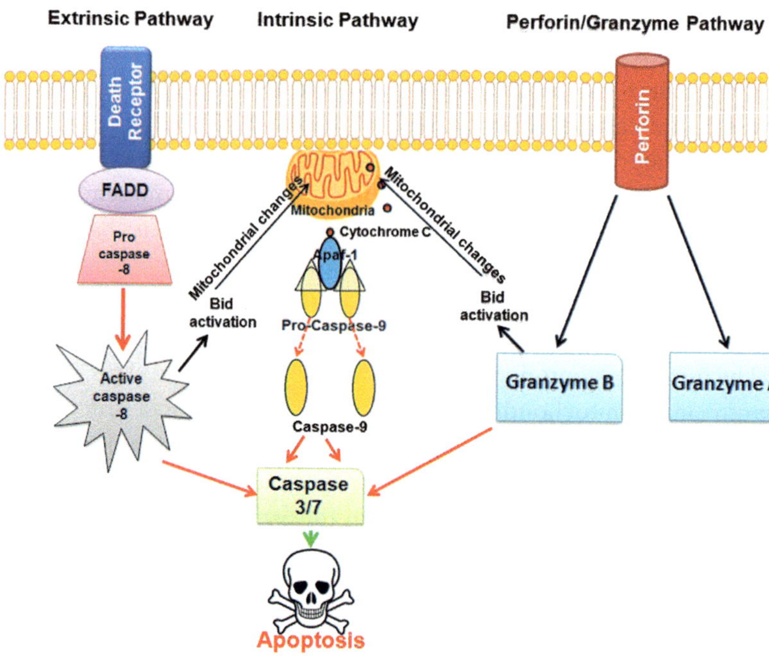

Fig. 1 Basic pathways of apoptosis. Induction of extrinsic, intrinsic and Perforin/ Granzyme pathways of apoptosis and their final execution of death phenomenon through activation of the caspase 3 protein.

Caspases (cysteine-dependent aspartate-specific proteases) are cysteine proteases that cleave substrates after specific aspartate residues. The specificity of target sites of these proteases is chiefly determined by a four-amino acid recognition motif, and also by other features of the three-dimensional structure of their target proteins. The caspases are synthesized in the cell in the form of inactive precursors are called procaspases. These procaspases are activated by proteolytic cleavage. Proteins such as *C. elegans* CED-4 can bind to pro-caspases and this multimerization in turn support cross-activation of adjacent caspase zymogens (Chang & Yang, 2006; Shi, 2004). Structurally, activated caspases consist of dimers of a large and a small subunit that, together, constitute the active site of the enzyme (Julien & Wells, 2017). Interestingly, the initiator procaspases have a long prodomain, which contains a caspase recruitment domain (CARD) that enables them to assemble with adaptor proteins into activation complexes when the cell receives a signal to undergo apoptosis.

2.3 Structural basis of apoptosis in protozoan parasites

Demonstration of apoptosis-like programmed cell death (PCD) in protozoan parasites including *Leishmania*, *Plasmodium*, and *Trypanosoma* initiated a great amount of interest in the molecular parasitology field and they have been found to share some common features with PCD in multicellular organisms, including reactive oxygen species (ROS) generation, phosphatidylserine (PS) exposure in the outer leaflet of the cell membrane, disruption of mitochondrial membrane potential, chromosome condensation, fragmentation of nuclear DNA, apoptotic bodies formation and cell shrinkage (Al-Olayan, Williams, & Hurd, 2002; Arambage, Grant, Pardo, Ranford-Cartwright, & Hurd, 2009). Although the molecular mechanisms and markers related to apoptosis-like programmed cell death (PCD) in parasitic organisms are limited, however; several studies have revealed the apoptotic markers in diverse parasitic taxa, including *Plasmodium*, *Leishmania*, *Trypanosomes* (Al-Olayan et al., 2002; Das, Mukherjee, & Shaha, 2001; Duszenko, Figarella, Macleod, & Welburn, 2006).

2.3.1 Programmed cell death (PCD) in Leishmania

Recently several proteins have been identified to be involved in Leishmania cell death. EndoG a potent cell death effector has been identified in two *Leishmania* species, LINF_100012900 in *L. infantum* and LdBPK_100600.1 in *L. donovani* showing a similar kind of nuclease activity like EndoG as in other organisms (Bose Dasgupta et al., 2008; Gannavaram, Vedvyas, & Debrabant, 2008). During the promastigote and amastigote stage, the enzyme localized within the mitochondrion and helps in their survival and its nuclease activity also remains inactive (Oliva et al., 2017; Rico et al., 2014). Upon encountering apoptotic stimuli, EndoG translocates to the nucleus and form a complex with nuclease TatD and Flap endonuclease-1 FEN-1 resulting in DNA degradation through exonuclease and endonuclease activity, acting as a cell death effector in the parasite (Bose Dasgupta et al., 2008; Gannavaram et al., 2008).

Further PCD in *Leishmania* has also been found to be associated with caspase-like activities and orthologs of caspases, i.e., metacaspases (MCs), which helps in this process (Lee, Gannavaram, Selvapandiyan, & Debrabant, 2007). MCs are cysteine proteases and despite their structural similarities with caspases, MCs possess distinct substrate specificity containing Lys/Arg from caspases (Asp). In *L. donovani* two MCs are involved in PCD in namely, LdMC1 and LdMC2, and its caspase-like activity do not require any proteolytic activation. Under normal conditions, they remain localized in acido calcisomes and after encountering apoptotic stimuli (H_2O_2) get released from it resulting in PCD (Lee et al., 2007). Another metacaspase termed LmjMCA has been found in *L. major* and in contrast to LdMCs it gets auto-processed upon encountering apoptotic signals and becomes 300 times more active than its full-length recombinant form. Furthermore, overexpression of LmjMCA induces *L. major* apoptosis after oxidative stress as well as the addition of anti-leishmanial drug curcumin or miltefosine, while its deletion inhibits curcumin or miltefosine induce apoptosis (Zalila et al., 2011). Recently three apoptotic pathways have been identified in *L. major*,

(i) Miltefosine induced apoptosis, involving LmjMCA activation,
(ii) Amphotericin B and H_2O_2 induced apoptosis involving LmjMCA inhibition, and,
(iii) LmjMCA independent apoptosis pathway induced by pentamidine (Basmaciyan, Azas, & Casanova, 2018).

Further another MC termed MCA has been studied in *L. mexicana*, mainly localized in the cytoplasm and helps in miltefosine induce apoptosis induction in *L. mexicana* (Castanys-Muñoz, Brown, Coombs, & Mottram, 2012). Despite all the findings, additional studies will be needed to understand the physiological substrates of *Leishmania* MCs and their various roles in *Leishmania*.

Recently BH3 domain-containing aquaporin protein Li-BH3AQP has been identified in *L. infantum* as being involved in cell death activity in *Leishmania* (Genes et al., 2016). Further, a cathepsin B-like protein cysteine proteinase C (CPC) has been found to acts as a cell death effector in Leishmania. CPC remains localized within the lysosome and after encountering apoptotic stimuli it releases from the lysosome and induces apoptosis in *Leishmania* (El-Fadili et al., 2010). Two additional proteins have also been identified in *L. major*, namely LmjHYD36 and LmjF.22.0600 that shows a potential role in the programmed cell death process of *Leishmania* (Basmaciyan, Azas, & Casanova, 2019; Basmaciyan, Jacquet, Azas, & Casanova, 2019).

2.3.2 Programmed cell death (PCD) in Plasmodium

Amongst the limited research works performed to identify the apoptotic markers in the malarial parasites majority was performed against the intra-erythrocytic asexual stages of *Plasmodium falciparum* and that too was in controlled laboratory conditions. However, observations of drug-induced

PCD mechanisms were not consistent in different *P. falciparum* strains therefore having controversial conclusions of cell death type (Jiménez-Ruiz et al., 2010). Additionally, markers of apoptosis-like PCD were also been observed in the motile zygote, ookinete in both in vivo and in vitro conditions (Jiménez-Ruiz et al., 2010). Various PCD pathways in *Plasmodium* sp. had been initiated by different extra and intracellular stimuli which include nutrient deprivation, heat shock, drug exposures, and existence of both oxidative and nitrosative stresses (Bruchhaus, Roeder, Rennenberg, & Heussler, 2007; Debrabant, Lee, Bertholet, Duncan, & Nakhasi, 2003; Duszenko et al., 2006). It was documented that both NO and/or RNS has the potential to induce apoptosis in *P. berghei* ookinetes (Ali, Al-Olayan, Lewis, Matthews, & Hurd, 2010). NO, RNS and ROS can be significant apoptosis inducers in the redox-active midgut lumen against *P. berghei* and *P. falciparum* ookinetes also undergo apoptosis in the midgut of *Anopheles stephensi*, its vector (Arambage et al., 2009).

Apoptosis was confirmed against erythrocyte stages of *P. falciparum* treated with both chloroquine (CQ) and atovaquone and the event was confirmed by the increased presence of JC-1 monomers (green staining) however, in the drug-resistant strains found to be was significantly lower (Meslin et al., 2007). Several triggers induce features of apoptosis in *P. falciparum* trophozoite stages after were exposed to CQ and staurosporine (ST) (Ch'ng, Kotturi, Chong, et al., 2010; Ch'ng, Liew, Goh, Sidhartha, & Tan, 2011). In addition to drug exposures, NO derivatives, oxidative stress (Ali et al., 2010; Kumar et al., 2008) and high parasite density can also prompt apoptotic death in *Plasmodium* sp. (Mutai & Waitumbi, 2010). Formation of DNA ladder in the CQ exposed malarial parasite was noted in the drug-sensitive *P. falciparum* 3D7 strain although the event was lacking in the CQ resistant *P. falciparum* Lili strain or *P. falciparum* F32 strain (Nyakeriga et al., 2006).

A combination of cell-based assays like flow cytometry and fluorescence microscopy detected apoptotic features like mitochondrial dysregulation, activation of cysteine proteases, and in situ DNA fragmentation in CQ and staurosporine (ST) exposed *Plasmodium* sp. (Ch'ng et al., 2010). It was inferred that the CQ-induced pathway is associated with the activation of cysteine protease thereafter mitochondrial dysregulation, amplified protease activity, and DNA fragmentation (Ch'ng et al., 2010). Novel synthetic coumarin-labeled CQ (CM-CQ) also showed that the apoptotic death of the exposed parasites was associated with concentration-dependent differences in drug localization (Ch'ng et al., 2010).

The malaria ookinete and the intra-erythrocytic stages were found to extremely susceptible to oxidative stress (Lanz-Mendoza et al., 2002). ROS generated through Selenium compounds exposure caused a marked collapse in $\Delta\Psi$m and DNA fragmentation in CQ-resistant *P. falciparum* strains, moreover, this also mediate ROS independent apoptosis in them (Suradji et al., 2011).

Bilirubin and actinomycin D also caused decrease in $\Delta\Psi$m, in *P. falciparum* NF-54 strain, following inhibition of hemozoin formation (Kumar et al., 2008). Astonishingly, a sixfold increase in the catalytic activity of a caspase-3-like protein was observed in the bilirubin-treated erythrocyte stages of *P. falciparum* (Kumar et al., 2008).

So far, few studies have been done to understand the molecular mechanism behind PCD in Plasmodium sp. and three metacaspases were identified to play a crucial role in this process. PfMCA1–3 in *P. falciparum*, PbMCA1–3 in *P. berghei* and PvMCA1–3 in *P. vivax* (Atkinson, Babbitt, & Sajid, 2009). They are mainly identified in the cytoplasm of the ookinete and exhibit caspase-like activity and that assist in the regulation of stress-dependent PCD and growth of the parasite. Metacaspase PfMCA1 has also been identified to possess an additional CARD domain and shows some percentage of similarity with human and mice caspase-1 (Meslin et al., 2007). However, the direct role of these MCAs is still unclear in the Plasmodium parasite and it needs further studies.

2.3.3 Programmed cell death (PCD) in Trypanosoma

In *Trypanosoma brucei* five metacaspases (MCs) have been identified that plays a role in Trypanosoma PCD namely TbMCA1, TbMCA2, TbMCA3, TbMCA4, and TbMCA5 of which TbMCA2–4 was found in the bloodstream form of the parasite life cycle, while TbMCA5 was expressed both in the bloodstream as well as procyclic forms of *Trypanosoma* life cycle. These MCs are mainly located within the endosomes (Meslin, Zalila, Fasel, Picot, & Bienvenu, 2011). Among these five metacaspases, TbMCA2, TbMCA3, and TbMCA-5 contain histidine/cysteine dyad and acts as active cysteine proteases while TbMCA1 and TbMCA4 lack this activity (Meslin et al., 2011). Two more metacaspases also had been identified In *Trypanosoma cruzi*, namely TcMCA-3 and TcMCA-5, homologous to *T. brucei* MCs TbMCA-3 and TbMCA-5. Both the MCs translocate from the cytoplasm to the nucleus upon encountering apoptotic stimulus, fresh human serum (FHS). Moreover, overexpression of TcMCA5 increases

T. cruzi sensitivity to FHS than the controls, clearly indicating the role of MCs in PCD of *Trypanosoma* (Proto, Coombs, & Mottram, 2013).

2.3.4 Apoptosis in other parasitic protozoans

Distinct morphological alterations that resemble apoptosis in metazoans were also noticed in nitric oxide (NO) donor sodium nitroprusside (SNP) exposed tachyzoites apicomplexan parasite *T. gondii* (Peng, Lin, Lin, Jiang, & Zhang, 2003). Externalization of PS to the outer leaflet of the cell membrane was also detected in *Toxoplasma* (Seabra, de Souza, & Damatta, 2004). However, there are still no reports of mitochondrial cytochrome-c release in *Plasmodium* or *Toxoplasma* (Jiménez-Ruiz et al., 2010). Besides the distinct other morphological apoptotic features nuclear pyknosis was also detected in *T. gondii* tachyzoites treated with SNP (Peng et al., 2003). Fragmentation of genomic DNA from SNP-treated *Toxoplasma* was monitored in extracellular tachyzoites (Peng et al., 2003). Additionally, TUNEL assay was also successfully employed to detect DNA strand breaks in *Toxoplasma* (Jiménez-Ruiz et al., 2010).

PCD had been confirmed in other unicellular organisms such as *Tetrahymena* (Davis, Ward, Herrick, & Allis, 1992) and *Dictyostelium discoideum* (Lee, Bertholet, Debrabant, et al., 2002). The occurrence of DNA strand breaks was confirmed in apoptosis-like death in organisms which a modified mitochondrion, namely *Blastocystis hominis* (Tan & Nasirudeen, 2005), and in anaerobic protozoa which is without any mitochondrion, as in *Trichomonas* (Chose, Sarde, Gerbod, Viscogliose, & Roseto, 2003; Mariante, Vancini, & Benchimol, 2005) and *Giardia* (Chose et al., 2003).

2.4 Structural basis of apoptosis activation in helminth parasites

Biochemical and cell biological analyses of EGL-1, CED-9, CED-4, and CED-3 have provided important insights on the mechanism of programmed cell death activation during *C. elegans* development (Conradt, Wu, & Xue, 2016). All these protein complexes are involved in the activation of CED-3. The immediate PCD of *C. elegans* is by proteolytic cleavage of pro-CED-3 to active CED-3. In this activation of pro-CED-3 along with CED-4 shape an oligomeric tetrameric complex known as apoptosome (Yuan & Akey, 2013).

This endogenous CED-9 and CED-4 are found in the mitochondrial surface in embryonic *C. elegans*. Not only the CED-9 protein, but also the CED-4 has been shown to interact directly with CED-3 in vitro.

Interestingly, CED-9 functions by inhibiting the translocation of CED-4 from mitochondria and by acting as a competitive inhibitor of CED-3 protein. Primarily, the pro-apoptotic protein EGL-1 binds with the CED-9 through its BH3 domain and induces its conformational changes, which eventually disrupts the CED-9–CED-4 interaction. CED-9 uses two different non-overlapping surface areas to bind to EGL-1 and CED-4. In the presence of EGL-1, CED-9 undergoes a series of conformational changes to create a hydrophobic cleft that can accommodate the extended BH3 α-helix of EGL-1 (Lee & Fairlie, 2012). At the same time, these conformational changes induce a rearrangement of the CED-4-binding surface on CED-9, which is likely to destabilize the CED-4–CED-9 interaction and allow the release of CED-4. Studies have shown that a mutated CED-9 (G169E) protein can constitutively bind to CED-4 with high affinity, but CED-3 cannot be activated, irrespective of the presence of EGL-1 (Parrish, Metters, Chen, et al., 2000; Yan, Chai, Lee, et al., 2005).

Recently in the Cestode parasite, *Hymenolepis diminuta* existence of the CED pathway way has been revealed (Roy, Joardar, Sinha Babu, & Lyndem, 2020). Though the obtained result is very preliminary however, the finding is worthy enough to state the previously mentioned CED pathway of apoptosis is conserved among different groups of helminth parasites.

Therefore, it is evident that induction of apoptosis is a useful strategy to combat different infectious diseases. Selective administration of compounds and/or drugs will eventually trigger an apoptotic signal that will ultimately kill the parasite (Figs. 2 and 3).

3. Therapeutic implications of apoptosis in anti-parasitic drug development

Induction of apoptosis through application of suitable therapeutic molecules inside the parasitic organisms residing inside the host body is now-a-day an exceptionally accepted strategy for combating such diseases. Likewise, in case of filarial nematodes drug induced oxidative stress often triggers apoptosis and autophagy. Taking under consideration of the vast importance of apoptosis this pathway has been investigated for selective chemotherapeutic interventions (Gucchait et al., 2018; Mukherjee et al., 2018; Mukherjee, Joardar, & Sinha Babu, 2018; Roy et al., 2019). Figs. 2 and 3 described that how apoptosis has been targeted for antiparasitic drug development.

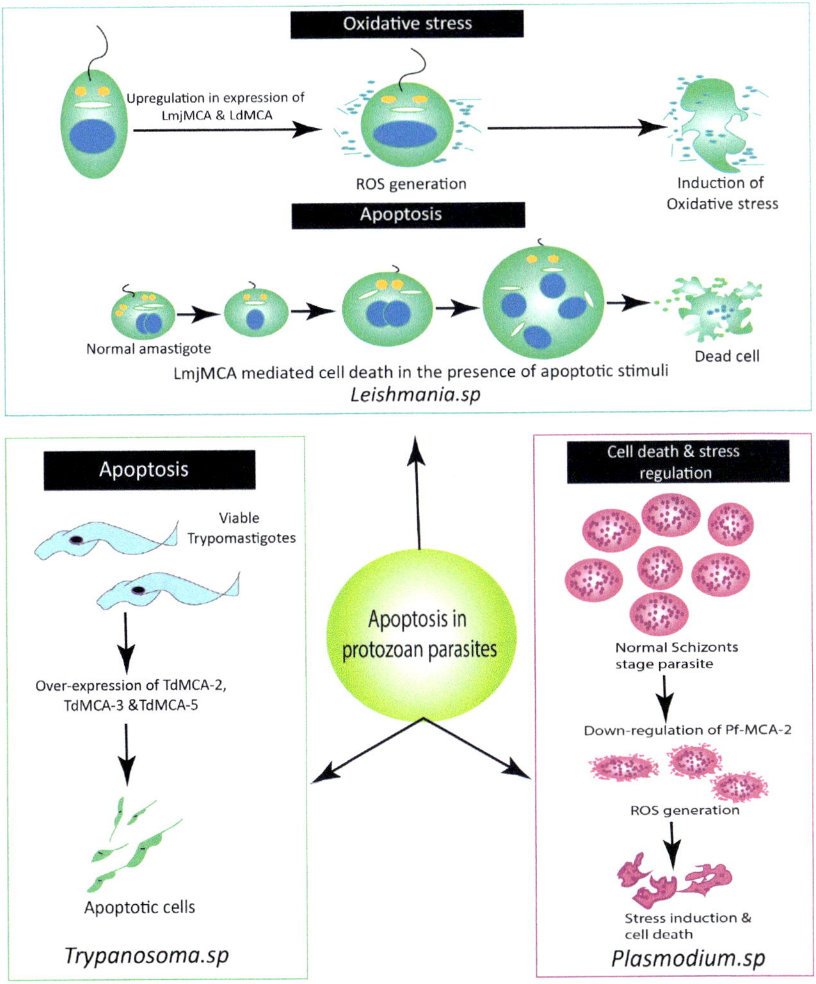

Fig. 2 Apoptosis induction in parasitic protozoans.

Several studies have identified that many virulence factors namely co-factors, enzymes and proteases benefits parasites to survive intracellular as well as extracellular host environment. For example, subtilisin-like serine protease found in Leishmania helps in differentiation of amastigotes to promastigotes (Da Silva-Lopez, Coelho, & De Simone, 2005). A cysteine protease, Cruzain of *T. cruzi* helps in immune evasion (Doyle et al., 2011). Further *Plasmodium* protease, Plasmepsin IX and X helps in parasitic invasion and egression, while protease Plasmepsin II and X causes breakdown of RBC etc. (Nasamu et al., 2017). Thus, parasitic proteases can be

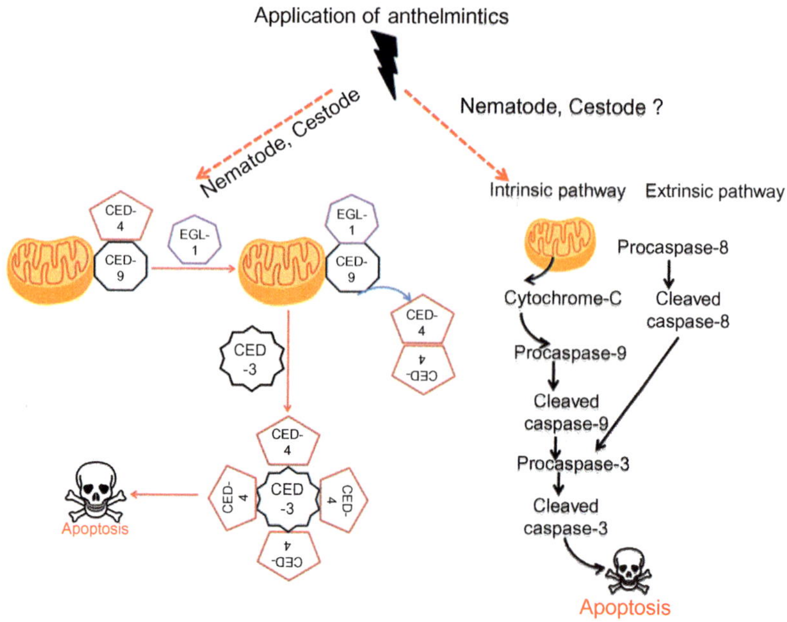

Fig. 3 Apoptosis induction in helminth parasites.

used as a potential drug targets against parasites based infectious diseases. In this context another group of parasitic proteases namely metacaspases, has also been reported to play a crucial role in maintaining the physiological conditions along with death and survival strategies of parasites. *Leishmania* and *Trypanosoma* metacaspases are rapidly being identified as imperative casts of programmed cell death, thus identifying the mechanism behind MCs mediated PCD provides opportunities to develop potential drugs that can target these proteases and help against parasitic infectious diseases. For example, miltefosine currently is being used to treat Leishmania that directly targets several metacaspases to induce PCD in different Leishmania species (Basmaciyan et al., 2018; Zalila et al., 2011). Therefore, identifying drugs like miltefosine that can targets metacaspases of *Plasmodium* as well as *Trypanosoma* parasites needed to be done. Further induction of parasitic apoptosis during their earlier stage of infection, in combination with inhibition of host cell apoptosis induced by a parasite can constitute a new therapeutic concept of future treatment that needs further investigation.

The apoptotic pathways in parasitic cestode and nematode can also be triggered through administration of several synthetic as well as natural compounds. Therefore, screening of authentic compounds having the potencies to trigger the downstream apoptotic pathway is highly demanded.

4. Conclusion

The elimination of unwanted cells by PCD is an important developmental and homeostatic process in a multicellular organism. Thus, dysfunction or deregulation of the apoptotic program can result in a variety of pathological conditions. Defects in apoptosis can result in cancer, autoimmune diseases, and pathogenesis of viral infections, whereas excessive apoptosis mediates neurodegenerative disorders, AIDS, and ischemic diseases (Fadeel et al., 1999). Triggering the downstream apoptotic pathway leading to death of the parasites is an effective way to combat infectious diseases. Maximally the administered compounds and/or drugs induce oxidative stress inside the parasitic body which in turn activate apoptosis (Gucchait et al., 2018; Mukherjee, Joardar, Mondal, et al., 2018; Mukherjee, Joardar, & Sinha Babu, 2018; Roy et al., 2019, 2020). Therefore, induction of PCD through application of therapeutic moieties has now become an effective strategy and hence applied widely. Initiation of apoptosis can therefore be an effective way for developing suitable therapeutics to combat these infectious organisms.

Acknowledgment

N.J. sincerely acknowledges the Council for Scientific and Industrial Research (CSIR), Govt. of India for providing the Senior Research Fellowship (Grant Number: 09/202/(0086) 2K19/EMR-I).

Conflict of interest

None.

Author contributions

N.J., N.M., S.H., K.J. and S.P.S. wrote the manuscript and critically reviewed the final draft before submission.

References

Ali, M., Al-Olayan, E. M., Lewis, S., Matthews, H., & Hurd, H. (2010). Naturally occurring triggers that induce apoptosis-like programmed cell death in *Plasmodium berghei* ookinetes. *PLoS One*, *5*, 12634.

Al-Olayan, E. M., Williams, G., & Hurd, H. (2002). Apoptosis in the malaria protozoan, Plasmodium berghei: a possible mechanism for limiting intensity of infection in the mosquito. *International Journal for Parasitology*, *32*, 1133–1143.

Ameisen, J. C. (2002). On the origin, evolution, and nature of programmed cell death: A timeline of four billion years. *Cell Death and Differentiation*, *9*, 367–393.

Arambage, S. C., Grant, K. M., Pardo, I., Ranford-Cartwright, L., & Hurd, H. (2009). Malaria ookinetes exhibit multiple markers for apoptosis-like programmed cell death *in vitro*. *Parasites & Vectors*, *2*, 32.

Atkinson, H. J., Babbitt, P. C., & Sajid, M. (2009). The global cysteine peptidase landscape in parasites. *Trends in Parasitology*, *25*, 573–581.

Baehrecke, E. H. (2002). How death shapes life during development. *Nature Reviews. Molecular Cell Biology*, *3*, 779–787.

Barcinski, M. A., & DosReis, G. A. (1999). Apoptosis in parasites and parasite induced apoptosis in the host immune system: A new approach to parasitic diseases. *Brazilian Journal of Medical and Biological Research*, *32*, 395–401.

Basmaciyan, L., Azas, N., & Casanova, M. (2018). Different apoptosis pathways in *Leishmania* parasites. *Cell Death & Disease*, *5*, 27.

Basmaciyan, L., Azas, N., & Casanova, M. (2019). A potential acetyltransferase involved in *Leishmania major* metacaspase dependent cell death. *Parasites & Vectors*, *12*, 266.

Basmaciyan, L., Jacquet, P., Azas, N., & Casanova, M. (2019). A novel hydrolase with a pro-death activity from the protozoan parasite *Leishmania major*. *Cell Death & Disease*, *5*, 99.

Birge, R. B., Boeltz, S., Kumar, S., et al. (2016). Phosphatidylserine is a global immunosuppressive signal in efferocytosis, infectious disease, and cancer. *Cell Death and Differentiation*, *23*, 962–978.

Borges, H. L., Linden, R., & Wang, J. Y. (2008). DNA damage-induced cell death: Lessons from the central nervous system. *Cell Research*, *18*, 17–26.

Bose Dasgupta, S., Das, B. B., Sengupta, S., Ganguly, A., Roy, A., Dey, S., et al. (2008). The caspase independent algorithm of programmed cell death in *Leishmania* induced by baicalein: The role of LdEndoG, LdFEN-1 andLdTatD as a DNA "degradesome". *Cell Death and Differentiation*, *15*, 1629–1640.

Bratton, S. B., & Salvesen, G. S. (2010). Regulation of the Apaf-1-caspase-9 apoptosome. *Journal of Cell Science*, *123*, 3209–3214.

Bruchhaus, I., Roeder, T., Rennenberg, A., & Heussler, V. T. (2007). Protozoan parasites: Programmed cell death as a mechanism of parasitism. *Trends in Parasitology*, *23*, 376–383.

Castanys-Muñoz, E., Brown, E., Coombs, G. H., & Mottram, J. C. (2012). *Leishmania mexicana* metacaspase is a negative regulator of amastigote proliferation in mammalian cells. *Cell Death & Disease*, *3*, e385.

Chang, D. W., Xing, Z., Capacio, V. L., Peter, M. E., & Yang, X. (2003). Interdimer processing mechanism of procaspase-8 activation. *The EMBO Journal*, *22*, 4132–4142.

Chang, H. Y., & Yang, X. (2006). Proteases for cell suicide: Functions and regulation of caspases. *Microbiology and Molecular Biology Reviews*, *64*, 821–846.

Ch'ng, J., Kotturi, S., Chong, A., et al. (2010). A programmed cell death pathway in the malaria parasite *Plasmodium falciparum* has general features of mammalian apoptosis but is mediated by clan CA cysteine proteases. *Cell Death & Disease*, *1*, e26.

Ch'ng, J. H., Liew, K., Goh, A. S., Sidhartha, E., & Tan, K. S. (2011). Drug-induced permeabilization of parasite's digestive vacuole is a key trigger of programmed cell death in *Plasmodium falciparum*. *Cell Death & Disease*, *2*, e216.

Chose, O., Sarde, C. O., Gerbod, D., Viscogliose, E., & Roseto, A. (2003). Programmed cell death in parasitic protozoans that lack mitochondria. *Trends in Parasitology*, *19*, 559–564.

Conradt, B., Wu, Y. C., & Xue, D. (2016). Programmed cell death during *Caenorhabditis elegans* development. *Genetics*, *203*, 1533–1562.

Da Silva-Lopez, R. E., Coelho, M. P., & De Simone, S. G. (2005). Characterization of an extracellular serine protease of *Leishmania (Leishmania) amazonensis*. *Parasitology*, *131*, 85–96.

D'Amelio, M., Tino, E., & Cecconi, F. (2008). The apoptosome: Emerging insights and new potential targets for drug design. *Pharmaceutical Research*, *25*, 740–751.

Das, M., Mukherjee, S. B., & Shaha, C. (2001). Hydrogen peroxide induces apoptosis-like death in *Leishmania donovani* promastigotes. *Journal of Cell Science*, *114*, 2461–2469.

Davis, M. C., Ward, J. G., Herrick, G., & Allis, C. D. (1992). Programmed nuclear death: Apoptotic-like degradation of specific nuclei in conjugating *Tetrahymena*. *Developmental Biology*, *154*, 419–432.

Debrabant, A., Lee, N., Bertholet, S., Duncan, R., & Nakhasi, H. L. (2003). Programmed cell death in trypanosomatids and other unicellular organisms. *International Journal for Parasitology*, *33*, 257–267.

Doyle, P. S., Zhou, Y. M., Hsieh, I., Greenbaum, D. C., McKerrow, J. H., & Engel, J. C. (2011). The *Trypanosoma cruzi* protease cruzain mediates immune evasion. *PLoS Pathogens*, 7, e1002139.

Duszenko, M., Figarella, K., Macleod, E. T., & Welburn, S. C. (2006). Death of a trypanosome: A selfish altruism. *Trends in Parasitology*, *22*, 536–542.

El-Fadili, A. K., Zangger, H., Desponds, C., Gonzalez, I. J., Zalila, H., Schaff, C., et al. (2010). Cathepsin B-like and cell death in the unicellular human pathogen *Leishmania*. *Cell Death & Disease*, *1*, e71.

Elmore, S. (2007). Apoptosis: A review of programmed cell death. *Toxicologic Pathology*, *35*, 495–516.

Fadeel, B., Gleiss, B., Hogstrand, K., Chandra, J., Wiedmer, T., Sims, P. J., et al. (1999). Phosphatidylserine exposure during apoptosis is a cell-type-specific event and does not correlate with plasma membrane phospholipid scramblase expression. *Biochemical and Biophysical Research Communications*, *266*, 504–511.

Fink, S. L., & Cookson, B. T. (2005). Apoptosis, pyroptosis, and necrosis: Mechanistic description of dead and dying eukaryotic cells. *Infection and Immunity*, *73*, 1907–1916.

Fuchs, Y., & Steller, H. (2011). Programmed cell death in animal development and disease. *Cell*, *147*, 742–758.

Gannavaram, S., Vedvyas, C., & Debrabant, A. (2008). Conservation of the pro-apoptotic nuclease activity of endonuclease G in unicellular trypanosomatid parasites. *Journal of Cell Science*, *121*, 99–109.

Genes, C. M., de Lucio, H., González, V. M., Sánchez-Murcia, P. A., Rico, E., Gago, F., et al. (2016). A functional BH3 domain in an aquaporin from *Leishmania infantum*. *Cell Death & Disease*, *2*, 16043.

Glücksmann, A. (1951). Cell deaths in normal vertebrate ontogeny. *Biological Reviews of the Cambridge Philosophical Society*, *26*, 59–86.

Green, D. R., & Llambi, F. (2015). Cell death Signaling. *Cold Spring Harbor Perspectives in Biology*, 7, a006080.

Gucchait, A., Joardar, N., Parida, P. K., Roy, P., Mukherjee, N., Yesuvadian, R., et al. (2018). Development of novel anti-filarial agents using carbamo(dithioperoxo)thioate derivatives. *European Journal of Medicinal Chemistry*, *143*, 598–610.

Hengartner, M. O. (2006). Roads to ruin: Apoptotic pathways in the nematode *Caenorhabditis elegans*. *Seikagaku*, *A50382*, S41–S43.

Ishikawa, K., Handa, N., & Kobayashi, I. (2009). Cleavage of a model DNA replication fork by a Type I restriction endonuclease. *Nucleic Acids Research*, *37*, 3531–3544.

Jiménez-Ruiz, A., Alzate, J. F., Macleod, E. T., Lüder, C. G., Fasel, N., & Hurd, H. (2010). Apoptotic markers in protozoan parasites. *Parasites & Vectors*, *3*, 104.

Julien, O., & Wells, J. A. (2017). Caspases and their substrates. *Cell Death and Differentiation*, *24*, 1380–1389.

Kay, J. G., & Grinstein, S. (2011). Sensing phosphatidylserine in cellular membranes. *Sensors (Basel)*, *11*, 1744–1755.

Kerr, J. F., Wyllie, A. H., & Currie, A. R. (1972). Apoptosis: A basic biological phenomenon with wide ranging implications in tissue kinetics. *British Journal of Cancer*, *26*, 239–257.

Kumar, S., Guha, M., Choubey, V., Maity, P., Srivastava, K., Puri, S. K., et al. (2008). Bilirubin inhibits *Plasmodium falciparum* growth through the generation of reactive oxygen species. *Free Radical Biology & Medicine*, *44*, 602–613.

Lanz-Mendoza, H., Hernandez-Martinez, S., Ku-Lopez, M., Rodriguez Mdel, C., Herrera-Ortiz, A., & Rodriguez, M. H. (2002). Superoxide anion in *Anopheles albimanus* hemolymph and midgut is toxic to *Plasmodium berghei* ookinetes. *The Journal of Parasitology*, *88*, 702–706.

Lawen, A. (2003). Apoptosis-an introduction. *BioEssays*, *25*, 888–896.

Lee, N., Bertholet, S., Debrabant, A., et al. (2002). Programmed cell death in the unicellular protozoan parasite *Leishmania*. *Cell Death and Differentiation*, *9*, 53–64.

Lee, E. F., & Fairlie, W. D. (2012). Structural biology of the intrinsic cell death pathway: What do we know and what is missing? *Computational and Structural Biotechnology Journal*, *1*. e201204007.

Lee, N., Gannavaram, S., Selvapandiyan, A., & Debrabant, A. (2007). Characterization of metacaspases with trypsin-like activity and their putative role in programmed cell death in the protozoan parasite *Leishmania*. *Eukaryotic Cell*, *6*, 1745–1757.

Leist, M., & Jaattela, M. (2001). Four deaths and a funeral: From caspases to alternative mechanisms. *Nature Reviews. Molecular Cell Biology*, *2*, 589–598.

Liang, Q., Li, W., Guo, N., et al. (2016). Identification and functional analysis of interleukin-1β in the Chinese soft-shelled turtle *Pelodiscus sinensis*. *Genes (Basel)*, *7*, 18.

Lockshin, R. A., & Zakeri, Z. (2001). Programmed cell death and apoptosis: Origins of the theory. *Nature Reviews. Molecular Cell Biology*, *2*, 545–550.

Malin, J. Z., & Shaham, S. (2015). Cell death in *C. elegans* development. *Current Topics in Developmental Biology*, *114*, 1–42.

Mariante, R. M., Vancini, R. G., & Benchimol, M. (2005). Cell death in trichomonads: New insights. *Histochemistry and Cell Biology*, *5*, 1–12.

Meier, P., Finch, A., & Evan, G. (2000). Apoptosis in development. *Nature*, *407*, 796–801.

Meslin, B., Barnadas, C., Boni, V., Latour, C., De Monbrison, F., Kaiser, K., et al. (2007). Features of apoptosis in *Plasmodium falciparum* erythrocytic stage through a putative role of PfMCA1 metacaspase-like protein. *The Journal of Infectious Diseases*, *195*, 1852–1859.

Meslin, B., Zalila, H., Fasel, N., Picot, S., & Bienvenu, A. L. (2011). Are protozoan metacaspases potential parasite killers? *Parasites & Vectors*, *4*, 26.

Mukherjee, S., Joardar, N., Mondal, S., Schiefer, A., Pfarr, K., Hoerauf, A., et al. (2018). Quinolone-fused cyclic sulfonamide as a novel benign antifilarial agent. *Scientific Reports*, *8*, 12073.

Mukherjee, N., Joardar, N., & Sinha Babu, S. P. (2018). Antifilarial activity of Azadirachtin fueled through reactive oxygen species induced apoptosis: A thorough molecular study on *Setaria cervi*. *Journal of Helminthology*, *93*, 519–528.

Mutai, B. K., & Waitumbi, J. N. (2010). Apoptosis stalks *Plasmodium falciparum* maintained in continuous culture condition. *Malaria Journal*, *9*, S6.

Nasamu, A. S., Glushakova, S., Russo, I., Vaupel, B., Oksman, A., Kim, A. S., et al. (2017). Plasmepsins IX and X are essential and druggable mediators of malaria parasite egress and invasion. *Science*, *358*, 518–522.

Nyakeriga, A. M., Perlmann, H., Hagstedt, M., Berzins, K., Troye-Blomberg, M., Zhivotovsky, B., et al. (2006). Drug-induced death of the asexual blood stages of *Plasmodium falciparum* occurs without typical signs of apoptosis. *Microbes and Infection*, *8*, 1560–1568.

Oliva, C., Sánchez-Murcia, P. A., Rico, E., Bravo, A., Menéndez, M., Gago, F., et al. (2017). Structure-based domain assignment in *Leishmania infantum* EndoG: Characterization of a pH-dependent regulatory switch and a C-terminal extension that largely dictates DNA substrate preferences. *Nucleic Acids Research*, *45*, 9030–9045.

Parrish, J., Metters, H., Chen, L., et al. (2000). Demonstration of the in vivo interaction of key cell death regulators by structure-based design of second-site suppressors. *Proceedings of the National Academy of Sciences*, *97*, 11916–11921.

Peng, B. W., Lin, J., Lin, J. Y., Jiang, M. S., & Zhang, T. (2003). Exogenous nitric oxide induces apoptosis in *Toxoplasma gondii* tachyzoites via a calcium signal transduction pathway. *Parasitology*, *126*, 541–550.

Proto, W. R., Coombs, G. H., & Mottram, J. C. (2013). Cell death in parasitic protozoa: Regulated or incidental? *Nature Reviews. Microbiology*, *11*, 58–66.

Rico, E., Oliva, C., Gutierrez, K. J., Alzate, J. F., Genes, C. M., Moreno, D., et al. (2014). *Leishmania infantum* EndoG is an endo/exo-nuclease essential for parasite survival. *PLoS One*, *9*, e89526.

Rock, K. L., & Kono, H. (2008). The inflammatory response to cell death. *Annual Review of Pathology*, *3*, 99–126.

Roy, S., Joardar, N., Sinha Babu, S. P., & Lyndem, L. M. (2020). *Senna* plant causes reactive oxygen species (ROS) mediated apoptosis in *Hymenolepis diminuta*. *Molecular and Biochemical Parasitology*, *238*, 111297.

Roy, P., Sengupta, A., Joardar, N., Bhattacharya, A., Saha, N. C., Misra, A. K., et al. (2019). Influence of autophagy, apoptosis and their interplay in filaricidal activity of C-cinnamoyl glycosides. *Parasitology*, *146*, 1451–1461.

Saraste, A., & Pulkki, K. (2000). Morphologic and biochemical hallmarks of apoptosis. *Cardiovascular Research*, *45*, 528–537.

Seabra, S. H., de Souza, W., & Damatta, R. A. (2004). *Toxoplasma gondii* exposes phosphatidylserine inducing a TGF-beta1 autocrine effect orchestrating macrophage evasion. *Biochemical and Biophysical Research Communications*, *324*, 744–752.

Shi, Y. (2004). Caspase activation, inhibition, and reactivation: A mechanistic view. *Protein Science*, *13*, 1979–1987.

Suradji, E. W., Hatabu, T., Kobayashi, K., Yamazaki, C., Abdulah, R., Nakazawa, M., et al. (2011). Selenium-induced apoptosis-like cell death in *Plasmodium falciparum*. *Parasitology*, *138*, 1–11.

Tan, K. S. W., & Nasirudeen, A. M. A. (2005). Protozoan programmed cell death—Insight from *Blastocystis* deathstyles. *Trends in Parasitology*, *21*, 547–550.

Tummers, B., & Green, D. R. (2017). Caspase-8: Regulating life and death. *Immunological Reviews*, *277*, 76–89.

Wang, C., & Youle, R. J. (2009). The role of mitochondria in apoptosis. *Annual Review of Genetics*, *43*, 95–118.

Wyllie, A. H. (1980). Glucocorticoid-induced thymocyte apoptosis is associated with endogenous endonuclease activation. *Nature*, *284*, 555–556.

Yan, N., Chai, J. J., Lee, E. S., et al. (2005). Structure of the CED-4/CED-9 complex reveals insights into programmed cell death in *Caenorhabditis elegans*. *Nature*, *437*, 831–837.

Yuan, S., & Akey, C. W. (2013). Apoptosome structure, assembly, and procaspase activation. *Structure*, *21*, 501–515.

Zalila, H., González, I. J., El-Fadili, A. K., Delgado, M. B., Desponds, C., Schaff, C., et al. (2011). Processing of metacaspase into a cytoplasmic catalytic domain mediating cell death in *Leishmania major*. *Molecular Microbiology*, *79*, 222–239.

Ziegler, U., & Groscurth, P. (2004). Morphological features of cell death. *News in Physiological Sciences*, *19*, 124–128.

CHAPTER EIGHT

Ferulic acid-mediated modulation of apoptotic signaling pathways in cancer

Ashutosh Gupta[a], Amit Kumar Singh[a], Mariam Loka[b], Abhay Kumar Pandey[a,*], and Anupam Bishayee[b,*]

[a]Department of Biochemistry, University of Allahabad, Prayagraj, Uttar Pradesh, India
[b]Lake Erie College of Osteopathic Medicine, Bradenton, FL, United States
*Corresponding authors: e-mail addresses: abishayee@lecom.edu; abishayee@gmail.com; akpandey@allduniv.ac.in

Contents

Abstract

Ferulic acid (4-hydroxy-3-methoxycinnamic acid, FA), a hydroxycinnamic acid derived from various seeds, nuts, leaves, and fruits, exists in a free form as well as is covalently conjugated with polysaccharides, glycoproteins, polyamines, lignin, and hydroxy fatty

ISSN 1876-1623
https://doi.org/10.1016/bs.apcsb.2020.12.005

acids of plant cell walls. It exhibits a variety of pharmacological effects, such as antioxidant, anti-inflammatory, vasodilatory, antithrombotic, antimicrobial, anti-allergic, antiviral, hepatoprotective, and anticancer activities. FA induces the expression of cell cycle-related proteins, such as p53 and p21, and reduces cyclin D1 and cyclin E levels. Moreover, FA triggers apoptosis and autophagic cell death depending on intracellular reactive oxygen species production in various cancer cell lines. The potential apoptotic action of FA is mediated by altered expression of procaspase-3, procaspase-8, procaspase-9, poly (ADP ribose) polymerase, Bcl-2, and Bax. It blocks the activation of both the canonical Smad and noncanonical extracellular-signal-regulated kinase/Akt (protein kinase B) pathways in various cancer cells. However, due to low solubility and permeability, its availability to biological systems is limited. Therefore, encapsulation of FA into chitosan tripolyphosphate nanoparticles may enhance its cytocompatibility, solubility, and anticancer potential. The nanohybrids of FA and double layered hydroxide exhibit cellular delivery properties of intercalated molecules on cancer cell lines. This chapter summarizes the anticancer efficacy of FA with an emphasis on the role of apoptosis, and underlying molecular mechanisms involving various signaling pathways in tumor cells.

1. Introduction

Cancer, a multistage disorder, is fundamentally categorized into the initiation, promotion, transformation, progression, invasion and metastatic stages. It is identified by unscheduled and unrestrained cellular proliferation in various tissues. Incidence of cancer in growing countries has been prevailed by types of tumor that are associated viral infections, genetic mutations, and bacterial contaminations (Jemal, Siegel, Xu, & Ward, 2010). According to the latest global cancer data, the projected global cancer burden has increased to 18.1 million, and there have been up to 9.6 million deaths in the year 2018. The United States accounts for 21% of cancer cases and 14.4% of cancer deaths globally. In contrast, the death ratio cause by cancer in Asia and Africa (57.3% and 7.3%, respectively) are higher than the extents of cancer frequency (48.4% and 5.8%, respectively) as compared to other regions of the world due to higher rates of certain types of cancer in these regions, which are related with poorer prognoses and higher death rates, in consort with limited resources to timely diagnoses and treatment (Gupta et al., 2019; Siegel, Miller, & Jemal, 2019). Moreover, literature survey displayed that lung cancer causes maximum deaths (18.4%) throughout the world due to the poor prognosis, followed by colorectal (9.2%), liver (8.2%), and stomach (8.2%) cancers. Breast cancer is the fifth major cause of cancer death (6.6%), which could be due to the favorable prognosis, at least in more developed countries (Siegel et al., 2019). Total cancer death is projected to rise to 20.3 million by the year 2026 because of transformation in size and arrangement of the population. Additionally, among the female population, cervix, breast, and ovarian cancers account for 34% of all deaths reported due to cancer.

Previous studies demonstrated that inadequate apoptosis can manifest cancer. The key characteristics of apoptosis involves morphologic changes, such as heterochromatin core mass, cell membrane shrinkage, and dislocation of organelles in cytoplasm (Safarzadeh, Sandoghchian Shotorbani, & Baradaran, 2014). Various stressful situations, such as precancerous lesions and activation of the DNA damage checkpoint, act through initiation of apoptosis to prevent carcinogenesis and maintain the genomic integrity (Kim, Kacew, & Lee, 2016; Vicente-Dueñas, Hauer, Cobaleda, Borkhardt, & Sánchez-García, 2018). Dysregulation of the apoptotic pathways may not only promote tumorigenesis but also render the cancer cell resistant to treatment. Therefore, the evasion of apoptosis is a critical hallmark of cancer (Antonarakis et al., 2019; Rayavarapu et al., 2017).

Natural products are essential and unremittable sources for drug development with negligible adverse effects. They have been ubiquitously present in fruits, vegetables, nuts, seeds, herbs, spices, stems and flower (Gupta et al., 2021; Kumar, Singh, Gupta, Bishayee, & Pandey, 2019). Previous studies documented that nearly 50% of all clinically used drugs are either developed directly from natural products, or from their derivatives, or their derivative's analogs (Gupta, Kumar, Bhattacharya, Bishayee, & Pandey, 2020; Singh, Kumar, & Pandey, 2018). Additionally, 74% of the commonly used drugs consist of plant-derived active ingredients (Gupta, Kumar, & Pandey, 2020). Approximately 60% of newly synthesized anticancer drugs are derived from natural products. Some popular chemotherapeutic agents of natural origin include vinca alkaloids (vinblastine, vincristine, vindesine, and vinorelbine), taxanes (paclitaxel and docetaxel); podophyllotoxin and its derivations (topothecan and irinothecan), and anthracyclines (doxorubicin, daunorubicin, epirubicin, and idarubicin) (Kumar, Kumar, Dwivedi, & Pandey, 2014; Sharma et al., 2017) (Table 1). Furthermore, traditional and recently emerging methods, including radiotherapy, chemotherapy, photodynamic therapy, and catalytic therapy, have not completely succeeded in countering the consequence of cancer to any significant extent (Kumar & Pandey, 2015). This limitation has drawn the attention of researchers to search alternative treatment options.

Table 1 Several common anticancer drugs derived from natural sources and their molecular targets.

Natural sources	Name of drugs	Molecular targets
Actinomyces spp.	Actinomycin D	DNA intercalation
Camptothececa	Camptothecin	Topoisomerase I
Cryptotheca crypta (sponge)	Cytarabine	DNA polymerizes
Chromobacterium spp.	FR901228 (depsipeptide)	HDAC
Dysoxylum binectariеferum	Flavopiridol	CDK9
Podophyllum spp.	Etoposide	Topoisomerase II
Myxobacterium sorangium	Epothilones	Microtubules and tubulin
Taxus brevifolia	Paclitaxel	
Vinca rosea	Vinblastine	
Streptomyces spp.	Mitomycin C	DNA alkylation
	Bleomycin	DNA breaks
	Pentostatin	Adenosine deaminase
	Lactacystin	Proteasome
	Geldanamycin	HSP90
	Staurosporine	Kinases
	Rapamycin	mTOR
	Doxorubicine	Topoisomerase II
	Daunomycin	

Ferulic acid (4-hydroxy-3-methoxycinnamic acid, FA) is a phenolic acid compound copiously present in a variety of fruits and vegetables. It has broad spectrum pharmacological properties, including antioxidant, hepatoprotective, antidiabetic, antithrombotic, antimicrobial, anti-allergic, antiviral, and antitumor activities (Kumar & Pruthi, 2014). In 1925, FA was synthesized chemically and structurally confirmed by spectroscopic technique. FA depicted the presence of an unsaturated side chain and the existence of both *cis* and *trans* isomeric forms (Nethaji, Pattabhi, & Desiraju, 1988). Many vital foods items, such as grain bran, whole grain foods, citrus fruits, banana, coffee, orange juice, eggplant, bamboo shoots, beetroot, cabbage, spinach and broccoli are rich sources of FA. It has a potent antioxidant agent that protects

DNA from oxidative damage and prevents lipid peroxidation by suppressing oxidative stress (Fang et al., 2016). Furthermore, it has been permitted as an additive antioxidant and food preservative in Japan (Graf, 1992; The Japan Food Chemical Research Faundation, 1996). Sodium ferulate, a salt of FA, is also used for treatment of cerebrovascular and cardiovascular diseases in China (Wang & Ou-Yang, 2005). FA can inhibit cancer proliferation in many ways, including altering the cancer cell cycle, inducing apoptosis, and regulating protein production. Since previously published reports on FA has received substantial attention for its anticancer properties, a systematic analysis regarding its antitumor potential and related mechanisms is desired. In this chapter, we discussed the efficacy of FA against various types of cancer with a focus on apoptotic mechanisms involving various signaling pathways in tumor cells.

2. Apoptosis and cancer therapy

Cellular homeostasis is maintained by cell proliferation and cell death. Apoptosis (programmed cell death) plays a central role to maintain this homeostasis. Evading apoptosis has been established as one of the key characteristic features of cancer cells (Hanahan & Weinberg, 2000). Apoptosis is initiated when cells sense irregularities, such as DNA damage, alterations in signaling by unusual stimulation of oncogenes, a lack of survival factors, or hypoxia. Thus, apoptosis of preneoplastic and malignant cells signifies a defensive mechanism against cancer development, and progression, because mechanistically apoptotic events can eliminate genetically damaged cells and stop them from proliferation under uncontrolled mitogenic stimuli (Gan et al., 2020; Kuno, Tsukamoto, Hara, & Tanaka, 2012). Free radicals, particularly reactive oxygen species (ROS), have been proposed as common mediators for apoptosis (Gupta & Pandey, 2020a; Haddad, 2004). An acute generation of ROS, or transient ROS burst, is frequently observed in apoptosis induced by various factors, such as stress, DNA fragmentation and chromatin condensation. Exposure of cancer cells to ROS generating anticancer agents exhaust the cellular antioxidant capacity, and a ROS level beyond a threshold leads to apoptosis (Kong, Beel, & Lillehei, 2000). Incubating cells with exogenous oxidants has been shown to cause apoptosis (Kannan & Jain, 2000; Kehrer, 2000). Many agents that induce apoptosis are either oxidants or stimulators of cellular oxidative metabolism.

Apoptosis is initiated through two major pathways: extrinsic (death receptor-mediated) and intrinsic (mitochondria-mediated) pathways.

In addition, mitogenic and stress-responsive pathways are involved in the regulation of apoptotic signaling (Khan, Adhami, & Mukhtar, 2008; Khan, Afaq, & Mukhtar, 2007; Kuno et al., 2012). Fine-tuning of the balance between the pro- and anti-apoptotic factors within each of these pathways in a cell leads to the final outcome of apoptosis or survival. One of the most important proapoptotic mediators involved in sensing DNA damage is the p53 protein, which is lost or functionally inactivated in more than 50% of human cancers (Hanahan & Weinberg, 2000). The activity of *TP53* (the gene encoding the p53 protein) is also controlled epigenetically: deacetylation of p53 through silent information regulator 1 (SIRT1), a member of the sirtuin histone deacetylase (HDAC) class III family, prevents *TP53*-mediated transactivation of the cell cycle inhibitor CDKN1A. CDKN1A encodes p21 and proapoptotic *BAX*, allowing for the promotion of cell survival after DNA damage and oncogenesis (Brooks & Gu, 2009). Therefore, inhibition of SIRT1 may lead to induction of apoptosis by counteracting the deacetylation of p53 and other key factors, including FOXO3a. Although SIRT1 can inactivate *TP53* and is up-regulated in several types of human cancers, SIRT1 has recently been reported to be a tumor suppressor (Herranz & Serrano, 2010).

It may be possible to prevent the initiation and progression steps of carcinogenesis and eliminate genetically damaged neoplastic cells by the initiation of apoptosis or cell cycle arrest using safe doses of dietary chemopreventive compounds. For example, it has been observed that resveratrol derived from wine, grape juice, peanuts, cocoa, and berries selectively targets tumor cells with presumably dysfunctional cell cycle checkpoints, while sparing normal tissue (Abotaleb et al., 2018; Gautam, Xu, Dumaguin, Janakiraman, & Chapman, 2000). These effects are dependent on the *TP53* status of the cells exposed to resveratrol. Epigallocatechin-3-gallate (EGCG) derived from tea affects a CDKN1C-associated survival pathway in normal epithelial cells, while inducing a proapoptotic pathway in oral cancer cells (Aggarwal et al., 2020). Ajoene extracted from garlic promotes apoptosis in human acute myeloid leukemia cells and peripheral blood mononuclear cells from patients with chronic leukemia, but not in normal and proliferating cells from healthy donors (Sakamoto, Lawson, & Milner, 1997). β-carotene, present in orange, carrots, sweet potatoes and apricots, can trigger apoptosis particularly in numerous human cancer cells, such as prostate cancer, colorectal cancer, breast cancer, and leukemia. In contrast, normal cells are extensively resistant to the induction of apoptosis by β-carotene (Palozza, 2005). In lung cells, quercetin from apples, black tea, berries, capers, red wine, and onions efficiently protects against H_2O_2-induced DNA damage, but this positive effect is counteracted by the reduction in glutathione (GSH) levels and an increase in leakage of lactate dehydrogenase and cytosolic-free calcium concentration (Mohammad et al., 2015). In general, polyphenols promote autophagic cell death in cancer cells through inhibition or activation of multiple signaling pathways that does not directly involve autophagy or apoptosis (Braicu et al., 2020).

3. FA: Chemistry and natural sources

FA ($C_{10}H_{10}O_4$) is a hydroxycinnamic acid having a molecular weight of 194.18g and a melting point of 172°C (Fig. 1). It is a phenolic acid ubiquitously present in the plant kingdom, which can be absorbed by the small intestine and excreted in the urine (Choudhury, Srai, Debnam, & Rice-Evans, 1999). FA and other hydroxycinnamic acids occur in plant tissues mainly as low molecular weight water-soluble conjugates present in the cytosol, and in bound forms esterified or etherified to cell wall polymers. Water-soluble conjugates are more readily absorbed as esters or aglycones in the upper digestive tract of nonruminants. Bound acids are primarily released by microbial action in the hindgut of nonruminants species, including humans (Chesson et al., 1999). The presence of double bond in the side chain is subjected to *cis-trans* isomerization, and the resonance stabilized phenoxy radical accounts for its effective antioxidant activity. Collectively with dihydroferulic acid, FA is the component of lignocelluloses, where it confers rigidity to the cell wall by crosslinking polysaccharides and lignin. FA is usually concentrated in the brans of grains, peels of fruits, and roots and peels of vegetables (Table 2) (Clifford, 1999; Herrmann, 1989; Mattila & Hellström, 2007). In most vegetables and fruits, FA is found in a conjugated form of hydroxyl acids, such as quinic acid (in coffee, cabbage, celery, and carrots), glucaric and galataric acids (in citrus), tartaric acid (in grape), and malic acid (in radish), or with mono/disaccharides, namely glucose (in apple and cabbage), digalactose (in spinach) and gentiobiose (in broccoli)

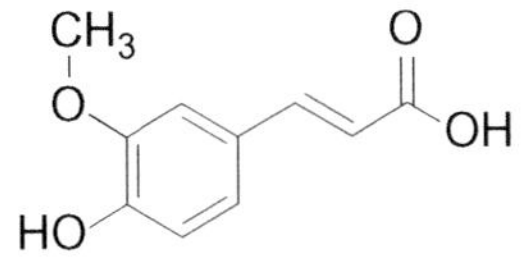

Fig. 1 Chemical structure of FA.

Table 2 Concentration of FA in fruits, grains, vegetables, and commercial foods products.

Dietary source	FA contents (mg/100 g)	Serving size (g)	mg/serving	References
Fruits				
Apples	0.27–0.85	125	0.3–1	Mattila, Hellstrom, and Torronen (2006)
Banana	5.4		6.75	
Berries	0.25–2.7		0.6–3.4	
Grapefruit	10.7–11.6		13.4–14.6	
Orange	9.2–9.9		11.5–12.5	
Plum, dark	1.47		1.8	
Rhubarb	2		2.5	
Grains				
Barley extract	1358–2293	5	68–115	Madhujith and Shahidi (2006)
Barley flour	25–34	30	7.5–10	Mattila, Pihlava, and Hellstrom (2005) and Nishizawa, Ohta, Egashira, and Sanada (1998)
Brown rice	42		8.7–12.6	Adom and Liu (2002) and Nishizawa et al. (1998)
Corn flour	38		11.4	Mattila et al. (2005)
Corn, dehulled kernels	174		52.2	Adom and Liu (2002)

Fine wheat bran	530–540	5	26–27	Andreasen, Kroon, Williamson, and Garcia-Conesa (2001) and Kroon, Faulds, Ryden, Robertson, and Williamson (1997)
Oat bran	33	–	–	Mattila et al. (2005)
Oats	25–35	30	10.5	Adom and Liu (2002) and Mattila et al. (2005)
Refined corn bran	2610–3300	5	130–151	Saulnier and Thibault (1999) and Zhao, Egashira, and Sanada (2005)
Rice endosperm cell wall	910		45	Shibuya (1984)
Rye bran	280		14	Andreasen et al. (2001) and Mattila et al. (2005)
Rye flour	86	30	25.8	Mattila et al. (2005)
Soft and hard wheat bran	1351–1456	5	68–73	Liyana-Pathirana and Shahidi (2006)
Wheat flour	89	30	26.7	Mattila et al. (2005)
Wheat kernels	64–127		19–38	Adom and Liu (2002) and Nishizawa et al. (1998)
Vegetables				
Avocado	1.1	200	2.2	Mattila and Hellström (2007)
Bamboo shoots	243.6	50	122	Nishizawa et al. (1998)
Broccoli	4.1	200	8.2	Mattila and Hellström (2007)

Continued

Table 2 Concentration of FA in fruits, grains, vegetables, and commercial foods products.—cont'd

Dietary source	FA contents (mg/100 g)	Serving size (g)	mg/serving	References
Burdock	7.3–19	100	7.3–19	Sakakibara, Honda, Nakagawa, Ashida, and Kanazawa (2003)
Carrot	1.2–2.8		1.2–2.8	Mattila and Hellström (2007)
Chinese cabbage	1.4	200	2.8	
Green bean/fresh	1.2		2.4	
Mizuna	1.4–1.8		2.8–3.6	Sakakibara et al. (2003)
Parsnip	2.2		4.4	Mattila and Hellström (2007)
Peanut	8.7	60	5.22	
Pot grown basil	1.5	200	3	
Pot grown lettuces	0.19–1.4		0.4–2.8	
Radish	4.6	100	4.6	
Red beet	25		25	
Red cabbages	6.3–6.5	200	12.6–13	
Soyabean	12	125	15	
Spinach/frozen	7.4	200	14.8	
Tomato	0.29–6		0.6–12	Bourne and Rice-Evans (1998) and Mattila and Hellström (2007)
Water dropwort/Eggplant	7.3–34		14.6–68	Sakakibara et al. (2003)

Commercial foods and beverages				
Bear	0.24–0.9	500	1–4.5	Bourne and Rice-Evans (1998) and Mattila et al. (2006)
Boiled spaghetti	13.6	100	13.6	Nishizawa et al. (1998)
Coffee	9.1–14.3	200	18.2–28.6	Mattila et al. (2006) and Nardini, Cirillo, Natella, and Scaccini (2002)
Orange juice	3–6.4		6–13.4	Mattila et al. (2006) and Rapisarda, Carollo, Tomaselli, and Maccarone (1998)
Pasta	12	100	12	Mattila et al. (2005)
Pickled red beet	39	25	9.75	Mattila and Hellström (2007)
Pickled red cabbage	1.5	50	0.5	
Popcorn	313	60	187.8	(Nishizawa et al. (1998))
Rice, brown, long grain parboiled	24	125	30	Mattila et al. (2005)
Sugar-beet pulp	800	10	80	Micard, Grabber, Ralph, Renard, and Thibault (1997)
Sweet corn	42	60	24	Nishizawa et al. (1998)
White wheat bread	8.2	35	2.87	Mattila et al. (2005)
Grain oat flakes	25–52		8.75–18	Mattila et al. (2005) and Nishizawa et al. (1998)
Grain rye bread	54		18.9	Mattila et al. (2005)

(Buchanan, Wallace, Fry, & Eastwood, 1996; Clifford, 1999; Herrmann, 1989). In roots of vegetables and grains, FA mostly occurs in pectic or hemicellulosic polysaccharides through its dimers (di-FA), and/or esterified with arabinose and galactose residues (Mathew & Abraham, 2004; Saulnier & Thibault, 1999). In grains, FA may also be esterified with sterols; one well-known example is FA-oryzanol which consists primarily of cycloartenol and 24-methylenecycloartanol. In several vegetables, such as burdock, water dropwort, and eggplant, free FA accounts for 50–90% of total of FA (Sakakibara et al., 2003). In contrast, the percentage of free FA in cereals only accounts for 0.1–0.5% (Adom & Liu, 2002).

3.1 Dietary intake of FA

Phenolic acids account for about one-third of the total dietary intake through fruits, vegetables and beverages. Although there is a remarkable variation in the dietary intake of phenolics throughout the different region of the world, it is calculated that the daily intake varies from about 20 mg to 1 g (Gupta & Pandey, 2020b; Hollman & Katan, 1998). Clifford (1999) suggested that individuals who frequently intake citrus fruits, coffee, cereal bran, and beer may consume 0.5–1 g caffeic acid and FA daily. A substantial intake of whole grains appreciably contributes to daily intake of FA up to 167 mg (35–89 mg total FA/100 g grain). Other dietary sources of FA may comprise orange juice, coffee, orange, cabbage, soybean, eggplant, water dropwort, banana, broccoli, radish, potato, spinach, tomato, grapefruit, and peanut supplying 5–70 mg FA/100 g (Table 2). Thus, one can measure that the total FA intake by consumption of cereals, vegetables, fruits, coffee, and juices may reach up to 150–250 mg/day. Furthermore, a study performed by Clifford (1999) reported that those who drink several cups of coffee per day (rich in bran) and citrus juices easily consume 500–1000 mg caffeic acid and FA (Ghatak & Panchal, 2010).

4. Pharmacokinetic properties of FA

The absorption, tissue distribution, metabolism and elimination pattern of FA has been extensively investigated in several animal studies. However, relevant human data are very limited.

4.1 Absorption

Several experimental conditions (*in situ* or *ex vivo*) revealed that FA can be absorbed from the stomach, jejunum, and ileum. After an incubation of half an hour, >70% of the FA disappeared from the stomach of a rat and appeared in the gastric mucosa, blood, bile and urine, showing a rapid-gastric absorption of FA (Zhao, Egashira, & Sanada, 2004). Also, FA rapidly moved from the jejunum to a significantly smaller amount from the ileum when it was perfused in an isolated rat intestine model (Spencer et al., 1999). Only 0.5–0.8% of administered FA was found in feces of rats, suggesting a very efficient absorption rate for FA (Zhao & Moghadasian, 2008). Numerous other *in vitro* studies documented the presence of H^+-driven transport system or a monocarboxylic acid transport (MCT) system for the uptake of cinnamic acid and structurally similar FA and q-coumaric acid (Itagaki et al., 2005; Konishi & Shimizu, 2003). Additionally, Wolffram, Weber, Grenacher, and Scharrer (1995) studied the presence of a Na^+-dependent saturable transport mechanism for mucosal uptake of radioactivity-labeled cinnamic acid and FA across the jejunal brush border membrane in rats (Wolffram et al., 1995). Similarly, studies also exhibited that MCT1, one of the best identified isoforms of MCTs did not participate in the MCT-mediated transport and absorption of FA in Caco-2 cells (Watanabe, Yashiro, Tohjo, & Konishi, 2006).

4.2 Distribution

Blood albumin is the chief carrier protein of FA (Kang et al., 2004). Zhao et al. (2004) evaluated that nearly 4%, 10%, and 53% of orally consumed FA can be found in the gastric mucosa, blood, and other tissues including liver and kidney, respectively. Further analysis in the rat intestine demonstrated that 49% of perfused FA might be distributed in the hepatic and peripheral tissues (Adam et al., 2002). Chang, Xu, Tao, and Feng (1993) demonstrated the amount of free FA recovered in the rat heart (14 μg/g), kidney (82 μg/g), spleen (22 μg/g), lung (34 μg/g), liver (28 μg/g), uterus (15 μg/g) and brain (2.6 μg/g) at approximately 30 min after an oral administration of 521 μmol/kg of FA. However, in most of the tissues the ratio of free FA decreased by 20% in kidneys, 80% in lungs, and 50% in other tissues after 60 min of administration (Chang et al., 1993). Another group of researchers noticed that <1% of the radioactivity of ^{14}C-FA remained in the rat liver, small intestine, cecum and skin up to 24 h following an intraperitoneal (i.p.) injection of FA (Teuchy & Sumere, 1971).

4.3 Metabolism

Several factors, such as dose, route of administration, and animal model, are accountable for metabolites of FA and their relative concentration.

Previous studies demonstrated that FA can be metabolized *in vivo* into various metabolites, including FA-sulfate, feruloylglycine, FA-glucuronide, vanillic acid, FA-diglucuronide, vanilloylglycine, *m*-hydroxyphenylpropionic acid, dihydroferulic acid, and FA-sulfoglucuronide (FA-diconjugate with sulfate and glucuronide) (Adam et al., 2002; Zhao & Moghadasian, 2008). In the presence of enzyme sulfotransferases and uridine diphosphate glucuronosyl transferases, conjugation reactions chiefly take place in the liver (Zhao et al., 2004). Intestinal mucosa and kidneys may also participate in this conjugation process, but to a lesser extent (Kern et al., 2003). Administration of FA through i.p. route leads to the excretion of 3-hydroxy phenyl propionic acid as a main urinary metabolite in rats (Booth, Emerson, Jones, & Deeds, 1957) (Fig. 2).

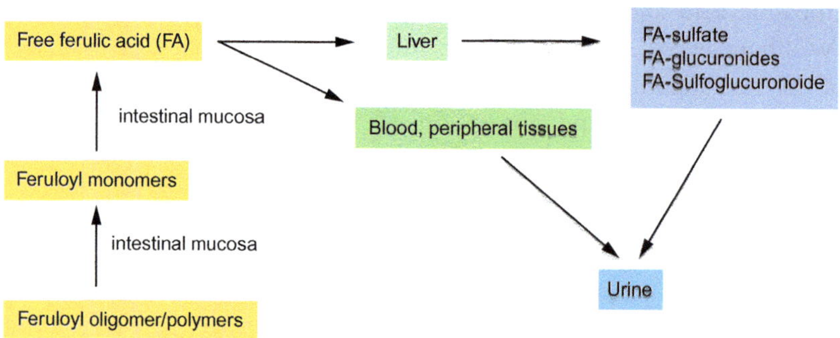

Fig. 2 Schematic representation of absorption and metabolism of FA.

4.4 Excretion

FA is predominantly excreted by urine in free and conjugated forms in rats (Rondini, Peyrat-Maillard, Marsset-Baglieri, & Berset, 2002). It is also eliminated through bile, which comprises approximately 4–6% of the oral prescription (Adam et al., 2002; Zhao et al., 2004). FA conjugates were observed in the bile only in the initial period after the i.p. administration (0–5 h) (Scheline, 1968), suggesting that augmented circulating levels of FA are required for biliary secretion. The bile excretion indicates the occurrence of FA and its derivatives in the feces of rats (Teuchy & Sumere, 1971). It is calculated that the half-life of FA can vary from 10 to 30 min in rats based on the concentration and the mode of administration (Zhao, Egashira, & Sanada, 2003a). The short half-life of FA may account for its low toxicity. The acute lethal doses (LD_{50}) of FA in female and male rats were 2.1 and 2.4 g/kg, respectively. No noteworthy subchronic toxicity were reported in male and female rats after long-term (13 weeks) consumption of dietary FA at 0.16 g/kg per day (Zhang et al., 2006).

4.5 Bioavailability

Due to the rapid conjugation process of FA in the liver its bioavailability is very low (Chang et al., 1993; Zhao et al., 2004). FA associated with arabinose or arabinoxylan from corn bran exhibited a lesser bioavailability than free FA. However, FA in wheat bran (96% bound with heteroxylans) displayed a remarkably higher bioavailability than free FA (Rondini et al., 2004). This variation may be due to the various existing forms of FA and the oral dosage, which alter the fate of dietary FA in the gastrointestinal tract and its systemic concentrations (van der Logt, Roelofs, Nagengast, & Peters, 2003). The collective urinary elimination of total FA may be used to measure the absorbability of dietary FA, because absorbed FA is eliminated chiefly by urine (Rondini et al., 2002). The urinary excretion in rats indicates the following sequence for absorbability of dietary FA: free FA followed by feruloyl mono-, di-saccharides and lastly feruloyl polysaccharides (Zhao et al., 2003a; Zhao, Egashira, & Sanada, 2003b). Feruloyl polysaccharides with complex structure may decrease the association between the polymers of FA and the hydrolyzing enzymes (ferulate esterases and xylanases), demonstrating a reduction in the release of FA (Zhao et al., 2005). Thus, the form of bound FA limits the degree of its absorbability; for example, feruloyl mono- and di-saccharide forms of FA in vegetables is a likely reason for the higher absorbability of FA from spinach as compared to feruloyl polysaccharide forms of FA in bran which has lower degree of absorbability (Adam et al., 2002; Zhao et al., 2005). Additionally, the absorbability of FA from corn bran is lower than that from the wheat bran because in corn bran FA binds with more complex heteroxylans than in wheat bran (Bunzel, Ralph, Marita, Hatfield, & Steinhart, 2001; Zhao et al., 2005).

5. FA-mediated proapoptotic and anticancer activities

5.1 Breast cancer

Breast cancer is the most common type of cancer and the second leading cause of death. This disease is the primary cause of mortality among women aged 45–55 years (Jemal et al., 2009). The incidence of breast cancer is almost 1-in-8 in women. Treatment options include surgery, chemotherapy, radiotherapy, and hormone therapy most of the time (Heravi Karimovi, Pourdehqan, Jadid Milani, Foroutan, & Aieen, 2006). The primary risk factors of breast cancer include age (Steiner, Klubert, & Knutson, 2008), high hormone levels (Yager & Davidson, 2006), race, economic status, and dietary iodine deficiency (Aceves, Anguiano, & Delgado, 2005; Stoddard II, Brooks, Eskin, & Johannes, 2008; Venturi, 2001). Chang et al. (2006) studied the antiproliferative activity of FA and its effect on human epidermal growth factor receptor 2 (HER2) expression in several human breast cancer cell lines (MCF-7, BT474, MDA-MB-231 and SKBR3). In breast cancer patients, HER2 gene overexpression is associated with carcinogenesis and poor prognosis. Moreover, the activation of estrogen receptor (ER) and abundant HER2 signaling pathways provide a strong stimulus for tumor progression and may contribute to hormonal therapy resistance in breast cancer (Osborne, Shou, Massarweh, & Schiff, 2005; Yarden & Sliwkowski, 2001). Treatment with FA up-regulated the HER2 gene expression in tested breast cancer lines, which could be completely abolished by the treatment with trastuzumab, a standard drug for breast cancer. Furthermore, FA function as an estrogenic agent, by triggering HER2 and its downstream Akt and extracellular signal-related kinase (ERK), promoting the transcriptional activity of ERα, and augmentation of pS2, ESR1 and HER2 genes along with protein synthesis of cell-cycle component cyclin D1, ERα and HER2. Consequently, these factors contribute to the cell proliferation in tested cell lines. Results were more pronounced in MCF-7 cells as compared to other cell lines. Collectively, FA might play an essential role in human breast cancer cell proliferation and in those patients with undiagnosed hormonal or resistance to trastuzumab therapy (Chang et al., 2006). Serafim et al. (2011) studied the effect of FA and its derivatives (hexyl ferulate and feruloylhexylamide) (75 μM) on cell proliferation, cell cycle progression, and apoptotic signaling in estrogen-sensitive (MCF-7) and insensitive (MDA-MB-231 and HS578T) human breast cancer cell lines. They observed that the test compounds can interact with ER or with downstream signaling pathways, contributing to the increased susceptibility of MCF-7 cells. Additionally, hexyl ferulate mostly acts through the mitochondrial pathway involving p53 and Bax. It is mediated by mitochondrial depolarization with consequent activation of caspase-9, which is more visible for 48 h, resulting in increased cell death (Serafim et al., 2011) (Table 3).

According to Zhang et al. (2016), metastasis is one of the leading causes of treatment failure in breast cancer. FA (10, 30 and 100 μM) prevented the migration and enhanced caspase-3 activity in a concentration-dependent

Table 3 *In vitro* anticancer effects of FA and underlying mechanisms of action.

Cancer type	Cell line	Conc.	Mechanisms	References
Breast	MDA-MB-231	3–100 μM	↑ Apoptosis ↓ cell viability, metastasis potential	Zhang et al. (2016)
	MCF7 and 4T1	5–1000 μM	↑ apoptosis ↓ cell viability, cell proliferation	Rezaei, Varshosaz, Fesharaki, Farhang, and Jafari (2019)
	MCF-7, MDA-MB-231 and HS578T	25–75 μM	↑ cytotoxicity, p53, Bax, caspase-9 ↓ cell proliferation,	Serafim et al. (2011)
	MCF7, BT474, MDAMB231 and SKBR3	10^{-8}–10^{-6} M	↑ HER2, ESR1, AKT, ERK1/2 and ERα signaling pathways, cyclin-D1	Chang et al. (2006)
	MDA-MB-231	10 and 100 μM	↑ PDI, IRE1α, PEPK and caspase-3 expression, Bax/Bcl-2 ratio, apoptosis ↓ invasion	Cheng, Zhang, Luo, Deng, and Jia (2020)
Cervical	HeLa and Caski	4–20 μM	↓ Cell viability, Bcl-2, Mcl-1 ↑ DNA condensation, apoptosis, pro-caspase PARP, Bax, ROS	Luo, Zhu, Tong, and Peng (2020)
	HeLa	1–50 μM	↓ Cell viability	Kim et al. (2013)
	HeLa and Caski	2.0 mM	↓ Cell invasion, MMP-9 expression, cyclin D1, cyclin E, LC3-II, Beclin1 Atg12-Atg5, arrest cell cycle ↑ p53, p21	Gao et al. (2018)
	HeLa and ME-180	1–40 μg/mL	↑ Radiation effects, lipid peroxidative markers, ROS, DNA damage, apoptosis	Karthikeyan, Kanimozhi, Prasad, and Mahalakshmi (2011)

Continued

Table 3 *In vitro* anticancer effects of FA and underlying mechanisms of action.—cont'd

Cancer type	Cell line	Conc.	Mechanisms	References
	ME-180	5–80 μM	↑ apoptosis, cytotoxicity	Panwar, Sharma, Kaloti, Dutt, and Pruthi (2016)
	HeLa	1–3 mM	↑ Histone H3 hyperacetylation, p53, inhibition of ERK1/2, S-phase arrest, apoptosis ↓ p21, Bcl-2, CDK-4 expression	Saenglee, Jogloy, Patanothai, Leid, and Senawong (2016)
Colon	HCT-15	100–250 μg/mL	↑ Cytotoxicity ↓ EGFR gene expression, cell proliferation	Roy et al. (2016)
	Caco-2	0.1–1500 μM	↑ S- and G2-phase ↓ cell proliferation, G1-phase, metabolic activity	Janicke, Onning, and Oredsson (2005)
	Caco-2	150 μM	↑ S- and G2-phase, CEP2, SMC1L1, WEE1 gene expression, centrosome associated protein ↓ CETN3, CCNB1 gene expression	Janicke et al. (2011)
	HT-29	0.1–100 μM	↑ Apoptosis ↓ cell viability, G0/G1-phase	Rosa et al. (2018)
Liver	HepG2	10–400 μM	↑ ROS, apoptosis, Bax/Bcl-2 ratio, p53 level ↓ cell cycle, GLUT-1 and GLUT-4, angiogenesis, metastasis protein expression	Das et al. (2019)
	HepG2	100–200 μg/mL	↑ Apoptosis, caspase-8 and caspase-9	ElKhazendar et al. (2019)
	Huh-7 and HepG2	4.1–7.2 μg/mL	↑ ROS, MMP, DNA damage, apoptosis, autophagy, cell cycle arrest ↓ cell viability,	Ezhuthupurakkal et al. (2018)

Lung	A549	50–1000 μM	↑ Superoxide anion ↓ cell migration, cell adhesion, DNA synthesis, ERK signaling pathway	Nasr Bouzaiene et al. (2015)
	H1299	0.06–0.6 mM	↑ ROS, Bax, apoptosis ↓ proliferation, survivin expression, MMP-2 and MMP-9 activity	Fong et al. (2016)
	NCI-H460		↑ p53, Bax, p21, GADD45A expression, apoptosis, cytotoxicity ↓ DNA repair inhibition, ATM gene expression, NF-κB	Bandugula and Rajendra Prasad (2013)
	A549	10–400 μM	↑ ROS, apoptosis, Bax/Bcl-2 ratio, p53 level ↓ cell cycle, GLUT-1 and GLUT-4, angiogenesis, metastasis protein expression	Das et al. (2019)
Pancreatic	MIA PaCa-2	150 μM–1 mM	↑ p53, Bax, PTEN caspase-3 and caspase-9 ↓ CCND1, CDK 4/6, Bcl-2 and caspase-8 and −10 expression, cell invasion and migration	Fahrioğlu et al. (2016)
	Panc-1	5–20 μM	↑ Apoptosis ↓ cell proliferation	Eitsuka et al. (2014)
	PaCa-2 and Panc-1	40–200 μM	↑ Apoptosis, ERK1/2 and RB proteins phosphorylation ↓ cell viability, PCNA and MKI67	Thakkar, Chenreddy, Wang, and Prabhu (2015)

Continued

Table 3 *In vitro* anticancer effects of FA and underlying mechanisms of action.—cont'd

Cancer type	Cell line	Conc.	Mechanisms	References
Prostate	PC-3 and LNCaP	20 μM–2 mM	↑ ATR, ATM, CDKN1A, CDKN1B, E2F4, RB1, and TP53 gene expression ↓ cell proliferation and invasion, CCND1, CCND2, CCND3, CDK-2,-4 and −6 gene expression	Eroğlu, Seçme, Bağcı, and Dodurga (2015)
	PC-3	1–900 μM	↑ CASP3, CASP7, CASP8, FAS, CYCS, TNF and PPARG genes	Eroglu et al. (2018)
Osteosarcoma	SaOS-2 and MG63	10–40 μM	↑ Apoptosis, caspase-3, Bax ↓ cell viability, Bcl-2, procaspase-3	Zhang, Wu, and Yang (2017)
	143B and MG63	10–150 μM	↑ Apoptosis, cell cycle arrest, Bax, caspase-3 ↓ CDK-2,-4 and −6, Bcl-2 expression, PI3K/Akt activation	Wang et al. (2016)

manner in breast cancer cells (MDA-MB-231) which confirmed the suppression of cancer cell proliferation and induction of apoptosis by FA. The expression of vimentin was expressively declined with noteworthy increase of E-cadherin after FA treatment. Epithelial-mesenchymal transition (EMT) occurred during the early stages of the transformation of a tumor into a malignant neoplasm and is required for breast cancer metastasis (Thiery, Acloque, Huang, & Nieto, 2009). FA treatment enhanced the expression of epithelial markers and suppressed mesenchymal markers in cancer cells. Additionally, oral administration of FA (100 mg/kg) on MDA-MB-231 xenograft mice remarkably inhibited proliferation and triggered apoptosis in tumor cells. Preventing EMT or inhibiting the function of EMT-related transcription factors may be other mechanisms involved in the antitumor activity of FA *in vivo* (Zhang et al., 2016) (Table 4). Similar findings have been reported by other investigators after the *in vitro* analysis of anticancer potential of FA (100 and 200 μg/mL) against MCF-7 breast cancer cell line. The 50% inhibitory concentration (IC_{50}) of cell proliferation values of FA for MCF-7 cells were 143.8, 75.4, and 85.6 μg/mL at 24, 48, and 72 h, respectively. Moreover, FA treatment appreciably enhanced apoptosis in MCF-7 cells as compared to the control. Caspase-8 and caspase-9 were both induced by FA, but its effect on caspase-8 was more significant than caspase-9 in MCF-7 cells (ElKhazendar et al., 2019).

Recently, Cheng et al. (2020) evaluated the synergetic activity of FA and epirubicin combination on MDA-MB-231 cells and revealed that the combination of these two test compounds (10 and 100 μM FA and 0.02–8 μM epirubicin) not only suppressed the invasion of MDA-MB-231 cells but also triggered the process of apoptosis. The synergism of both compounds (10 μM FA and 1 μM epirubicin) enhanced the expression of the endoplasmic reticulum (ER)-stress associated proteins, such as protein disulfide isomerase (PDI), inositol-requiring transmembrane kinase/endoribonuclease 1α (IRE1α), and PEPK. Moreover, the combination of test compounds enhanced the Bax/Bcl-2 ratio and the expression level of caspase-3, which indicated that the Bcl-2 family is intricate in the initiation of epirubicin/FA-mediated apoptosis by Bax/Bcl-2/caspase-3 pathway in MDA-MB-231 cells. The ER stress increases PERK activity, which phosphorylates eukaryotic initiation factor 2α (eIF2α) to reduce protein translation, thus enhance the sensitivity of MDA-MB-231 cells (Cheng et al., 2020). Cyclodextrin nanosponges (CD-NSs) are a novel group of crosslinked CD derivatives which can be used to enhance the solubility of low-soluble bioactive compounds. During *in vitro* study, the cytotoxic effect of FA and FA-NSs was demonstrated

Table 4 *In vivo* anticancer effects of FA and underlying mechanisms of action.

Cancer Type	Model	Treatment duration	Dose (mg/kg)	Mechanisms	Reference
Breast	BALB/c nude mice	6 weeks	100	↑ Apoptosis ↓ tumor volumes and weight, proliferation, metastatic potential	Zhang et al. (2016)
Colon	F334 rats	3 weeks	15	FA prevented the development of ACF, suppression of metabolic activation and enhancement of detoxification	Kawabata et al. (2000)
Liver	Rats	–	3.6 μg/mL	↓ Liver damage, GST-P expression, Ki-67, 8-OHdG-positive cells	Ezhuthupurakkal et al. (2018)
	Mice	30 days	50	↑ Caspase-3, apoptosis ↓ tumor volume (1.2-fold), tumor growth (3.5-fold)	Das et al. (2019)
Pancreatic	SCID mice	5 weeks	75	↑ p-ERK1/2 and p-RB ↓ growth of pancreatic tumor, tumor volume (45%), PCNA (46%) and MKI67 (43%) expression, proliferation	Thakkar et al. (2015)

against 4T1 (mouse breast cancer) and MCF-7 (human breast cancer) cell lines. FA-NSs (5–80 μM) did not have any significant effect on the viability of both the tested cells. The IC_{50} of cell proliferation for FA-NS in both MCF7 and 4T1 cell lines has been found to be 250 μM, while this value is 750 μM in case of free FA. Additionally, the cancer cell lines treated with FA-NSs (250 μM) showed remarkable structural changes compared with other formulations and also induced 50% cell death. Conclusively, encapsulation of FA into the NSs leads to a higher solubility and stability of FA and may have contributed to higher anticancer properties of FA (Rezaei et al., 2019).

5.2 Cervical cancer

Cervical cancer is one of the most common forms of malignancies in the female population. The frequency ranges from 8 to 30 newly reported cases per 100,000 women per year, based on the country and region (Denny, 2012). Approximately, 6% of cervical cancers are adenocarcinomas and less than 2% are adenosquamous (mixed) carcinomas (Katanyoo, Sanguanrungsirikul, & Manusirivithaya, 2012). Several factors have been recognized to trigger the malignancy to a greater or lesser extent: early coitus (before the age of 20), promiscuity (multiple sexual partners), pregnancy in younger age, higher number of births, infection with herpes simplex virus and human papilloma virus, poor socioeconomic status of women, and others (Dillner et al., 2008). On the basis of degree of differentiation, cervical cancers are divided into grades such as good, medium, poor, and undifferentiated grades. Gao et al. (2018) investigated the therapeutic potential of FA on HeLa and Caski cervical carcinoma cell lines by analyzing cell proliferation, invasion, apoptosis, and autophagy. They reported that FA (1.0, 2.0, and 4.0 mM) promoted apoptosis and arrest cell cycle at G_0/G_1 phase by triggering the expression of p53 and p21 protein with a reduction in the levels of cyclin D1 and cyclin E in both cell lines. FA suppressed the autophagy-related proteins, such as Beclin1, LC3-II and Atg12-Atg5, in a concentration-dependent manner and prevented the invasion in cervical cancer cells (HeLa and Caski) by reducing matrix metallopeptidase-9 (MMP-9) mRNA expression (Gao et al., 2018). A separate study on HeLa and Caski cells revealed that FA treatment caused DNA damage, induced apoptosis through caspases activation [procaspase-3, procaspase-8, procaspase-9, and poly (ADP-ribose) polymerase (PARP)], and promoted ROS formation in a concentration-dependent manner (2–25 μM). Additionally, FA treatment remarkably reduced the expression of Bcl-2 and Mcl-1 proteins, trigged the pro-apoptotic protein level (Bax) and down-regulated the phosphoinositide 3-kinases (PI3K)/Akt signaling pathway in Caski cells (Luo et al., 2020).

Another group of researchers evaluated the radio-sensitizing potential of FA against HeLa and ME-180 cells (Karthikeyan et al., 2011). They reported that FA (10 μg/mL) in combination with radiation Cobalt 60, 1.6 Gy/min) remarkably prevented the cell growth and gave additive effects. Reduced GSH:oxidized glutathione (GSSG) ratio indicated that FA depleted GSH and prevented the recycling of GSH in HeLa and ME-180 cells in a concentration-dependent manner (1–40 μg/mL). Furthermore, it is implicated that FA interact with inhibitors of *de novo* synthesis of GSH to sensitize tumor cells to radiation. The inhibition of cell growth, decreased antioxidant status (superoxide dismutase, catalase, and glutathione peroxidase), increased lipid peroxidation, and DNA damage were associated with profound induction of cell apoptosis in HeLa and ME-180 cells. FA influenced the mitochondrial activity of tested cancer cells, which may account for enhanced cytotoxicity of FA plus radiation treated cells (Karthikeyan et al., 2011). Panwar et al. (2016) used chitosan-tripolyphosphate pentasodium (CS-TPP) nanoparticles (NPs) to increase the bioavailability of FA and evaluated the therapeutic potential of these FA-loaded NPs against human cervical cancer cell line (ME-180). During analysis they observed that 40 μM FA/CS-TPP NPs caused maximum inhibition (57%) of cell proliferation as compared to other formulations (20 μM native FA, 40 μM native FA, unloaded CS-TPP NPs and 20 μM FA/CS-TPP). Similarly, treatment of ME-180 with 40 μM FA/CS-TPP NPs for 24 h exhibited morphological characteristics, including irregularity in shape, cytoplasmic condensation, and higher cell shrinkage, which indicated that FA-loaded CS-NPs trigger apoptotic cell death in ME-180 cells. On the basis of experimental findings, they suggested that FA/CS-TPP NPs would have bound to the tumor cell membranes *via* electron interactions, entered the cells through endocytosis, and release the encapsulated FA into cytosol (endosomal burst). Due to enhanced bioavailability, FA exhibited its antitumor effect by altering the mitochondrial activity and antioxidant status of the cell, thus disrupting the cell's organelles which would eventually lead to cell death (Panwar et al., 2016).

Previous studies on cancer revealed that epigenetic histone modifications are a promising avenue for cancer preventive and therapeutic strategies. To analyze this aspect of treatment, Saenglee et al. (2016) evaluated the *in vitro* antiproliferative and HDACs inhibitory efficacy of FA against HeLa cells and concluded that FA treatment (1, 2 and 3 mM) resulted in histone H3 hyperacetylation in both type of cervical cell lines in a concentration-dependent manner therefore exhibiting potential for HDACs inhibition. HDACs inhibitors are supposed to be therapeutic candidates for cancer by virtue of their capability to trigger expression of tumor suppressor genes, repress oncogenes expression, and stimulate apoptosis by ROS formation and DNA damage (Conley, Wright, & Kummar, 2006; Robert & Rassool, 2012). On the other hand, in HeLa cells FA decreased the level of anti-apoptotic Bcl-2 protein, arrested G1-phase, induced upregulation of p53 and down-regulation of p21. Blockage of cell cycle progression in HeLa cells by FA was most likely due to suppression of CDK4 expression and inhibition of ERK1/2 phosphorylation (Saenglee et al., 2016).

5.3 Colon cancer

Every year, approximately 1.36 million new colon cancer patients are reported, with above half a million mortality. This accounts for about 40% of the total cancer cases detected annually (Banerjee et al., 2017). Delivery of colon targeting drugs aim direct action at affected sites, avoid the pre-systemic metabolism, and lower absorption of drugs in the small intestine; therefore, augmenting therapeutic potential and declining destructive systemic side effects (Lu et al., 2016). Kawabata, Yoshioka, and Terao (2019) evaluated the therapeutic activity of FA against azoxymethane (AOM)-triggered colon carcinogenesis in male rats. They conducted the study in to three phases with different parameters and FA concentrations and suggested that FA (15 mg/kg) administration inhibited the aberrant crypt foci (ACF) formation. It prevented the advancement of colonic ACF and suppressed progression of preneoplasia to malignant neoplasia. Furthermore, FA (250 and 500 ppm) had an inhibitory action on intestinal neoplasms (23% and 27%, respectively) by exposure in the initial phase. FA expressively improved the detoxification enzymes activities (glutathione S-transferase and quinone reductase) in the liver and/or colonic mucosa in a dose-dependent manner (Kawabata et al., 2019). A study on human colonic endothelial tumor cell line (Caco-2) suggested that FA treatment (1500 μM) suppressed the metabolic activity with reduced number of Caco-2 cells to 43–75% of the control. FA also decreased the proportion of cells in the G1 phase and increased the proportion of cells in the S and G2 phases. Researchers reported that FA inhibited cell proliferation by presumably affecting different cell cycle phases. One of the mechanisms of such chemopreventive action of FA on colon carcinogenesis is the suppression of metabolic activation and enhancement of detoxification (Janicke et al., 2005).

A further study on Caco-2 cells by Janicke et al. (2011) reported that FA (150 μM) significantly altered the mRNA expression of numerous genes of

Caco-2 cells related to cell cycle and proliferation. Treatment with FA up-regulated the mRNA expressions of centrosomal protein 2 (CEP2), RAB GTPase activating protein 1 (RABGAP1) (both CEP2 and RABGAP1 genes are involved in centrosome assembly), structural maintenance of chromosome 1-like 1 protein (SMC1L1, a S-phase checkpoint protein), and Wee1. The upregulation seen may indicate that the Caco-2 cells were inhibited to enter mitosis by this mechanism when treated with FA. FA also down-regulated centrin 3 (CETN3), polymerase (DNA directed) sigma (PLOS), and G2/mitotic-specific cyclin B1 (CCNB1) mRNA expressions. Additionally, FA triggered a delay in the S-phase and altered genes that regulate centrosome assembly and S-phase DNA damage checkpoints (Janicke et al., 2011). Furthermore, a separate group of researchers explored the cytotoxic effect of FA (100–250 μg/mL) on the human colorectal cancer cell line (HCT-15) and also investigated its impact on human epidermal growth factor receptor (EGFR) gene expression. FA treatment appreciably decreased the viability of HCT-15 cells in a concentration dependent manner (IC_{50} 154 μg/mL). Similarly, FA (175 μg/mL) remarkably reduced the expression of the target gene EGFR up to 4.28-fold as compared to the control and exhibited noteworthy binding ability with the EGFR gene at molecular level *in silico* (Roy et al., 2016). Rosa et al. (2018) evaluated the impact of FA on cell viability, cell cycle progression, and rate of apoptosis in human colon adenocarcinoma cells (HT-29). They revealed that FA (5 μM) promoted a remarkable reduction (63%) in cell viability and exhibited a decline in the proportion of cells in the G0/G1 phase (10 and 100 μM) during cell cycle analysis. Moreover, FA (10 μM) suppressed the percentage of viable cells, stimulated S-phase elongation and reduced G2/M phase of the cell cycle, with an increase in the percentage of apoptotic cells (Rosa et al., 2018).

5.4 Hepatic cancer

Liver cancer is one of the most common forms of malignancy and the third foremost cause of cancer-associated death worldwide (Yu & Schwabe, 2017). Due to the lack of specific symptoms in initial stages, the absence of initial diagnostic markers, and the low percentage (10–20%) of radical resection hepatocellular carcinoma (HCC) on diagnosis, most HCC patients are frequently spotted in an advanced stage with poor prognosis (Kobayashi et al., 2017). Even if sorafenib and regorafenib are prescribed in HCC, the overall outcome of HCC remains insufficient with the median survival in early and advanced HCC of about 6–9 months and 1–2 months, respectively

(Ren, Ma, Duan, & Chen, 2020). ElKhazendar et al. (2019) explored the therapeutic effect of FA against a liver cancer cell line (HepG2) and revealed that FA (100 and 200 μg/mL) exerted cytotoxic effects on HepG2 cells with IC_{50} values 150.7, 81.38, and 210.4 μg/mL at 24, 48, and 72 h, respectively. Additionally, caspases-8 and caspase-9 were both induced by FA in a concentration-dependent manner, but its action on caspase-8 was more pronounced than caspase-9 suggesting that both initiators of apoptosis are involved, but primarily through caspase-8 extrinsic pathway (ElKhazendar et al., 2019).

Separately, *in vitro* anticancer potential of FA with zinc oxide nanoparticles (ZnONPs) on Huh-7 (8.2 μg/mL) and HepG2 cells (7.2 μg/mL) demonstrated a noteworthy reduction in the cell viability than that of ZnONP- and FA-treated cells. Furthermore, excessive ROS production was observed in ZnONPs-FA treated cells as compared to individual treatment which, in turn, lead to DNA damage and altered mitochondrial membrane potential. This may be due to the combined action of intracellular ROS production by ZnONPs and the prooxidative mechanism of FA in the presence of Zn ions in ZnONPs-FA treated cells. Analysis of alterations in Ki67 and γH2AX gene expression also confirmed the cell cycle arrest followed by DNA damage and induced apoptosis due to ZnONPs-FA (Ezhuthupurakkal et al., 2018). A similar outcome has also been observed in diethylnitrosamine (DEN)-induced HCC in rats. ZnONPs-FA treatment reduced the hepatic damages and protected its normal architecture in HCC rats. Moreover, ZnONPs-FA suppressed the expression of GST-P (preneoplastic marker of hyperplastic nodule), Ki67 (proliferative marker) and 8-OHdG (indicates the ROS generation) in DEN-induced HCC (Ezhuthupurakkal et al., 2018).

Radiotherapy increases ROS level beyond the threshold limit to promote cellular death. Das et al. (2019) demonstrated that FA (10–400 μM) enhanced the radiation sensitivity of HepG2 cells through collapsing redox homeostasis. In their experiment, they observed that FA pretreatment eliminated the intracellular ROS initially in HepG2 cancer cells. It suppressed the surface level and expression of glucose transporters (GLUT) such as GLUT1 and GLUT4 in the carcinoma cells and therefore, hindered the uptake of glucose. Moreover, combined treatment of FA and ionizing radiation prominently decreased the expression of angiogenesis and metastatic proteins. This combined therapy triggered ROS formation which prevented Akt activation followed by reduced the GLUT1 and GLUT4 expression with suppression of invasion markers, such as MMP-9, VEGF, PDGFRβ and PECAM1 expressions. This mechanistically showed that combination

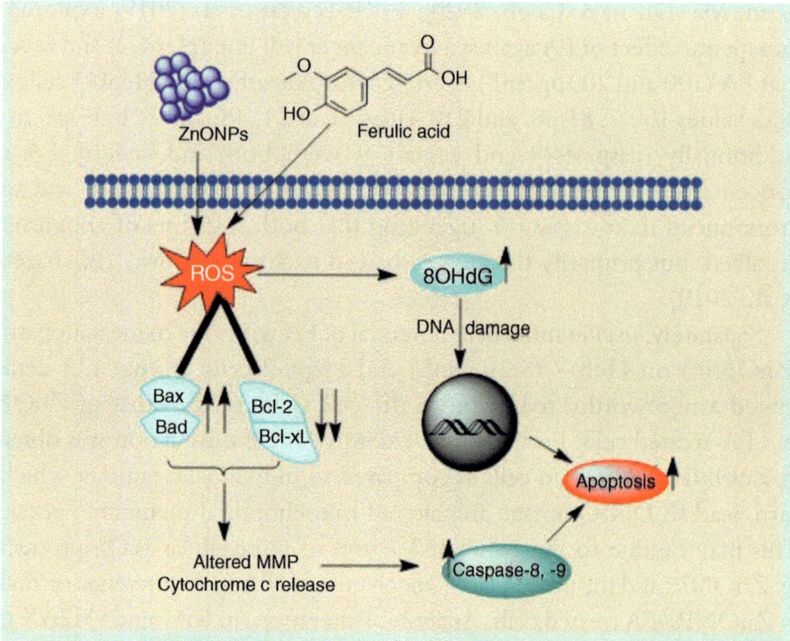

Fig. 3 Anticancer effect of FA on hepatocellular carcinoma. FA encapsulated in zinc oxide nanoparticle (ZnONPs) triggered the intracellular ROS concentration that promoted DNA damage and finally lead to apoptosis.

treatment effectively disrupted the ROS-Akt-GLUT-glycolysis network. Furthermore, combination treatment activated ROS-p53-Bax-cytochrome c-intrinsic caspases-DNA damage axis in both carcinoma cells thereby programming them for apoptosis (Das et al., 2019) (Fig. 3).

5.5 Lung cancer

Lung cancer is the third most common cancer after breast and prostate but has the largest proportion of all cancer-related deaths (22%). Approximately 80% of human lung cancers are non-small cell lung cancer (NSCLC), and their development involves multiple genetic abnormalities that lead to malignant transformation of the bronchial epithelial cells, followed by invasion in lymph nodes and distant metastases. A previous report suggested that 2-deoxy-D-glucose (2DG), a glycolytic inhibitor, inhibits growth and induces apoptosis in a number of cancer cell lines (Zhang et al., 2006). Bandugula and Rajendra Prasad (2013) revealed that combination therapy of 2DG (4 mM), FA (53.8 μM), and ionizing radiation strengthened the potential of radiation by preventing of DNA repair, enhancing the expression of proapoptotic protein p53 and Bax, arresting the cell cycle and overcoming of apoptosis resistance in NCI-H460 NSCLC cells. ATM expression was down-regulated by 2DG and FA, which ultimately resulted in the decreased expression of nuclear factor kappa light chain enhancer of activated B cells (NF-κB). Combination of 2DG and FA also increased the levels of p21 and GADD45A in NCI-H460 cells. DNA repair inhibition is expected to be a possible mechanism for the radio-sensitization observed, which is apparent by the down-regulation of radiation-induced ataxia-telangiectasia mutated gene expression upon treatment with 2DG and/or FA. Moreover, Western blot analysis of NF-κB and caspase-3 revealed the involvement of apoptotic signals in the cytotoxicity exhibited by the combination of 2DG and FA. Additionally, an increased fraction of cells was observed in G0 phase upon treatment with 2DG and FA compared with 2DG and FA alone (Bandugula & Rajendra Prasad, 2013).

Separately, FA (50–1000 μM) inhibited the proliferation of human lung (A549) cell line by the suppression of DNA synthesis. A moderate increase in ROS can promote cell proliferation and differentiation, whereas excessive amounts of ROS can cause oxidative damages. FA scavenged ROS produced in a concentration-dependent manner and reduced cell adhesion in A549 cells by 77.9% at the higher tested concentration of FA (200 μM). Furthermore, it has been observed that the inhibitory effect of FA on cell migration is likely due to the reduced attachment to extracellular matrix proteins (Nasr Bouzaiene et al., 2015).

Trans-FA treatment did not alter the cellular morphology and proliferation of lung cancer cell (H1299) until after the initial 48 h. Although, long-term medication suppressed colony formation and anchorage-independent growth in H1299 cells. *Trans*-FA exhibited inhibitory effects on cellular proliferation of lung cancer cells H1299 and lung fibroblast cells (HEL-299) suggesting the efficacy of *trans*-FA for selectively inhibiting lung cancer. Furthermore, *trans*-FA (0.6 mM) caused an accumulation of the G0/G1 population and triggered a moderate rise in the apoptotic population (Fong et al., 2016). *Trans*-FA treatment promoted the phosphorylation of β-catenin at residues Thr41 and Ser45 and led to the proteasomal degradation of cytoplasmic β-catenin, causing the down-regulation of β-catenin protein levels. *Trans*-FA treatment inhibited the migration and invasion of lung cancer cells and concurrently attenuated the activities of both MMP-2 and MMP-9. The remarkable rise in ROS concentration and noteworthy alterations in Bax and survivin expression occurred alongside the antiproliferative effects observed

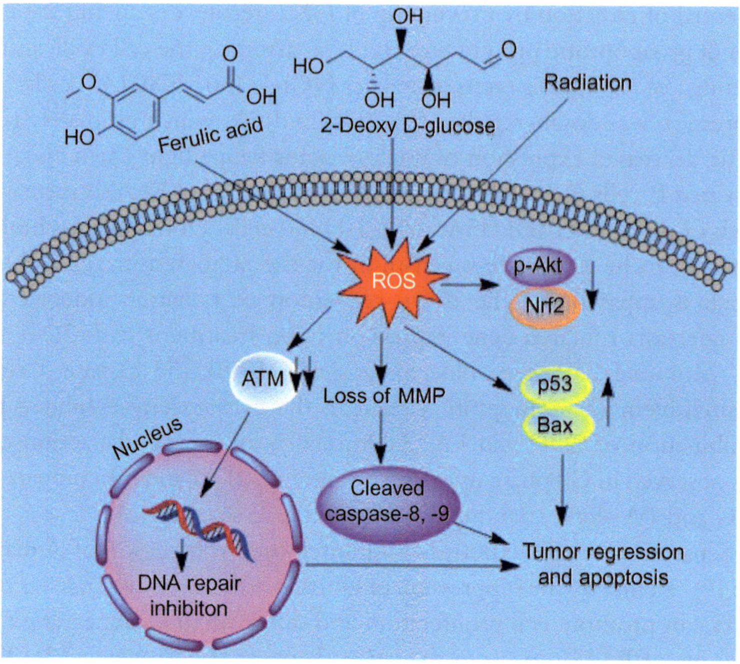

Fig. 4 Mechanism of action of FA against lung cancer cells. FA together with radiation stimulate ROS formation and proapoptotic gene (p53 and Bax) expression. It also suppressed ATM, p-Akt and Nrf2 gene expression which altered DNA repair mechanism and promoted apoptosis in cancer cells.

following *trans*-FA treatment (Fong et al., 2016). Similar results have been reported by Das et al. (2019) on a lung cancer cell line (A549). They suggested that combination of irradiation (cobalt-60) and FA activated the mitochondrial apoptotic pathway and concomitantly inhibited the cell cycle progression and survival pathway. Additionally, the combination medication displayed tremendous tumor regression, activation of caspase-3, and nuclear fragmentation in tested tumor cells (Fig. 4).

5.6 Pancreatic cancer

Pancreatic cancer has been one of the deadliest cancers with a higher mortality rate of about 3% of all cancers with an estimated five-year survival rate (Xu & Yang, 2016). The projected number of individuals predicted to be identified with pancreatic cancer in 2018 is about 55,440 (29,200 men and 26,240 women), whereas, approximately 44,330 individuals (21,310 women and 23,020 men) die of pancreatic cancer (Cancer Facts & Figures, 2017).

An *in vitro* study on human pancreatic cancer cells (MIA PaCa-2) reported the effects of FA on gene expression, viability, colony formation and migration/invasion. FA (150–750 μM and 1 mM) caused a remarkable reduction in the expression of CCND1, CDK 4/6, Bcl-2, caspase-8 and caspase-10, while causing a rise in the p53, Bax, PTEN, caspase-3 and caspase-9 expression. FA also decreased colony formation and prevented cell invasion and migration (Fahrioğlu, Dodurga, Elmas, & Seçme, 2016). CCND1 regulates the cell cycle at the G1-S checkpoint and can cause cell arrest in G1 phase. Amplification or upregulation of CCND1 has been reported to play a role in tumor formation in different tissues like the breast (Çıtışlı et al., 2015). In FA-treated cells, the mRNA expression of CCND1, CDK4, and CDK6 were significantly reduced indicating that FA may play a role in the prevention of MIA PaCa-2 cell proliferation by declining expression of these genes and perhaps by stimulating apoptotic mechanisms *via* p53 and Bax pathway, not the FADD, Akt, caspase-8 and caspase-9 pathway. Moreover, FA can down-regulate the PARP expression in the MIA PaCa-2 cells which may contribute to decreased cellular viability (Fahrioğlu et al., 2016).

Thakkar et al. (2015) evaluated the effect of chitosan-coated solid-lipid nanoparticles (c-SLN) of FA and aspirin (ASP) on human pancreatic cancer cells (MIA PaCa-2 and Panc-1) and demonstrated that combinations of low concentrations of free FA (200 μM) and ASP (1 mM) significantly reduced cell viability by 45% and 60% in MIA PaCa-2 and Panc-1, respectively. However, when encapsulated within c-SLNs, a 5-fold and 40-fold decline in FA (40 μM) and ASP (25 μM) concentration was observed. Separately, they displayed little or no decrease in the cell viability, but when combined, a significant reduction by 70% was observed in MIA PaCa-2 and Panc-1 cells. Furthermore, oral administration of combinations of FA (75 mg/kg) and ASP (25 mg/kg) c-SLNs to MIA PaCa-2 pancreatic tumor xenograft mouse model (severe combined immunodeficient mice) repressed the growth of the tumor by 45% compared to the control. However, immunohistochemical examination of tumor tissues exhibited remarkable reduction in proliferative proteins PCNA and MKI67 expression and enhanced expression of p-RB, p21, and p-ERK1/2 apoptotic proteins suggesting the proapoptotic role of the regimen (Thakkar et al., 2015). Synergistic inhibitory effect of FA (5–20 μM) and δ-tocotrienol (δ-T3) (10–12.5 μM) against pancreatic cancer cells (PANC-1) demonstrated that combined treatment noticeably reduced proliferation of PANC-1 cells relative to δ-T3 alone, although 20 μM FA alone showed no inhibitory effect. The combination of δ-T3 (12.5 μM) and FA (20 μM) inhibited the proliferation of

PANC-1 cells by 32%. Although δ-T3 induced G1 arrest by upregulating p21 in PANC-1 cells, more cells accumulated in G1-phase with the combination of d-T3 and FA. This synergistic effect was attributed to an increase in the cellular concentration of δ-T3 by FA (Eitsuka et al., 2014).

5.7 Prostate cancer

Prostate cancer is the most prevalent type of cancer in men with the highest mortality rate, especially in developed countries (Siegel, Miller, & Jemal, 2016). When it is diagnosed at an early stage, such as stage 1 or stage 2, the 5-year survival rate can reach 90%, while the likelihood of successful treatment at advanced stages such as stage 3 and 4 is low (Trewartha & Carter, 2013). Eroğlu et al. (2015) demonstrated the cytotoxic effects of FA in human prostate cancer cell lines (PC-3 and LNCaP) by analyzing 84 key genes expression for both cell cycle regulation and apoptosis. FA treatment (20 μM–2 mM) enhanced the expressions of ATM, ATR, RB1 and p53 while decreased the expression of CCND1, CCND2, and CCND3 genes in PC-3 cell line compared to the control. However, no such alteration in the gene expression was observed in the LNCaP cell line. Furthermore, FA in PC-3 and LNCaP cell lines suppressed the invasion in the ratios of $33.6 \pm 0.5\%$ and $23 \pm 1.9\%$, respectively, while it suppressed the colony formation in the ratios of $38 \pm 0.82\%$ and $40 \pm 1.7\%$, respectively. This indicated that FA strengthened cellular response due to the amplified gene expression leading to antiproliferative action and blocked cell cycle progression in tested prostate cancer cells (Eroğlu et al., 2015).

Furthermore, synergistic effect of FA (30–900 μM) and gemcitabine (5–150 μM) on apoptosis and metastasis in PC-3 cells revealed that combined treatment of test compounds significantly increased the expressions of *CASP3, CASP7, CASP8, FAS, CYCS, TNF* and *PPARG* genes compared with the control. Likewise, a significant increase in the expression of *TIMP-1, TIMP-2* and *CDH1* genes confirmed the antimetastatic effect of FA and gemcitabine combination (Eroglu et al., 2018). Similarly, Eitsuka et al. (2014) reported that combined treatment of FA (5–20 μM) and δ-tocotrienol (δ-T3) (10–12.5 μM) markedly reduced proliferation (59%) and arrested cell cycle at G1 phase by upregulating p21 gene expression in DU-145 cells.

5.8 Osteosarcoma

Osteosarcoma is a primary malignant bone tumor with a worldwide incidence of 3.4 per million people per year. For most of the 20th century,

5-year survival rates for classic osteosarcoma were 20% (Misaghi, Goldin, Awad, & Kulidjian, 2018). An *in vitro* study reported that FA inhibited proliferation and induced apoptosis in both human osteosarcoma cells (143B and MG63) in a concentration-dependent manner. Aberration of PI3K/Akt pathway is a very common mechanism for many human cancers including osteosarcoma since it can mediate survival pathway and apoptosis pathway (Nouri et al., 2020). Treatment with FA (30, 100, and 150 μM) down-regulated the expression of p-Akt, Bcl-2 CDK2, CDK4, and CDK6 and up-regulated Bax expression. This indicates that FA inhibited PI3K/Akt pathway, enhanced apoptosis of osteosarcoma cells, and acted as an antiproliferative agent (Wang et al., 2016). A separate study on SaOS-2 and MG63 osteosarcoma cell lines displayed that FA (10–40 μM) critically declined the cell viabilities and promoted apoptosis through caspase-3 activation in a concentration-dependent manner. In the FA treated SaOS-2 and MG63 cells, procaspase-3, Bcl-2 mRNA, and proteins were decreased, while Bax was simultaneously increased compared to the control group (Zhang et al., 2017).

6. Conclusion and future perspectives

FA is a dietary compound present in various plant seeds and leaves, both in free form and covalently conjugated to polysaccharides, glycoproteins, polyamines, lignin and hydroxy fatty acids of plant cell walls. FA exhibits a wide variety of pharmacological activities, such as antioxidant, anti-inflammatory, antimicrobial, antiallergic, hepatoprotective, anticancer, antithrombotic effects, increased sperm viability, antiviral, vasodilatory actions, metal chelation, modulation of enzyme activity, activation of transcriptional factors, gene expression and signal transduction. Furthermore, recent *in vitro* and *in vivo* studies revealed that it has potential for prevention and treatment of numerous cancers, such as breast, lung, cervical, prostrate, hepatic, colon and pancreatic cancer. FA exhibits potential anticancer effect by inhibiting the growth of cancer cells through suppression of proliferation, initiation of apoptosis and cell cycle arrest as well as modulating various signaling pathways. Most of the cancer related studies have been performed *in vitro* with limited *in vivo* results, therefore the observations on the cancer prevention and treatment needs further investigation. It is essential to explore the pharmacological attributes and the mechanism of actions of metabolites for a better interpretation of FA as an anticancer agent. Still, the mechanisms underlying the antitumor potential of FA are not completely understood and more *in vivo* analysis is required. Therefore, further clinical studies on anticancer efficacy of FA are also warranted to investigate the systemic bioavailability of FA from dietary sources or specific drug formulations, optimum dose, and duration. In addition, the limited bioavailability of FA and its metabolites pose a challenge for their clinical development. To resolve these issues, there is a need to explore advanced drug delivery systems as viable options for delivering remedial concentrations of FA into the systemic circulation as well as target tissues.

Declarations of competing interest

The authors declare that there are no conflicts of interests.

References

Abotaleb, M., Samuel, S. M., Varghese, E., Varghese, S., Kubatka, P., Liskova, A., et al. (2018). Flavonoids in cancer and apoptosis. *Cancers, 11*(1). https://doi.org/10.3390/cancers11010028.

Aceves, C., Anguiano, B., & Delgado, G. (2005). Is iodine a gatekeeper of the integrity of the mammary gland? *Journal of Mammary Gland Biology and Neoplasia, 10*(2), 189–196. https://doi.org/10.1007/s10911-005-5401-5.

Adam, A., Crespy, V., Levrat-Verny, M.-A., Leenhardt, F., Leuillet, M., Demigné, C., et al. (2002). The bioavailability of ferulic acid is governed primarily by the food matrix rather than its metabolism in intestine and liver in rats. *The Journal of Nutrition, 132*(7), 1962–1968. https://doi.org/10.1093/jn/132.7.1962.

Adom, K. K., & Liu, R. H. (2002). Antioxidant activity of grains. *Journal of Agricultural and Food Chemistry, 50*(21), 6182–6187. https://doi.org/10.1021/jf0205099.

Aggarwal, V., Tuli, H. S., Tania, M., Srivastava, S., Ritzer, E. E., Pandey, A., et al. (2020). Molecular mechanisms of action of epigallocatechin gallate in cancer: Recent trends and advancement. *Seminars in Cancer Biology*. https://doi.org/10.1016/j.semcancer.2020.05.011.

Andreasen, M. F., Kroon, P. A., Williamson, G., & Garcia-Conesa, M. T. (2001). Esterase activity able to hydrolyze dietary antioxidant hydroxycinnamates is distributed along the intestine of mammals. *Journal of Agricultural and Food Chemistry, 49*(11), 5679–5684.

Antonarakis, E. S., Shaukat, F., Isaacsson Velho, P., Kaur, H., Shenderov, E., Pardoll, D. M., et al. (2019). Clinical features and therapeutic outcomes in men with advanced prostate cancer and DNA mismatch repair gene mutations. *European Urology, 75*(3), 378–382. https://doi.org/10.1016/j.eururo.2018.10.009.

Bandugula, V. R., & Rajendra Prasad, N. (2013). 2-Deoxy-D-glucose and ferulic acid modulates radiation response signaling in non-small cell lung cancer cells. *Tumour Biology, 34*(1), 251–259. https://doi.org/10.1007/s13277-012-0545-6.

Banerjee, A., Pathak, S., Subramanium, V. D., Dharanivasan, G., Murugesan, R., & Verma, R. S. (2017). Strategies for targeted drug delivery in treatment of colon cancer: Current trends and future perspectives. *Drug Discovery Today, 22*(8), 1224–1232. https://doi.org/10.1016/j.drudis.2017.05.006.

Booth, A. N., Emerson, O. H., Jones, F. T., & Deeds, F. (1957). Urinary metabolites of caffeic and chlorogenic acids. *The Journal of Biological Chemistry, 229*(1), 51–59.

Bourne, L. C., & Rice-Evans, C. (1998). Bioavailability of ferulic acid. *Biochemical and Biophysical Research Communications, 253*(2), 222–227. https://doi.org/10.1006/bbrc.1998.9681.

Braicu, C., Zanoaga, O., Zimta, A.-A., Tigu, A. B., Kilpatrick, K. L., Bishayee, A., et al. (2020). Natural compounds modulate the crosstalk between apoptosis- and autophagy-regulated signaling pathways: Controlling the uncontrolled expansion of tumor cells. *Seminars in Cancer Biology*. https://doi.org/10.1016/j.semcancer.2020.05.015.

Brooks, C. L., & Gu, W. (2009). How does SIRT1 affect metabolism, senescence and cancer? *Nature Reviews Cancer, 9*(2), 123–128. https://doi.org/10.1038/nrc2562.

Buchanan, C. J., Wallace, G., Fry, S. C., & Eastwood, M. A. (1996). In vivo release of 14C-labelled phenolic groups from intact dietary spinach cell walls during passage through the rat intestine. *Journal of the Science of Food and Agriculture, 71*(4), 459–469. https://doi.org/10.1002/(SICI)1097-0010(199608)71:4<459::AID-JSFA602>3.0.CO;2-H.

Bunzel, M., Ralph, J., Marita, J. M., Hatfield, R. D., & Steinhart, H. (2001). Diferulates as structural components in soluble and insoluble cereal dietary fibre. *Journal of the Science of Food and Agriculture, 81*(7), 653–660. https://doi.org/10.1002/jsfa.861.

Cancer Statistics, Cancer Facts & Figures, (2017). Retrieved September 14, 2020, from https://www.cancer.org/research/cancer-facts-statistics/all-cancer-facts-figures/cancer-facts-figures-2017.html.

Chang, C. J., Chiu, J. H., Tseng, L. M., Chang, C. H., Chien, T. M., Wu, C. W., et al. (2006). Modulation of HER2 expression by ferulic acid on human breast cancer MCF7 cells. *European Journal of Clinical Investigation, 36*(8), 588–596. https://doi.org/10.1111/j.1365-2362.2006.01676.x.

Chang, M. X., Xu, L. Y., Tao, J. S., & Feng, Y. (1993). Metabolism and pharmacokinetics of ferulic acid in rats. *Zhongguo Zhong Yao Za Zhi, 18*(5), 300–302. 319.

Cheng, W.-J., Zhang, P.-P., Luo, Q.-Q., Deng, S.-M., & Jia, A.-Q. (2020). The chemosensitizer ferulic acid enhances epirubicin-induced apoptosis in MDA-MB-231 cells. *Journal of Functional Foods, 73*, 104130. https://doi.org/10.1016/j.jff.2020.104130.

Chesson, A., Provan, G. J., Russell, W. R., Scobbie, L., Richardson, A. J., & Stewart, C. (1999). Hydroxycinnamic acids in the digestive tract of livestock and humans. *Journal of the Science of Food and Agriculture, 79*(3), 373–378. https://doi.org/10.1002/(SICI)1097-0010(19990301)79:3<373::AID-JSFA257>3.0.CO;2-6.

Choudhury, R., Srai, S. K., Debnam, E., & Rice-Evans, C. A. (1999). Urinary excretion of hydroxycinnamates and flavonoids after oral and intravenous administration. *Free Radical Biology & Medicine, 27*(3–4), 278–286. https://doi.org/10.1016/s0891-5849(99)00054-4.

Çıtışlı, V., Dodurga, Y., Eroğlu, C., Seçme, M., Avcı, Ç. B., & Şatıroğlu-Tufan, N. L. (2015). Temozolomide may induce cell cycle arrest by interacting with URG4/URGCP in SH-SY5Y neuroblastoma cells. *Tumour Biology: The Journal of the International Society for Oncodevelopmental Biology and Medicine, 36*(9), 6765–6772. https://doi.org/10.1007/s13277-015-3373-7.

Clifford, M. N. (1999). Chlorogenic acids and other cinnamates—Nature, occurrence and dietary burden. *Journal of the Science of Food and Agriculture, 79*(3), 362–372. https://doi.org/10.1002/(SICI)1097-0010(19990301)79:3<362::AID-JSFA256>3.0.CO;2-D.

Conley, B. A., Wright, J. J., & Kummar, S. (2006). Targeting epigenetic abnormalities with histone deacetylase inhibitors. *Cancer, 107*(4), 832–840. https://doi.org/10.1002/cncr.22064.

Das, U., Manna, K., Adhikary, A., Mishra, S., Saha, K. D., Sharma, R. D., et al. (2019). Ferulic acid enhances the radiation sensitivity of lung and liver carcinoma cells by collapsing redox homeostasis: Mechanistic involvement of Akt/p38 MAPK signalling pathway. *Free Radical Research, 53*(9–10), 944–967. https://doi.org/10.1080/10715762.2019.1655559.

Denny, L. (2012). Cervical cancer prevention: New opportunities for primary and secondary prevention in the 21st century. *International Journal of Gynaecology and Obstetrics: The Official Organ of the International Federation of Gynaecology and Obstetrics, 119*(Suppl. 1), S80–S84. https://doi.org/10.1016/j.ijgo.2012.03.023.

Dillner, J., Rebolj, M., Birembaut, P., Petry, K.-U., Szarewski, A., Munk, C., et al. (2008). Long term predictive values of cytology and human papillomavirus testing in cervical cancer screening: Joint European cohort study. *BMJ*, *337*, a1754. https://doi.org/10.1136/bmj.a1754.

Eitsuka, T., Tatewaki, N., Nishida, H., Kurata, T., Nakagawa, K., & Miyazawa, T. (2014). Synergistic inhibition of cancer cell proliferation with a combination of δ-tocotrienol and ferulic acid. *Biochemical and Biophysical Research Communications*, *453*(3), 606–611. https://doi.org/10.1016/j.bbrc.2014.09.126.

ElKhazendar, M., Chalak, J., El-Huneidi, W., Vinod, A., Abdel-Rahman, W. M., & Abu-Gharbieh, E. (2019). Antiproliferative and proapoptotic activities of ferulic acid in breast and liver cancer cell lines. *Tropical Journal of Pharmaceutical Research*, *18*(12), 6. https://doi.org/10.4314/tjpr.v18i12.16.

Eroglu, C., Avci, E., Secme, M., Dodurga, Y., Vural, H., & Kurar, E. (2018). The combination effect of ferulic acid and gemcitabine on expression of genes related apoptosis and metastasis in PC-3 prostate cancer cells. *European Journal of Cell Biology*, 77(1), 32–37. https://doi.org/10.26650/EurJBiol.2018.0003.

Eroğlu, C., Seçme, M., Bağcı, G., & Dodurga, Y. (2015). Assessment of the anticancer mechanism of ferulic acid via cell cycle and apoptotic pathways in human prostate cancer cell lines. *Tumour Biology: The Journal of the International Society for Oncodevelopmental Biology and Medicine*, *36*(12), 9437–9446. https://doi.org/10.1007/s13277-015-3689-3.

Ezhuthupurakkal, P. B., Ariraman, S., Arumugam, S., Subramaniyan, N., Muthuvel, S. K., Kumpati, P., et al. (2018). Anticancer potential of ZnO nanoparticle-ferulic acid conjugate on Huh-7 and HepG2 cells and diethyl nitrosamine induced hepatocellular cancer on Wistar albino rat. *Nanomedicine: Nanotechnology, Biology, and Medicine*, *14*(2), 415–428. https://doi.org/10.1016/j.nano.2017.11.003.

Fahrioğlu, U., Dodurga, Y., Elmas, L., & Seçme, M. (2016). Ferulic acid decreases cell viability and colony formation while inhibiting migration of MIA PaCa-2 human pancreatic cancer cells in vitro. *Gene*, *576*(1 Pt. 3), 476–482. https://doi.org/10.1016/j.gene.2015.10.061.

Fang, J., Pang, X., Yan, R., Lian, W., Li, C., Wang, Q., et al. (2016). Discovery of neuroprotective compounds by machine learning approaches. *RSC Advances*, *6*(12), 9857–9871. https://doi.org/10.1039/C5RA23035G.

Fong, Y., Tang, C.-C., Hu, H.-T., Fang, H.-Y., Chen, B.-H., Wu, C.-Y., et al. (2016). Inhibitory effect of trans-ferulic acid on proliferation and migration of human lung cancer cells accompanied with increased endogenous reactive oxygen species and β-catenin instability. *Chinese Medicine*, *11*, 45. https://doi.org/10.1186/s13020-016-0116-7.

Gan, Q., Wang, J., Hu, J., Lou, G., Xiong, H., Peng, C., et al. (2020). Modulation of apoptosis by plant polysaccharides for exerting anti-cancer effects: A review. *Frontiers in Pharmacology*, *11*, 792. https://doi.org/10.3389/fphar.2020.00792.

Gao, J., Yu, H., Guo, W., Kong, Y., Gu, l., Li, Q., et al. (2018). The anticancer effects of ferulic acid is associated with induction of cell cycle arrest and autophagy in cervical cancer cells. *Cancer Cell International*, *18*, 102. https://doi.org/10.1186/s12935-018-0595-y.

Gautam, S. C., Xu, Y. X., Dumaguin, M., Janakiraman, N., & Chapman, R. A. (2000). Resveratrol selectively inhibits leukemia cells: A prospective agent for ex vivo bone marrow purging. *Bone Marrow Transplantation*, *25*(6), 639–645. https://doi.org/10.1038/sj.bmt.1702189.

Ghatak, S. B., & Panchal, S. J. (2010). Ferulic acid—An insight into its current research and future prospects. *Trends in Food Science & Technology*. https://doi.org/10.1016/j.tifs.2010.11.004.

Graf, E. (1992). Antioxidant potential of ferulic acid. *Free Radical Biology and Medicine*, *13*(4), 435–448. https://doi.org/10.1016/0891-5849(92)90184-I.

Gupta, A., Kumar, R., Bhattacharya, P., Bishayee, A., & Pandey, A. K. (2020). *Terminalia bellirica* (Gaertn.) roxb. (Bahera) in health and disease: A systematic and comprehensive review. *Phytomedicine*, 77, 153278. https://doi.org/10.1016/j.phymed.2020.153278.

Gupta, A., Kumar, R., Ganguly, R., Singh, A. K., Rana, H. K., & Pandey, A. K. (2021). Antioxidant, anti-inflammatory and hepatoprotective activities of *Terminalia bellirica* and its bioactive component ellagic acid against diclofenac induced oxidative stress and hepatotoxicity. *Toxicology Reports*, *8*, 44–52. https://doi.org/10.1016/j.toxrep.2020.12.010.

Gupta, A., Kumar, R., & Pandey, A. K. (2020). Antioxidant and antidiabetic activities of *Terminalia bellirica* fruit in alloxan induced diabetic rats. *South African Journal of Botany*, *130*, 308–3015. https://doi.org/10.1016/j.sajb.2019.12.010.

Gupta, A., & Pandey, A. K. (2020a). Saffron and its active ingredients: A natural product with potent anticancer property. In K. Saravanan, C. Egbuna, H. I. Averal, S. Kannan, & S. Elavarasi (Eds.), *Drug development for cancer and diabetes* (pp. 199–215). Apple Academic Press.

Gupta, A., & Pandey, A. K. (2020b). Aceclofenac-induced hepatotoxicity: An ameliorative effect of *Terminalia bellirica* fruit and ellagic acid. *World Journal of Hepatology*, *12*(11), 949–964. https://doi.org/10.4254/wjh.v12.i11.949.

Gupta, A., Singh, A. K., Kumar, R., Ganguly, R., Rana, H. K., Pandey, P. K., et al. (2019). Corilagin in cancer: A critical evaluation of anticancer activities and molecular mechanisms. *Molecules (Basel, Switzerland)*, *24*(18), 3399. https://doi.org/10.3390/molecules24183399.

Haddad, J. J. (2004). Redox and oxidant-mediated regulation of apoptosis signaling pathways: Immuno-pharmaco-redox conception of oxidative siege versus cell death commitment. *International Immunopharmacology*, *4*(4), 475–493. https://doi.org/10.1016/j.intimp.2004.02.002.

Hanahan, D., & Weinberg, R. A. (2000). The hallmarks of cancer. *Cell*, *100*(1), 57–70. https://doi.org/10.1016/S0092-8674(00)81683-9.

Heravi Karimovi, M., Pourdehqan, M., Jadid Milani, M., Foroutan, S. K., & Aieen, F. (2006). Study of the effects of group counseling on quality of sexual life of patients with breast cancer under chemotherapy at imam Khomeini hospital. *Journal of Mazandaran University of Medical Sciences*, *16*(54), 43–51.

Herranz, D., & Serrano, M. (2010). Sirt1: Recent lessons from mouse models. *Nature Reviews. Cancer*, *10*(12), 819–823. https://doi.org/10.1038/nrc2962.

Herrmann, K. (1989). Occurrence and content of hydroxycinnamic and hydroxybenzoic acid compounds in foods. *Critical Reviews in Food Science and Nutrition*, *28*(4), 315–347. https://doi.org/10.1080/10408398909527504.

Hollman, P. C., & Katan, M. B. (1998). Bioavailability and health effects of dietary flavonols in man. *Archives of Toxicology. Supplement*, *20*, 237–248. https://doi.org/10.1007/978-3-642-46856-8_21.

Itagaki, S., Kobayashi, Y., Otsuka, Y., Kubo, S., Kobayashi, M., Hirano, T., et al. (2005). Food – drug interaction between ferulic acid and nateglinide involving the fluorescein/H+ cotransport system. *Journal of Agricultural and Food Chemistry*, *53*(7), 2499–2502. https://doi.org/10.1021/jf047990i.

Janicke, B., Hegardt, C., Krogh, M., Onning, G., Akesson, B., Cirenajwis, H. M., et al. (2011). The antiproliferative effect of dietary fiber phenolic compounds ferulic acid and p-coumaric acid on the cell cycle of Caco-2 cells. *Nutrition and Cancer*, *63*(4), 611–622. https://doi.org/10.1080/01635581.2011.538486.

Janicke, B., Onning, G., & Oredsson, S. M. (2005). Differential effects of ferulic acid and p-coumaric acid on S phase distribution and length of S phase in the human colonic cell line Caco-2. *Journal of Agricultural and Food Chemistry*, *53*(17), 6658–6665. https://doi.org/10.1021/jf050489l.

Jemal, A., Siegel, R., Ward, E., Hao, Y., Xu, J., & Thun, M. J. (2009). Cancer statistics, 2009. *CA: A Cancer Journal for Clinicians*, *59*(4), 225–249. https://doi.org/10.3322/caac.20006.

Jemal, A., Siegel, R., Xu, J., & Ward, E. (2010). Cancer statistics, 2010. *CA: a Cancer Journal for Clinicians*, *60*(5), 277–300. https://doi.org/10.3322/caac.20073.

Kang, J., Liu, Y., Xie, M.-X., Li, S., Jiang, M., & Wang, Y.-D. (2004). Interactions of human serum albumin with chlorogenic acid and ferulic acid. *Biochimica et Biophysica Acta*, *1674*(2), 205–214. https://doi.org/10.1016/j.bbagen.2004.06.021.

Kannan, K., & Jain, S. K. (2000). Oxidative stress and apoptosis. *Pathophysiology*, 7(3), 153–163. https://doi.org/10.1016/S0928-4680(00)00053-5.

Karthikeyan, S., Kanimozhi, G., Prasad, N. R., & Mahalakshmi, R. (2011). Radiosensitizing effect of ferulic acid on human cervical carcinoma cells in vitro. *Toxicology In Vitro*, *25*(7), 1366–1375. https://doi.org/10.1016/j.tiv.2011.05.007.

Katanyoo, K., Sanguanrungsirikul, S., & Manusirivithaya, S. (2012). Comparison of treatment outcomes between squamous cell carcinoma and adenocarcinoma in locally advanced cervical cancer. *Gynecologic Oncology*, *125*(2), 292–296. https://doi.org/10.1016/j.ygyno.2012.01.034.

Kawabata, K., Yamamoto, T., Hara, A., Shimizu, M., Yamada, Y., Matsunaga, K., et al. (2000). Modifying effects of ferulicacid on azoxymethane-induced colon carcinogenesis in F344 rats. *Cancer Letters*, *157*(1), 15–21. https://doi.org/10.1016/s0304-3835(00)00461-4.

Kawabata, K., Yoshioka, Y., & Terao, J. (2019). Role of intestinal microbiota in the bioavailability and physiological functions of dietary polyphenols. *Molecules*, *24*(2), 370. https://doi.org/10.3390/molecules24020370.

Kehrer, J. P. (2000). Cause-effect of oxidative stress and apoptosis. *Teratology*, *62*(4), 235–236. https://doi.org/10.1002/1096-9926(200010)62:4<235::AID-TERA11>3.0.CO;2-3.

Kern, S. M., Bennett, R. N., Needs, P. W., Mellon, F. A., Kroon, P. A., & Garcia-Conesa, M.-T. (2003). Characterization of metabolites of hydroxycinnamates in the in vitro model of human small intestinal epithelium caco-2 cells. *Journal of Agricultural and Food Chemistry*, *51*(27), 7884–7891. https://doi.org/10.1021/jf030470n.

Khan, N., Adhami, V. M., & Mukhtar, H. (2008). Apoptosis by dietary agents for prevention and treatment of cancer. *Biochemical Pharmacology*, *76*(11), 1333–1339. https://doi.org/10.1016/j.bcp.2008.07.015.

Khan, N., Afaq, F., & Mukhtar, H. (2007). Apoptosis by dietary factors: The suicide solution for delaying cancer growth. *Carcinogenesis*, *28*(2), 233–239. https://doi.org/10.1093/carcin/bgl243.

Kim, H. S., Kacew, S., & Lee, B. M. (2016). Genetic and epigenetic cancer chemoprevention on molecular targets during multistage carcinogenesis. *Archives of Toxicology*, *90*(10), 2389–2404. https://doi.org/10.1007/s00204-016-1813-9.

Kim, H.-J., Ryu, K., Kang, J.-H., Choi, A.-J., Kim, T.-I., & Oh, J.-M. (2013). Anticancer activity of ferulic acid-inorganic nanohybrids synthesized via two different hybridization routes, reconstruction and exfoliation-reassembly. *The Scientific World Journal*, *2013*(3). https://doi.org/10.1155/2013/421967.

Kobayashi, T., Aikata, H., Kobayashi, T., Ohdan, H., Arihiro, K., & Chayama, K. (2017). Patients with early recurrence of hepatocellular carcinoma have poor prognosis. *Hepatobiliary & Pancreatic Diseases International*, *16*(3), 279–288. https://doi.org/10.1016/S1499-3872(16)60181-9.

Kong, Q., Beel, J. A., & Lillehei, K. O. (2000). A threshold concept for cancer therapy. *Medical Hypotheses*, *55*(1), 29–35. https://doi.org/10.1054/mehy.1999.0982.

Konishi, Y., & Shimizu, M. (2003). Transepithelial transport of Ferulic acid by monocarboxylic acid transporter in Caco-2 cell monolayers. *Bioscience, Biotechnology, and Biochemistry*, *67*(4), 856–862. https://doi.org/10.1271/bbb.67.856.

Kroon, P. A., Faulds, C. B., Ryden, P., Robertson, J. A., & Williamson, G. (1997). Release of covalently bound ferulic acid from fiber in the human colon. *Journal of Agricultural and Food Chemistry*, *45*(3), 661–667. https://doi.org/10.1021/jf9604403.

Kumar, S., Kumar, R., Dwivedi, A., & Pandey, A. K. (2014). *In vitro* antioxidant, antibacterial, and cytotoxic activity and *in vivo* effect of *Syngonium podophyllum* and *Eichhornia crassipes* leaf extracts on isoniazid induced oxidative stress and hepatic markers. *BioMed Research International*, *2014*, 1–11. https://doi.org/10.1155/2014/459452.

Kumar, S., & Pandey, A. K. (2015). Free radicals: Health implications and their mitigation by herbals. *Journal of Advances in Medicine and Medical Research*, 7, 438–457. https://doi.org/10.9734/BJMMR/2015/16284.

Kumar, N., & Pruthi, V. (2014). Potential applications of ferulic acid from natural sources. *Biotechnology Reports*, *4*, 86–93. https://doi.org/10.1016/j.btre.2014.09.002.

Kumar, R., Singh, A. K., Gupta, A., Bishayee, A., & Pandey, A. K. (2019). Therapeutic potential of Aloe vera—A miracle gift of nature. *Phytomedicine*, *60*, 152996. https://doi.org/10.1016/j.phymed.2019.152996.

Kuno, T., Tsukamoto, T., Hara, A., & Tanaka, T. (2012). Cancer chemoprevention through the induction of apoptosis by natural compounds. *Journal of Biophysical Chemistry*, *3*(2), 720–726. https://doi.org/10.4236/jbpc.2012.32018.

Liyana-Pathirana, C. M., & Shahidi, F. (2006). Importance of insoluble bound phenolics to antioxidant properties of wheat. *Journal of Agricultural and Food Chemistry*, *54*, 1256–1264. https://doi.org/10.1021/jf052556h.

Lu, L., Chen, G., Qiu, Y., Li, M., Liu, D., Hu, D., et al. (2016). Nanoparticle-based oral delivery systems for colon targeting: Principles and design strategies. *Science Bulletin*, *61*(9), 670–681. https://doi.org/10.1007/s11434-016-1056-4.

Luo, L., Zhu, S., Tong, Y., & Peng, S. (2020). Ferulic acid induces apoptosis of HeLa and caski cervical carcinoma cells by down-regulating the phosphatidylinositol 3-kinase (PI3K)/Akt signaling pathway. *Medical Science Monitor : International Medical Journal of Experimental and Clinical Research*, *26*, e920095-1–e920095-7. https://doi.org/10.12659/MSM.920095.

Madhujith, T., & Shahidi, F. (2006). Optimization of the extraction of antioxidative constituents of six barley cultivars and their antioxidant properties. *Journal of Agricultural and Food Chemistry*, *54*(21), 8048–8057. https://doi.org/10.1021/jf061558e.

Mathew, S., & Abraham, T. E. (2004). Ferulic acid: An antioxidant found naturally in plant cell walls and feruloyl esterases involved in its release and their applications. *Critical Reviews in Biotechnology*, *24*(2–3), 59–83. https://doi.org/10.1080/07388550490491467.

Mattila, P., & Hellström, J. (2007). Phenolic acids in potatoes, vegetables, and some of their products. *Journal of Food Composition and Analysis*, *20*(3), 152–160. https://doi.org/10.1016/j.jfca.2006.05.007.

Mattila, P., Hellstrom, J., & Torronen, R. (2006). Phenolic acids in berries, fruits, and beverages. *Journal of Agricultural and Food Chemistry*, *54*(19), 7193–7199. https://doi.org/10.1021/jf0615247.

Mattila, P., Pihlava, J. M., & Hellstrom, J. (2005). Contents of phenolic acids, alkyl- and alkenylresorcinols, and avenanthramides in commercial grain products. *Journal of Agricultural and Food Chemistry*, *53*(21), 8290–8295. https://doi.org/10.1021/jf051437z.

Micard, V., Grabber, J. H., Ralph, J., Renard, C. M. G. C., & Thibault, J. F. (1997). Dehydrodiferulic acids from sugar-beet pulp. *Phytochemistry*, *44*(7), 1365–1368. https://doi.org/10.1016/S0031-9422(96)00699-1.

Misaghi, A., Goldin, A., Awad, M., & Kulidjian, A. A. (2018). Osteosarcoma: A comprehensive review. *SICOT-J*, *4*, 12. https://doi.org/10.1051/sicotj/2017028.

Mohammad, R. M., Muqbil, I., Lowe, L., Yedjou, C., Hsu, H.-Y., Lin, L.-T., et al. (2015). Broad targeting of resistance to apoptosis in cancer. *Seminars in Cancer Biology*, *35*(0), S78–S103. https://doi.org/10.1016/j.semcancer.2015.03.001.

Nardini, M., Cirillo, E., Natella, F., & Scaccini, C. (2002). Absorption of phenolic acids in humans after coffee consumption. *Journal of Agricultural and Food Chemistry*, *50*(20), 5735–5741. https://doi.org/10.1021/jf0257547.

Nasr Bouzaiene, N., Kilani Jaziri, S., Kovacic, H., Chekir-Ghedira, L., Ghedira, K., & Luis, J. (2015). The effects of caffeic, coumaric and ferulic acids on proliferation, superoxide production, adhesion and migration of human tumor cells in vitro. *European Journal of Pharmacology*, *766*, 99–105. https://doi.org/10.1016/j.ejphar.2015.09.044.

Nethaji, M., Pattabhi, V., & Desiraju, G. R. (1988). Structure of 3-(4-hydroxy-3-methoxyphenyl)-2-propenoic acid (ferulic acid). *Acta Crystallographica Section C, 44*(2), 275–277. https://doi.org/10.1107/S0108270187009211.

Nishizawa, C., Ohta, T., Egashira, Y., & Sanada, H. (1998). Ferulic acid contents in typical cereals (in Japanese with abstract in English). *Nippon Shokuhin Kagaku Kogaku Kaishi, 45*(8), 499–503. https://doi.org/10.3136/nskkk.45.499.

Nouri, Z., Fakhri, S., Nouri, K., Wallace, C. E., Farzaei, M. H., & Bishayee, A. (2020). Targeting multiple signaling pathways in cancer: The rutin therapeutic approach. *Cancers, 12*(8), 2276. https://doi.org/10.3390/cancers12082276.

Osborne, C. K., Shou, J., Massarweh, S., & Schiff, R. (2005). Crosstalk between estrogen receptor and growth factor receptor pathways as a cause for endocrine therapy resistance in breast cancer. *Clinical Cancer Research, 11*(2), 865s–870s.

Palozza, P. (2005). Can β-carotene regulate cell growth by a redox mechanism? An answer from cultured cells. *Biochimica et Biophysica Acta (BBA)—Molecular Basis of Disease, 1740*(2), 215–221. https://doi.org/10.1016/j.bbadis.2004.12.008.

Panwar, R., Sharma, A. K., Kaloti, M., Dutt, D., & Pruthi, V. (2016). Characterization and anticancer potential of ferulic acid-loaded chitosan nanoparticles against ME-180 human cervical cancer cell lines. *Applied Nanoscience, 6*(6), 803–813. https://doi.org/10.1007/s13204-015-0502-y.

Rapisarda, P., Carollo, G., Tomaselli, F., & Maccarone, E. (1998). Hydroxycinnamic acids as markers of Italian blood orange juices. *Journal of Agricultural and Food Chemistry, 46*, 464–470. https://doi.org/10.1021/jf9603700.

Rayavarapu, S., Yarla, N. S., Kadiri, S. K., Bishayee, A., Vidavalur, S., Tadikonda, R., et al. (2017). Synthesis of Saccharumoside-B analogue with potential of antiproliferative and pro-apoptotic activities. *Scientific Reports*, 7(1), 8309. https://doi.org/10.1038/s41598-017-05832-w.

Ren, Z., Ma, X., Duan, Z., & Chen, X. (2020). Diagnosis, therapy, and prognosis for hepatocellular carcinoma. *Analytical Cellular Pathology, 2020*, 1–2. https://doi.org/10.1155/2020/8157406.

Rezaei, A., Varshosaz, J., Fesharaki, M., Farhang, A., & Jafari, S. M. (2019). Improving the solubility and in vitro cytotoxicity (anticancer activity) of ferulic acid by loading it into cyclodextrin nanosponges. *International Journal of Nanomedicine, 14*, 4589–4599. https://doi.org/10.2147/IJN.S206350.

Robert, C., & Rassool, F. V. (2012). HDAC inhibitors: roles of DNA damage and repair. *Advances in Cancer Research, 116*, 87–129. https://doi.org/10.1016/B978-0-12-394387-3.00003-3.

Rondini, L., Peyrat-Maillard, M.-N., Marsset-Baglieri, A., & Berset, C. (2002). Sulfated ferulic acid is the main in vivo metabolite found after short-term ingestion of free ferulic acid in rats. *Journal of Agricultural and Food Chemistry, 50*(10), 3037–3041. https://doi.org/10.1021/jf011295i.

Rondini, L., Peyrat-Maillard, M.-N., Marsset-Baglieri, A., Fromentin, G., Durand, P., Tomé, D., et al. (2004). Bound ferulic acid from bran is more bioavailable than the free compound in rat. *Journal of Agricultural and Food Chemistry, 52*(13), 4338–4343. https://doi.org/10.1021/jf0348323.

Rosa, L.d. S., Jordão, N. A., Soares, N.d. C. P., de Mesquita, J. F., Monteiro, M., & Teodoro, A. J. (2018). Pharmacokinetic, antiproliferative and apoptotic effects of phenolic acids in human colon adenocarcinoma cells using in vitro and in silico approaches. *Molecules : A Journal of Synthetic Chemistry and Natural Product Chemistry, 23*(10), 2569. https://doi.org/10.3390/molecules23102569.

Roy, N., Narayanankutty, A., Nazeem, P. A., Valsalan, R., Babu, T. D., & Mathew, D. (2016). Plant phenolics ferulic acid and P-coumaric acid inhibit colorectal cancer cell proliferation through EGFR down-regulation. *Asian Pacific Journal of Cancer Prevention: APJCP, 17*(8), 4019–4023.

Saenglee, S., Jogloy, S., Patanothai, A., Leid, M., & Senawong, T. (2016). Cytotoxic effects of peanut phenolics possessing histone deacetylase inhibitory activity in breast and cervical cancer cell lines. *Pharmacological Reports: PR*, *68*(6), 1102–1110. https://doi.org/10.1016/j.pharep.2016.06.017.

Safarzadeh, E., Sandoghchian Shotorbani, S., & Baradaran, B. (2014). Herbal medicine as inducers of apoptosis in cancer treatment. *Advanced Pharmaceutical Bulletin*, *4*(Suppl. 1), 421–427. https://doi.org/10.5681/apb.2014.062.

Sakakibara, H., Honda, Y., Nakagawa, S., Ashida, H., & Kanazawa, K. (2003). Simultaneous determination of all polyphenols in vegetables, fruits, and teas. *Journal of Agricultural and Food Chemistry*, *51*(3), 571–581. https://doi.org/10.1021/jf020926l.

Sakamoto, K., Lawson, L. D., & Milner, J. A. (1997). Allyl sulfides from garlic suppress the in vitro proliferation of human A549 lung tumor cells. *Nutrition and Cancer*, *29*(2), 152–156. https://doi.org/10.1080/01635589709514617.

Saulnier, L., & Thibault, J.-F. (1999). Ferulic acid and diferulic acids as components of sugar-beet pectins and maize bran heteroxylans. *Journal of the Science of Food and Agriculture*, *79*(3), 396–402. https://doi.org/10.1002/(SICI)1097-0010(19990301)79:3<396::AID-JSFA262>3.0.CO;2-B.

Scheline, R. R. (1968). Metabolism of phenolic acids by the rat intestinal microflora. *Acta Pharmacologica et Toxicologica*, *26*(2), 189–205. https://doi.org/10.1111/j.1600-0773.1968.tb00437.x.

Serafim, T. L., Carvalho, F. S., Marques, M. P. M., Calheiros, R., Silva, T., Garrido, J., et al. (2011). Lipophilic caffeic and ferulic acid derivatives presenting cytotoxicity against human breast cancer cells. *Chemical Research in Toxicology*, *24*(5), 763–774. https://doi.org/10.1021/tx200126r.

Sharma, U. K., Sharma, A. K., Gupta, A., Kumar, R., Pandey, A., & Pandey, A. K. (2017). Pharmacological activities of cinnamaldehyde and eugenol: Antioxidant, cytotoxic and anti-leishmanial studies. *Cellular and Molecular Biology (Noisy-le-Grand, France)*, *63*(6), 73–78. https://doi.org/10.14715/cmb/2017.63.6.15.

Shibuya, N. (1984). Phenolic-acids and their carbohydrate esters in rice endosperm cell-walls. *Phytochemistry*, *23*(10), 2233–2237. https://doi.org/10.1016/S0031-9422(00)80526-9.

Siegel, R. L., Miller, K. D., & Jemal, A. (2016). Cancer statistics, 2016. *CA: a Cancer Journal for Clinicians*, *66*(1), 7–30. https://doi.org/10.3322/caac.21332.

Siegel, R. L., Miller, K. D., & Jemal, A. (2019). Cancer statistics, 2019. *CA: a Cancer Journal for Clinicians*, *69*(1), 7–34. https://doi.org/10.3322/caac.21551.

Singh, A. K., Kumar, R., & Pandey, A. K. (2018). Hepatocellular carcinoma: Causes, mechanism of progression and biomarkers. *Current Chemical Genomics and Translational Medicine*, *12*, 9–26. https://doi.org/10.2174/2213988501812010009.

Spencer, J. P. E., Chowrimootoo, G., Choudhury, R., Debnam, E. S., Srai, S. K., & Rice-Evans, C. (1999). The small intestine can both absorb and glucuronidate luminal flavonoids. *FEBS Letters*, *458*(2), 224–230. https://doi.org/10.1016/S0014-5793(99)01160-6.

Steiner, E., Klubert, D., & Knutson, D. (2008). Assessing breast Cancer risk in women. *American Family Physician*, *78*(12), 1361–1366.

Stoddard, F. R., II, Brooks, A. D., Eskin, B. A., & Johannes, G. J. (2008). Iodine alters gene expression in the MCF7 breast cancer cell line: Evidence for an anti-estrogen effect of iodine. *International Journal of Medical Sciences*, *5*(4), 189–196.

Teuchy, H., & Sumere, C. F. V. (1971). The metabolism of [1-14C] phenylalanine, [3-14C] cinnamic acid and [2-14C] ferulic acid in the rat. *Archives Internationales de Physiologie et de Biochimie*, *79*(3), 589–618. https://doi.org/10.3109/13813457109085341.

Thakkar, A., Chenreddy, S., Wang, J., & Prabhu, S. (2015). Ferulic acid combined with aspirin demonstrates chemopreventive potential towards pancreatic cancer when delivered using chitosan-coated solid-lipid nanoparticles. *Cell & Bioscience*, *5*, 46. https://doi.org/10.1186/s13578-015-0041-y.

The Japan Food chemical Research Faundation. (1996). Retrieved September 14, 2020, from https://www.ffcr.or.jp/en/tenka/list-of-existing-food-additives/list-of-existing-food-additives.html.

Thiery, J. P., Acloque, H., Huang, R. Y. J., & Nieto, M. A. (2009). Epithelial-mesenchymal transitions in development and disease. *Cell, 139*(5), 871–890. https://doi.org/10.1016/j.cell.2009.11.007.

Trewartha, D., & Carter, K. (2013). Advances in prostate cancer treatment. *Nature Reviews Drug Discovery, 12*(11), 823–824. https://doi.org/10.1038/nrd4068.

van der Logt, E. M. J., Roelofs, H. M. J., Nagengast, F. M., & Peters, W. H. M. (2003). Induction of rat hepatic and intestinal UDP-glucuronosyltransferases by naturally occurring dietary anticarcinogens. *Carcinogenesis, 24*(10), 1651–1656. https://doi.org/10.1093/carcin/bgg117.

Venturi, S. (2001). Is there a role for iodine in breast diseases? *Breast (Edinburgh, Scotland), 10*(5), 379–382. https://doi.org/10.1054/brst.2000.0267.

Vicente-Dueñas, C., Hauer, J., Cobaleda, C., Borkhardt, A., & Sánchez-García, I. (2018). Epigenetic priming in cancer initiation. *Trends in Cancer, 4*(6), 408–417. https://doi.org/10.1016/j.trecan.2018.04.007.

Wang, T., Gong, X., Jiang, R., Li, H., Du, W., & Kuang, G. (2016). Ferulic acid inhibits proliferation and promotes apoptosis via blockage of PI3K/Akt pathway in osteosarcoma cell. *American Journal of Translational Research, 8*(2), 968–980.

Wang, B.-H., & Ou-Yang, J.-P. (2005). Pharmacological actions of sodium ferulate in cardiovascular system. *Cardiovascular Drug Reviews, 23*(2), 161–172. https://doi.org/10.1111/j.1527-3466.2005.tb00163.x.

Watanabe, H., Yashiro, T., Tohjo, Y., & Konishi, Y. (2006). Non-involvement of the human monocarboxylic acid transporter 1 (MCT1) in the transport of phenolic acid. *Bioscience, Biotechnology, and Biochemistry, 70*(8), 1928–1933. https://doi.org/10.1271/bbb.60093.

Wolffram, S., Weber, T., Grenacher, B., & Scharrer, E. (1995). A Na(+)-dependent mechanism is involved in mucosal uptake of cinnamic acid across the jejunal brush border in rats. *The Journal of Nutrition, 125*(5), 1300–1308. https://doi.org/10.1093/jn/125.5.1300.

Xu, Y.-P., & Yang, M. (2016). Advancement in treatment and diagnosis of pancreatic cancer with radiopharmaceuticals. *World Journal of Gastrointestinal Oncology, 8*(2), 165–172. https://doi.org/10.4251/wjgo.v8.i2.165.

Yager, J. D., & Davidson, N. E. (2006). Estrogen carcinogenesis in breast Cancer. *The New England Journal of Medicine, 354*(3), 270–282. https://doi.org/10.1056/NEJMra050776.

Yarden, Y., & Sliwkowski, M. X. (2001). Untangling the ErbB signalling network. *Nature Reviews Molecular Cell Biology, 2*(2), 127–137. https://doi.org/10.1038/35052073.

Yu, L.-X., & Schwabe, R. F. (2017). The gut microbiome and liver cancer: Mechanisms and clinical translation. *Nature Reviews Gastroenterology & Hepatology, 14*(9), 527–539. https://doi.org/10.1038/nrgastro.2017.72.

Zhang, X. D., Deslandes, E., Villedieu, M., Poulain, L., Duval, M., Gauduchon, P., et al. (2006). Effect of 2-deoxy-D-glucose on various malignant cell lines in vitro. *Anticancer Research, 26*(5A), 3561–3566.

Zhang, X., Lin, D., Jiang, R., Li, H., Wan, J., & Li, H. (2016). Ferulic acid exerts antitumor activity and inhibits metastasis in breast cancer cells by regulating epithelial to mesenchymal transition. *Oncology Reports, 36*(1), 271–278. https://doi.org/10.3892/or.2016.4804.

Zhang, X.-d., Wu, Q., & Yang, S. (2017). Ferulic acid promoting apoptosis in human osteosarcoma cell lines. *Pakistan Journal of Medical Sciences, 33*(1), 127–131. https://doi.org/10.12669/pjms.331.12066.

Zhao, Z., Egashira, Y., & Sanada, H. (2003a). Ferulic acid sugar esters are recovered in rat plasma and urine mainly as the sulfoglucuronide of ferulic acid. *The Journal of Nutrition, 133*(5), 1355–1361. https://doi.org/10.1093/jn/133.5.1355.

Zhao, Z., Egashira, Y., & Sanada, H. (2003b). Digestion and absorption of ferulic acid sugar esters in rat gastrointestinal tract. *Journal of Agricultural and Food Chemistry*, *51*(18), 5534–5539. https://doi.org/10.1021/jf034455u.

Zhao, Z., Egashira, Y., & Sanada, H. (2004). Ferulic acid is quickly absorbed from rat stomach as the free form and then conjugated mainly in liver. *The Journal of Nutrition*, *134*(11), 3083–3088. https://doi.org/10.1093/jn/134.11.3083.

Zhao, Z., Egashira, Y., & Sanada, H. (2005). Phenolic antioxidants richly contained in corn bran are slightly bioavailable in rats. *Journal of Agricultural and Food Chemistry*, *53*(12), 5030–5035. https://doi.org/10.1021/jf050111n.

Zhao, Z., & Moghadasian, M. H. (2008). Chemistry, natural sources, dietary intake and pharmacokinetic properties of ferulic acid: A review. *Food Chemistry*, *109*(4), 691–702. https://doi.org/10.1016/j.foodchem.2008.02.039.

CHAPTER NINE

S1PR1 signaling in cancer: A current perspective

B. Anu†, N.N. Namitha†, and K.B. Harikumar*

Cancer Research Program, Rajiv Gandhi Centre for Biotechnology (RGCB), Thiruvananthapuram, KL, India
*Corresponding author: e-mail address: harikumar@rgcb.res.in

Contents

Abstract

Sphingosine-1-phosphate receptor 1 (S1PR1) is a G-protein coupled receptor for the bioactive lysosphingolipid sphingosine 1-phosphate (S1P). S1PR1 belongs to the sphingosine-1-phosphate receptor subfamily comprising five members (S1PR1–5). It has prominent roles in regulating endothelial cell cytoskeletal structure, cell migration, immunomodulation, vasculogenesis during embryogenesis, T cell egress and Multiple sclerosis. This review is addressing the role of S1PR1 in tumorigenesis and therapeutic opportunities to target S1PR1 in cancer.

† These authors contributed equally.

Advances in Protein Chemistry and Structural Biology, Volume 125
ISSN 1876-1623
https://doi.org/10.1016/bs.apcsb.2020.12.006

1. Introduction

Sphingolipids are essential components of the cellular membrane. It consists of a sphingoid base, fatty amino alcohol of typically 18 carbons called sphingosine (van Meer, Voelker, & Feigenson, 2008). Ceramidase is the enzyme that converts ceramide to sphingosine, which in turn gets phosphorylated by two isoenzyme specific kinases, namely Sphingosine kinase (SphK) 1& 2 to form a vital lipid mediator known as sphingosine 1-phosphate (S1P) (Spiegel & Milstien, 2011). The levels of S1P are tightly controlled in the cells where it can be dephosphorylated by phosphatases (SPPs) or degraded by S1P lyase. S1P can act either in a receptor-dependent manner through five G-protein coupled receptors termed as S1PR 1–5 or receptor-independent manner (Figs. 1 and 2) (Cartier & Hla, 2019; Hannun & Obeid, 2018; Maceyka & Spiegel, 2014; Mohammed & Harikumar, 2017; Nagahashi et al., 2014).

S1PR1 is the first identified and characterized receptor by Hla and Macaig through the differential display method (Hla & Maciag, 1990). Initially, the receptors were called endothelial differentiation gene-1 (EDG-1) and later renamed S1PR1. The other four receptors were characterized later. S1PR1 couples with Gαi/o initiate downstream signaling pathways, which involve GTPase Rac, cAMP, MAPK, etc. S1PR1 is essential for vascular morphogenesis and complete knockout of S1PR1 is embryonically lethal between D12.5 and 14.5 (Liu et al., 2000). Gaengel et al. showed that S1PR1 is a central component of a biological program that regulate angiogenesis in vessels during developmental stage (Gaengel et al., 2012). S1PR1 is also requiring for the transfer of immature B cells from the bone marrow to the blood (Allende et al., 2010). Studies using different genetic approaches indicated that S1PR1 is an important molecule which regulates T cell trafficking; because T cells from S1PR1-deficient mice failed to egress from the thymus and other lymphoid organs confirming that S1PR1 regulate T cell egress (Allende, Dreier, Mandala, & Proia, 2004; Matloubian et al., 2004). Several agonists and functional agonists of S1PR1 were reported by various investigators and a summary of known S1PR1 modulators are given in Table 1.

2. Structure of S1PR1

S1PR1 is a 382 amino acid protein consisting of three extracellular N-terminus loops termed as ECL1–ECL3, seven transmembrane α-helices spanning lipid bilayer (TM1–TM7) and three intracellular C terminus loops (ICL1–ICL3). The structure of S1PR1 was solved using a comparison study

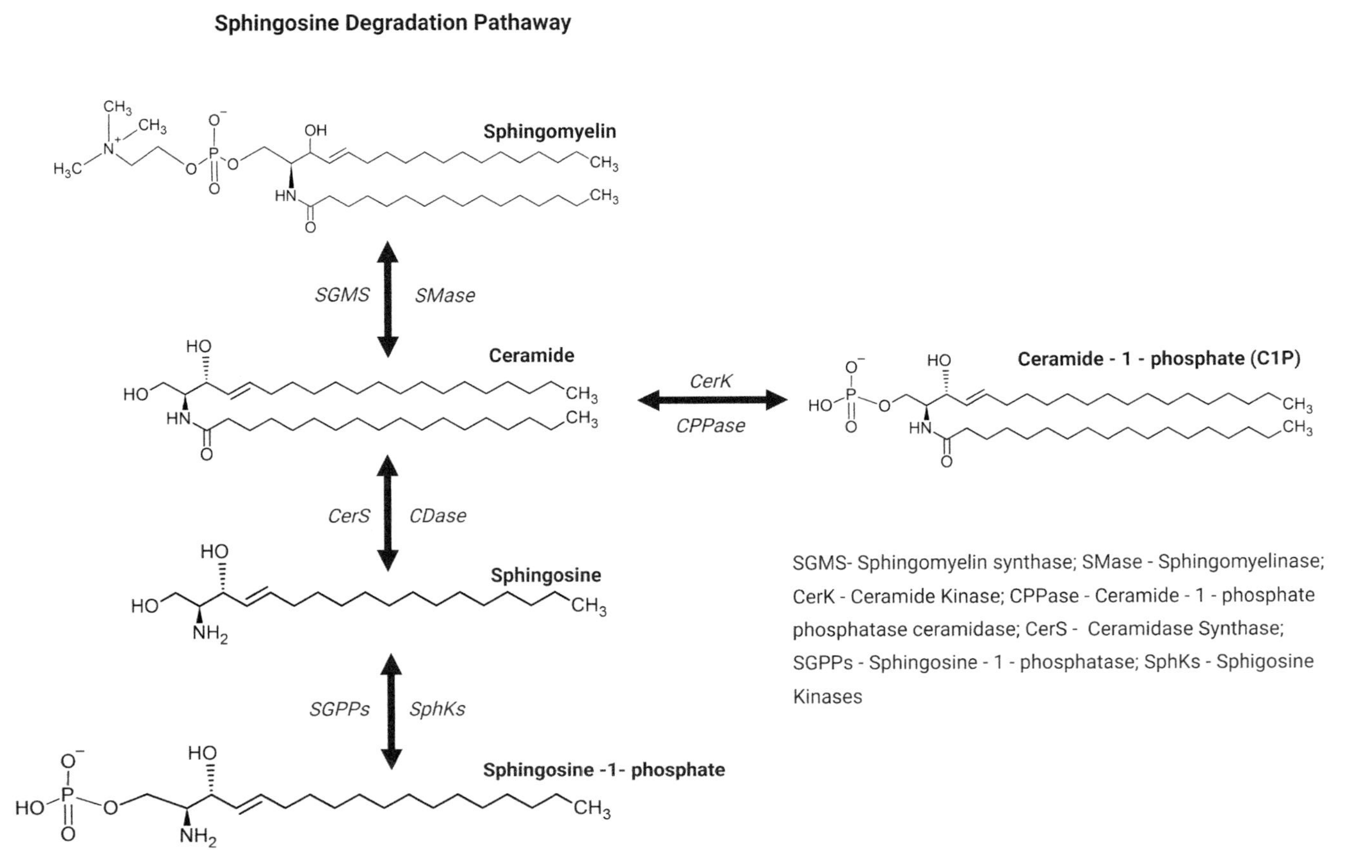

Fig. 1 The Sphingolipid pathway: It starts from Sphingomyelin and various steps and enzymes involved in each conversion step is described.

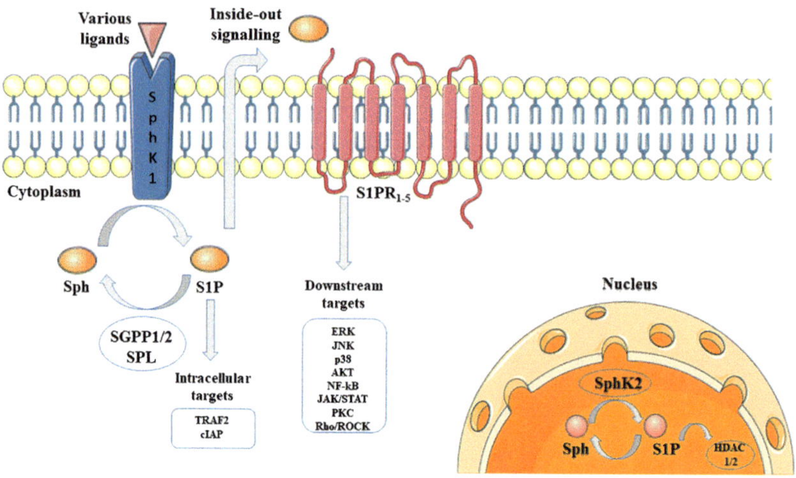

Fig. 2 The signaling pathways mediated by S1P: The S1P is produced by phosphorylation of sphingosine by SphK1 or SphK2. S1P can be dephosphorylated by SGPPs or irreversibly cleaved by S1P lyase. The S1P can act intracellularly or extracellularly.

Table 1 List of S1PR1 agonist and antagonists.

Sl.No.	Molecule	Agonist/Antagonist
1	Fingolimod (FTY720)	Agonist and functional antagonist to S1PR1 Agonist for S1PR3–5
2	Ponesimod (ACT-128800)	Agonist
3	Siponimod (BAF312)	Dual agonist to S1PR1 and S1PR5
4	Ozanimod (RPC1063)	Agonist
5	Ceralifimod (ONO-4641)	Agonist
6	Cenerimod (ACT-334441)	Agonist
7	Amiselimod (MT-1303)	Functional antagonist
8	CYM5442	Agonist
9	S1PR1-antibody	Antagonist
10	W146	Antagonist
11	W-061	Agonist
12	SEW2871	Agonist
13	VPC23019	Antagonist to S1PR1/S1PR3
14	VPC44116	Antagonist to S1PR1/S1PR3

Table 1 List of S1PR1 agonist and antagonists.—cont'd

Sl.No.	Molecule	Agonist/Antagonist
15	AUY954	Selective agonist to S1PR1
16	ONO-4641	Agonist
17	CS-2100	Agonist
18	IMMH001 (SYL930)	Agonist
19	SAR247799	Agonist
20	Sonepcizumab	Antagonist (Monoclonal antibody to S1PR1)
21	NIBR-0213	Selective Antagonist to S1PR1
22	KRP-203	Selective agonist to S1PR1
23	CS-0777	S1PR1 agonist and functional antagonist
24	AAL(R)	Agonist for S1P1/2/3/4/5
25	TASP0277308	Antagonist
26	RP-002	Agonist

with bovine rhodopsin (Okada et al., 2004; Schürer et al., 2008). The X-ray diffraction and micro diffraction studies were used in the structure elucidation. The ECL1 and ECL3 blocks the ligand binding with the receptor, while ECL2 and ECL3 maintained the interloop disulfide bonds. The binding of S1P is not a direct process, and a putative mechanism involving the entry between TM7 and TM1 followed by lateral diffusion into the binding pocket (Hanson et al., 2012; O'Sullivan & Dev, 2013; Parrill, Lima, & Spiegel, 2012). The structural study implicated that S1PR1 is a highly conserved protein.

3. S1PR1 in cancer

3.1 S1PR1 and breast cancer

It has been shown that the level of SphK1 is elevated in breast cancer as compared to normal breast tissue. A microarray analysis using 1269 breast tumor samples reported a worse outcome for patients with high SphK1 expression. Previously it was shown that estradiol (E2) induces the export of S1P from breast cancer cells through ABCC1 and ABCG2 drug transporters. Liu et al. demonstrated that loss of S1PR1 leads to the generation of vasculogenic mimicry (VM). This is a tumor-specific blood supply system which occurs independently of angiogenesis or endothelial cells. Analysis of S1PR1 in human breast cancer tissue samples revealed that expression is correlated with clinical stage and lymphatic metastasis and VM. The study did not observe any difference in S1PR1 expression between hormone receptor-positive and negative samples. The KM (Kaplan–Meier) survival analysis indicated that in hormone-positive breast cancer cases, the expression of S1PR1 was correlated with poor patient survival (Liu et al., 2019). Yet another study demonstrated that a significant positive correlation of S1PR1 expression was observed with levels of infiltrating both CD4+ and $CD8^+$ T cells, macrophages neutrophils, and dendritic cells, indicating the role of this receptor in immune cell infiltration (Zhong et al., 2020). IL-22 is an effector cytokine that plays a crucial role in several diseases, including cancer. Gene expression analysis of breast cancer patients samples with bone or brain metastasis showed a positive correlation between S1PR1 and IL-22R1. Both receptors' relative expression was less in breast cancer cell lines with low metastatic potential (MCF-7, T47D, (Zhong et al., 2020) MDA-MB453, and SKBR3). Relatively higher expression was observed in breast cancer cell lines with higher metastatic potential (HCC1954, MDA-MB231, and MDA-MB157). Further, IL-22 increased S1P levels. This study clearly showed IL-22, together with S1PR1, involved in breast cancer's bone or brain metastasis (Kim et al., 2020). Using an in vitro system Xiao and Yang showed that the S1PR1 antibody potentiated the effect of carboplatin in triple-negative breast cancer cell lines; however, the authors did not provide a conclusive mechanism for the observed effect (Xiao & Yang, 2019). Silencing both S1PR1 and S1PR3 decreased the proliferation of MDA-MB231 cells but not in MCF-7 cells, and authors argue that observed effects might be due to the presence or absence of estrogen receptor (Calis, Cosan, & Mutlu, 2017). Brune et al. reported that specific silencing of S1PR1 in macrophages inhibited the lung metastasis of breast cancer cells. However, the silencing of this receptor did not influence the tumor immune cell infiltration and the formation of the premetastatic niche but had a profound influence on lymphangiogenesis. They observed that the effects of S1PR1 were mediated through global gene expression through NLRP3 and IL-1b. S1PR1 regulated the production of IL-1 in the macrophages and promoted the process of lymphangiogenesis in vivo. The study provides strong evidence to target S1PR1/NLRP3/IL-1β axis to inhibit breast cancer metastasis (Weichand et al., 2017).

Most of the chemotherapeutic drugs used in TNBC treatment and management are known to affect peripheral sensory nerves and induce neurotoxic side effects, especially painful peripheral neuropathy (CIPN or Chemotherapy-induced peripheral neuropathy). But it is not well understood how each drug causes CIPN. The signals received by sensory neurons then relay these signals specifically to the dorsal horn of the spinal cord (DHSC). It was observed that the development of CIPN is associated with of S1PR1 activation. Administration of the specific antagonists or functional antagonists of S1PR1 inhibited the activation of neuroinflammatory processes neuropathic pain without compromising chemotherapeutic drugs' anticancer properties (paclitaxel and oxaliplatin). The S1PR1 modulators are already in use to treat multiple sclerosis, and this study further confirmed that the spinal cord is the site of action of S1PR1 modulators. The study also demonstrated that inhibiting S1PR1 signaling subsequently block the downstream NF-κB and MAPKs pathways, thereby changing the spinal cord cytokine environment to more anti-inflammatory in nature (Janes et al., 2014). Cancer-induced bone pain, CIBP) is reported by almost 30%–50% of all cancer patients. But therapeutic strategies for the management of CIBP are multimodal, mostly targeted at relieving pain and quality of life. Using a murine breast cancer model of CIBP, it was observed that the level of both S1P and dihydro S1P was elevated in the spinal cord while ceramides levels were decreased and no significant changes in the sphingomyelin. Intrathecal injection of TASP0277308 or FTY720 inhibited the occurrence of CIPB as observed by flinching and guarding behavior. The cytokine measurement indicated that inhibiting the S1PR1 signaling leads to the production of IL-10, which plays a significant role in relieving the CIBP. The study showed that S1PR1 has a significant role in CIBP (Grenald et al., 2017). In some breast cancer patients, chemotherapy induces a state known as chemo-brain, which is characterized by cognitive impairment and memory dysfunction (Hermelink, 2015). p-FTY720 was shown to inhibit HDAC and improve memory and learning in mice, and it also raises the possibility that S1PR1 modulators can be used in the management of chemo-brain (Hait et al., 2014).

FTY720 potentiated the anti-proliferative effects of radiation in vitro in TNBC cell lines by modulating PI3K/AKT and MAPK signaling (Marvaso et al., 2014). SphK2 is the enzyme responsible for the phosphorylation of FTY in vivo and enhances histone acetylations, and regulates expression of a subset of gene independently of S1PRs. Using spontaneous breast cancer transgenic mouse model (MMTV-PyMT), Hait et al. showed that FTY inhibited HFD driven tumor formation and increased the expression of both ER and PR. Further same study also provides convincing evidence to prove that FTY reactivates ERα expression and sensitizes TNBC cells to TAM therapy (Hait et al., 2015). Obesity is also a major risk factor for breast cancer and associated metastasis. S1P levels were higher in obese breast cancer patients as well as obese mice bearing mammary tumors. Targeting the S1P/S1PR1 axis with FTY720 in syngeneic mouse models inhibited obesity driven tumorigenesis and pulmonary metastasis and improved tumor-bearing mice (Nagahashi et al., 2018). Rupp et al. also showed that FTY720 inhibited the growth of TNBC tumor growth in both immunocompetent and immunodeficient mouse models, and synergism was observed with the combination of Cisplatin and FTY720. Interestingly in a more clinically relevant setting, FTY was found to decrease tumor recurrence after the surgical removal of the primary tumor (Rupp et al., 2021). Martin et al. also reported that a combination of gefitinib (an EGFR inhibitor)-FTY720 was effective in treating basal-like breast cancers and improved the survival of tumor-bearing mice (Martin et al., 2017). There were few attempts to make nanoparticle-based delivery of FTY either alone or in combination with other chemotherapeutic drugs. One such study using breast cancer as a model system demonstrated the efficacy of a combination of docetaxel and FTY and found that the combination enhanced anticancer effects and reduced systemic toxicity in pre-clinical mouse models (Alshaker et al., 2017).

3.2 S1PR1 and colorectal cancer

We have shown that the S1P/S1PR1 axis is involved in amplifying NF-κB/IL-6/Stat3 loop in inflammation-driven colon cancer (Liang et al., 2013). FTY720 treatment decreased the severity of colitis and tumor incidence in colitis-associated colon cancer model in mice. More importantly, FTY720 was given after the initiation of tumor formation, and still, this compound was able to reduce the progression of the tumors. Others had also shown that FTY720 was effective in preventing the severity associated with colitis. FTY720 also modulates the immune responses in colitis. In a TNBS colitis model, fingolimod increased the expression of CD25 and FoxP3 (Daniel et al., 2007). Using the co-culture experiment, Liu et al. also confirmed this observation (Liu, Jiang, et al., 2012).

3.3 S1PR1 and brain cancer

The role of S1PR1 in brain cancer is slightly controversial, owing to opposite results reported by different investigators. In one of the initial studies,

S1PRs (1–3 and 5) were overexpressed in U-118 MG cells, and S1PR1 overexpression had a profound effect on cell proliferation. S1P enhances glioma cell proliferation through an ERK-dependent pathway. S1PR1 overexpression increased the S1P dependent ERK activation and cell proliferation (Young & Van Brocklyn, 2007). Using human glioma samples, Yoshida et al. showed that S1PR1 expression is decreased in glioma cases both at mRNA and protein levels.

Further, it was shown that there was a significant correlation between S1PR1 expression and PFS and OS in glioma cases. Patients with higher and lower levels of S1PR1 had a PFS rate of 42.8% and 13.3% while an OS rate of 77.7% and 36.8%, respectively, at 1 year. Downregulation of S1PR1 significantly increased cell proliferation, while overexpression decreased the effect. U87 cells overexpressing S1PR1 were injected intracerebrally to mice, and tumor growth was delayed in these mice compared to vector controls. The authors are proposing S1PR1 as a possible prognostic marker in glioma (Yoshida et al., 2010). S1P/S1PR1 axis also regulates the expression of urokinase plasminogen activator (uPA) and its receptor uPAR; both play a vital role in the process of invasion (Ref). uPAR levels were elevated in glioma patients (Ref). Both S1PR1 and 2 signaling increased expression of a secreted matricellular protein, namely CCN1, in glioma. Young et al. showed that S1PR1 regulates both uPA and CCN1 pathways in glioma cell invasion (Young, Pearl, & Van Brocklyn, 2009). Rauch et al. reported an almost fourfold increase in S1PR1 expression in GBM as compared to control brains. Kaplan-Meier analyses in GBM cases showed a correlation between the expression of both S1PR1 and 2 to patients' survival. Patients with higher S1PR1 expression survived the most (Bien-Möller et al., 2016). FTY720 inhibited the proliferation and induced apoptosis in brain tumor stem cells (Estrada-Bernal, Palanichamy, Ray Chaudhury, & Van Brocklyn, 2012). FTY720 also inhibited an intracranial xenograft in SCID mice induced by brain tumor stem cells and improved tumor-bearing mice's survival, and a combination of FTY and TMZ had synergetic action in prolonging the survival of mice (Estrada-Bernal et al., 2012; Stessin et al., 2017). Another study showed that the sensitization of TMZ by FTY is dependent on the NRF2-ARE pathway (Zhang & Wang, 2017). Garner et al. using medulloblastoma patient-derived xenografts (PDXs), reported that FTY decreased the tumor formation. In in vitro system, FTY also inhibited migration and invasion of PDX derived cells and arrested the cell cycle in the G1 phase (Garner et al., 2018). Though the authors did not provide a detailed mechanism for observed beneficial effects of FTY, this study is more relevant in the clinical settings. Studies using human glioma cell lines showed that FTY inhibited the migration and invasion of the cells by downregulating the expression of MMP-2 and 9 while upregulating the expression of TIMP-1 (Zhang, Wang, Zhu, Ding, & Xu, 2014). Perla et al. showed that FTY also increased the histone H3 acetylation in medulloblastoma cell lines and proposed that this could be one of the mechanisms for FTY action (Perla et al., 2020). Dimethyl fumarate (DMF) is currently under clinical trial for newly diagnosed cases of GBM. DMF is rapidly metabolized in the body to monomethyl-fumarate (MMF), and this molecule is used in the treatment of multiple sclerosis. A combination of MMF and FTY was found to have better efficacy to induce apoptosis in GBM PDX lines and sensitize temozolomide and ionizing radiation. The suggested molecular mechanism involved the degradation of Ras proteins and inhibition of the YAP/TAZ pathway (Dent, Booth, Roberts, Poklepovic, & Hancock, 2020). FTY also inhibited the growth of tumors in a glioma allograft model in rats. This S1PR1 functional antagonist could block the recruitment of glioma-associated macrophages through the internalization of CXCR4 and a decrease in the production of IL-6. This is the first study depicting the role of FTY in modulating glioma tumor microenvironment, which affects the therapeutic outcome (Guo et al., 2020). Similarly, other groups also confirmed the observation that FTY suppresses the growth of glioma cells (Guo et al., 2020; Kolodziej et al., 2020). A much broader understanding is required of S1P signaling in GBM pathogenesis.

3.4 S1PR1 and leukemia

Immunohistochemical analysis showed that S1PR1 expression is upregulated in the surface of mantle lymphoma cells. Further mRNA analysis also supported the findings (Nishimura et al., 2010). Chronic lymphocytic leukemia (CLL) is mostly affected the B cell populations. The BCR signaling represses the S1PR1 expression, which leads to the retention of CLL cells in lymph nodes. Three different inhibitors of BCR were used, and only idelalisib (which targets PI3Kδ) increased expression of S1PR1 and stimulated S1P driven migration. Reversing the BCR mediated repression of S1PR1 resulted in the enhanced egress of tumor cells from affected lymph nodes (Till, Pettitt, & Slupsky, 2015). CLL also express abnormally higher surface expression of CXCR4 and CCR7. A similar trend was observed in CLL lymph node biopsies, and S1PR1 was negative. Ibrutinib, a Btk inhibitor, could suppress both BCR and chemokine receptor signaling in leukemia. Cells from patients were analyzed before the start of Ibrutinib treatment and subsequently, after 4 months and found that S1PR1 expression was significantly elevated with a modest decrease in CCR7. However, no change in CXCR4 expression was observed. Increasing the S1PR1 expression has a favorable clinical outcome (Patrussi et al., 2015).

Ribosomal proteins have been shown to involve in the development of diseases. Rlp22 is a component of the 60S ribosome, and loss of Rlp22 was observed in 10%–15% cases of T-cell acute lymphoblastic leukemia/lymphoma (T-ALL). Interestingly the loss is correlated with poor patient survival. Rpl22 deficiency resulted in the formation of thymic lymphoma in MyrAkt2 transgenic mice. Similar results were observed in Pten$^{-/-}$ *Rpl22*$^{-/-}$ mice, which showed an increased incidence of thymic lymphomas. Rpl22 deficient lymphomas had a decreased S1PR1 expression. When S1PR1 expression was restored in Rpl22-deficient thymic lymphoma cells, they regained their ability to migrate toward the S1P gradient. KLF2 is a known transcription factor that regulates S1PR1 expression. In RPl22 deficient cells, there was a marked decrease in the expression of KLF2 both at mRNA and protein levels. Ectopic expression of RPl22 restored both Klf2 and S1PR1 expression (Rao et al., 2016). In human acute myeloid leukemia cells silencing S1PR1 expression resulted in decreased cell proliferation, increased apoptosis, and activation of JNK signaling (Xu et al., 2016). Hodkin/Reed-Sternberg cells (HRS) are often considered as the hallmark of Hodgkin lymphoma. S1PR1 is overexpressed in HRS cells and regulates their migratory capacity (Kluk et al., 2013). S1P signaling through S1PR1 increased thePI3K pathway in HRS cells. Further gene expression analysis identified a feed-forward loop comprising S1PR1 and BATF3, which is involved in aberrant transcription of genes. This study implicates inhibiting S1P signaling could also affect the oncogenic BATF3 signaling in Hodgkin lymphoma (Vrzalikova et al., 2018). Koresawa reported that S1PR1 could be used as a prognostic marker in diffuse large B-cell lymphoma (DLBCL). More importantly, STAT3 levels were also higher in these samples suggesting the existence of the S1PR1/STAT3 loop (Koresawa et al., 2016). Siponimod inhibited the tumor volume and angiogenesis in a mouse model of DLBCL, together with a smaller number of CD31+ cells in tumors. These studies pointed out the possible use of S1PR1 in the management of leukemia.

3.5 S1PR1 and pancreatic cancer

Pancreatic cancer is one of the deadliest forms of cancer, with an average 5% survival rate of less than 6%. We reported that treatment of murine and human pancreatic cancer cell lines with FT720 resulted in decreased cell proliferation, increased apoptosis, and cell cycle arrest. Importantly FTY targets S1PR1/STAT3 loop in pancreatic cancer both in vitro and in vivo and improved the efficacy of gemcitabine in pre-clinical models. The combination effectively decreased Sonic-Hedgehog signaling, epithelial-mesenchymal transition, and desmoplasia, which led to the abrogation of the critical pathways responsible for drug resistance and metastasis. During the study, we also noticed that the activity and function of the tumor suppressor PP2A were restored by FTY720. Our results clearly indicate the therapeutic benefits of FTY720 in the management of pancreatic cancer (Lankadasari et al., 2018).

3.6 S1PR1 and ovarian cancer

S1PR1 expression was elevated in ovarian cancer tissue samples compared to adjacent normal and cell lines. microRNA 148b is shown as a negative regulator of S1PR1 (Wen et al., 2015). p53 null SKOV3 cells were overexpressed with mutant p53 (R248), and microarray was performed to understand the differential gene expression. S1PR1, EDN2, THBS1, and HB-EGF were the top four genes that affect the invasion and migration properties. The knockdown of p53 abrogates the expression above four gene panel, confirming the microarray data. Further, it was identified that rad51 is one of the transcription factors which can regulate the expression of these genes. The silencing of Rad21 decreased the expression of S1PR1 and THBS1. However, a detailed mechanism of p53 regulation of S1PR1 in ovarian cancer is yet to elucidate (Ahn, Kim, Lee, & Choi, 2017). Knockdown of S1PR1 significantly decreased S1P stimulated secretion of VEGF, IL-8, and IL-6. Both S1PR1 and 3 are responsible for the angiogenic potential of ovarian cancer cells. Using a mouse model of intraperitoneal ovarian cancer, it was shown that treatment with VPC23019 decreased tumor number and volume. Immunohistochemical analysis showed that the VPC compound also reduced the levels of VEGF, IL-8, and IL-6 in tissue samples. This study showed the importance of the SphK1/S1P/S1PR1/3 axis in ovarian cancer (Ahn et al., 2017).

4. Conclusion

Available evidence strongly suggests that S1PR1 plays a pivotal role in tumor development, affecting different tumorigenesis stages and modulating the significant pathways involved. At present, FTY720, a functional antagonist of S1PR1, is approved for clinical use. However, numerous studies, including studies from our laboratory, reported the pre-clinical efficacy of FTY720, but still, clinical trials using FTY720 in cancer are yet to start except for a safety study in newly diagnosed high-grade glioma (NCT02490930). Several other molecules are also been developed/characterized which act as either agonist or antagonist of S1PR1 are in different stages preclinical and clinical development. We propose that S1PR1 modulators will be a better chemosensitizer in cancer therapy.

Acknowledgments

The work in KBH laboratory was supported by a research grant from Department of Biotechnology (No.BT/PR6965/BRB/10/1145/2012) and DBT-Ramalingaswami fellowship (No.BT/RLF/*Re*-entry/38/2011).

References

Ahn, J.-H., Kim, T. J., Lee, J. H., & Choi, J.-H. (2017). Mutant p53 stimulates cell invasion through an interaction with Rad21 in human ovarian cancer cells. *Scientific Reports*, *7*, 9076.

Allende, M. L., Dreier, J. L., Mandala, S., & Proia, R. L. (2004). Expression of the sphingosine 1-phosphate receptor, S1P1, on T-cells controls thymic emigration. *The Journal of Biological Chemistry*, *279*, 15396–15401.

Allende, M. L., Tuymetova, G., Lee, B. G., Bonifacino, E., Wu, Y.-P., & Proia, R. L. (2010). S1P1 receptor directs the release of immature B cells from bone marrow into blood. *The Journal of Experimental Medicine*, *207*, 1113–1124.

Alshaker, H., Wang, Q., Srivats, S., Chao, Y., Cooper, C., & Pchejetski, D. (2017). New FTY720-docetaxel nanoparticle therapy overcomes FTY720-induced lymphopenia and inhibits metastatic breast tumour growth. *Breast Cancer Research and Treatment*, *165*, 531–543.

Bien-Möller, S., Lange, S., Holm, T., Böhm, A., Paland, H., Küpper, J., et al. (2016). Expression of S1P metabolizing enzymes and receptors correlate with survival time and regulate cell migration in glioblastoma multiforme. *Oncotarget*, *7*, 13031–13046.

Calis, I. U., Cosan, D. T., & Mutlu, F. (2017). Effects of S1P1 and S1P3 in ER+ and ER − breast cancer cells. *Anticancer Research*, *37*, 5469–5475.

Cartier, A., & Hla, T. (2019). Sphingosine 1-phosphate: Lipid signaling in pathology and therapy. *Science (New York, N.Y.)*, *366*(6463), 1–12. eaar5551.

Daniel, C., Sartory, N., Zahn, N., Geisslinger, G., Radeke, H. H., & Stein, J. M. (2007). FTY720 ameliorates Th1-mediated colitis in mice by directly affecting the functional activity of CD4 ⁺ CD25 ⁺ regulatory T cells. *The Journal of Immunology*, *178*, 2458–2468.

Dent, P., Booth, L., Roberts, J. L., Poklepovic, A., & Hancock, J. F. (2020). Fingolimod augments monomethylfumarate killing of GBM cells. *Frontiers in Oncology*, *10*, 22.

Estrada-Bernal, A., Palanichamy, K., Ray Chaudhury, A., & Van Brocklyn, J. R. (2012). Induction of brain tumor stem cell apoptosis by FTY720: A potential therapeutic agent for glioblastoma. *Neuro-Oncology*, *14*, 405–415.

Gaengel, K., Niaudet, C., Hagikura, K., Laviña, B., Siemsen, B. L., Muhl, L., et al. (2012). The sphingosine-1-phosphate receptor S1PR1 restricts sprouting angiogenesis by regulating the interplay between VE-cadherin and VEGFR2. *Developmental Cell*, *23*, 587–599.

Garner, E. F., Williams, A. P., Stafman, L. L., Aye, J. M., Mroczek-Musulman, E., Moore, B. P., et al. (2018). FTY720 decreases tumorigenesis in group 3 medulloblastoma patient-derived xenografts. *Scientific Reports*, *8*, 6913.

Grenald, S. A., Doyle, T. M., Zhang, H., Slosky, L. M., Chen, Z., Largent-Milnes, T. M., et al. (2017). Targeting the S1P/S1PR1 axis mitigates cancer-induced bone pain and neuroinflammation. *Pain*, *158*, 1733–1742.

Guo, X.-D., Ji, J., Xue, T.-F., Sun, Y.-Q., Guo, R.-B., Cheng, H., et al. (2020). FTY720 exerts anti-glioma effects by regulating the glioma microenvironment through increased CXCR4 internalization by glioma-associated microglia. *Frontiers in Immunology*, *11*, 178.

Hait, N. C., Avni, D., Yamada, A., Nagahashi, M., Aoyagi, T., Aoki, H., et al. (2015). The phosphorylated prodrug FTY720 is a histone deacetylase inhibitor that reactivates ERα expression and enhances hormonal therapy for breast cancer. *Oncogenesis*, *4*, e156.

Hait, N. C., Wise, L. E., Allegood, J. C., O'Brien, M., Avni, D., Reeves, T. M., et al. (2014). Active, phosphorylated fingolimod inhibits histone deacetylases and facilitates fear extinction memory. *Nature Neuroscience*, *17*, 971–980.

Hannun, Y. A., & Obeid, L. M. (2018). Sphingolipids and their metabolism in physiology and disease. *Nature Reviews Molecular Cell Biology*, *19*, 175–191.

Hanson, M. A., Roth, C. B., Jo, E., Griffith, M. T., Scott, F. L., Reinhart, G., et al. (2012). Crystal structure of a lipid G protein-coupled receptor. *Science (New York, N.Y.)*, *335*, 851–855.

Hermelink, K. (2015). Chemotherapy and cognitive function in breast cancer patients: The so-called chemo brain. *JNCI Monographs*, *2015*, 67–69.

Hla, T., & Maciag, T. (1990). An abundant transcript induced in differentiating human endothelial cells encodes a polypeptide with structural similarities to G-protein-coupled receptors. *The Journal of Biological Chemistry*, *265*, 9308–9313.

Janes, K., Little, J. W., Li, C., Bryant, L., Chen, C., Chen, Z., et al. (2014). The development and maintenance of paclitaxel-induced neuropathic pain require activation of the sphingosine 1-phosphate receptor subtype 1. *Journal of Biological Chemistry*, *289*, 21082–21097.

Kim, E.-Y., Choi, B., Kim, J.-E., Park, S.-O., Kim, S.-M., & Chang, E.-J. (2020). Interleukin-22 mediates the chemotactic migration of breast cancer cells and macrophage infiltration of the bone microenvironment by potentiating S1P/SIPR signaling. *Cell*, *9*, 131.

Kluk, M. J., Ryan, K. P., Wang, B., Zhang, G., Rodig, S. J., & Sanchez, T. (2013). Sphingosine-1-phosphate receptor 1 in classical Hodgkin lymphoma: Assessment of expression and role in cell migration. *Laboratory Investigation*, *93*, 462–471.

Kolodziej, M., Al Barim, B., Nagl, J., Weigand, M., Uhl, E., Uhle, F., et al. (2020). Sphingosine-1-phosphate analogue FTY720 exhibits apotent anti-proliferative effect on glioblastoma cells. *International Journal of Oncology*, *57*, 1039–1046.

Koresawa, R., Yamazaki, K., Oka, D., Fujiwara, H., Nishimura, H., Akiyama, T., et al. (2016). Sphingosine-1-phosphate receptor 1 as a prognostic biomarker and therapeutic target for patients with primary testicular diffuse large B-cell lymphoma. *British Journal of Haematology*, *174*, 264–274.

Lankadasari, M. B., Aparna, J. S., Mohammed, S., James, S., Aoki, K., Binu, V. S., et al. (2018). Targeting S1PR1/STAT3 loop abrogates desmoplasia and chemosensitizes pancreatic cancer to gemcitabine. *Theranostics*, *8*, 3824–3840.

Liang, J., Nagahashi, M., Kim, E. Y., Harikumar, K. B., Yamada, A., Huang, W.-C., et al. (2013). Sphingosine-1-phosphate links persistent STAT3 activation, chronic intestinal inflammation, and development of colitis-associated cancer. *Cancer Cell*, *23*, 107–120.

Liu, Y., Jiang, J., et al. (2012). The sphingosine-1-phosphate receptor agonist FTY720 and its phosphorylated form affect the function of CD4 + CD25 + T cells in vitro. *International Journal of Molecular Medicine*, *30*, 211–219.

Liu, S., Ni, C., Zhang, D., Sun, H., Dong, X., Che, N., et al. (2019). S1PR1 regulates the switch of two angiogenic modes by VE-cadherin phosphorylation in breast cancer. *Cell Death & Disease*, *10*, 200.

Liu, Y., Wada, R., Yamashita, T., Mi, Y., Deng, C. X., Hobson, J. P., et al. (2000). Edg-1, the G protein-coupled receptor for sphingosine-1-phosphate, is essential for vascular maturation. *The Journal of Clinical Investigation*, *106*, 951–961.

Maceyka, M., & Spiegel, S. (2014). Sphingolipid metabolites in inflammatory disease. *Nature*, *510*, 58–67.

Martin, J. L., Julovi, S. M., Lin, M. Z., de Silva, H. C., Boyle, F. M., & Baxter, R. C. (2017). Inhibition of basal-like breast cancer growth by FTY720 in combination with epidermal growth factor receptor kinase blockade. *Breast Cancer Research*, *19*, 90.

Marvaso, G., Barone, A., Amodio, N., Raimondi, L., Agosti, V., Altomare, E., et al. (2014). Sphingosine analog fingolimod (FTY720) increases radiation sensitivity of human breast cancer cells in vitro. *Cancer Biology & Therapy*, *15*, 797–805.

Matloubian, M., Lo, C. G., Cinamon, G., Lesneski, M. J., Xu, Y., Brinkmann, V., et al. (2004). Lymphocyte egress from thymus and peripheral lymphoid organs is dependent on S1P receptor 1. *Nature*, *427*, 355–360.

Mohammed, S., & Harikumar, K. B. (2017). Sphingosine 1-phosphate: A novel target for lung disorders. *Frontiers in Immunology*, *8*, 296.

Nagahashi, M., Hait, N. C., Maceyka, M., Avni, D., Takabe, K., Milstien, S., et al. (2014). Sphingosine-1-phosphate in chronic intestinal inflammation and cancer. *Advances in Biological Regulation*, *54*, 112–120.

Nagahashi, M., Yamada, A., Katsuta, E., Aoyagi, T., Huang, W.-C., Terracina, K. P., et al. (2018). Targeting the SphK1/S1P/S1PR1 axis that links obesity, chronic inflammation, and breast cancer metastasis. *Cancer Research*, *78*, 1713–1725.

Nishimura, H., Akiyama, T., Monobe, Y., Matsubara, K., Igarashi, Y., Abe, M., et al. (2010). Expression of sphingosine-1-phosphate receptor 1 in mantle cell lymphoma. *Modern Pathology*, *23*, 439–449.

Okada, T., Sugihara, M., Bondar, A.-N., Elstner, M., Entel, P., & Buss, V. (2004). The retinal conformation and its environment in rhodopsin in light of a new 2.2 A crystal structure. *Journal of Molecular Biology*, *342*, 571–583.

O'Sullivan, C., & Dev, K. K. (2013). The structure and function of the S1P1 receptor. *Trends in Pharmacological Sciences*, *34*, 401–412.

Parrill, A. L., Lima, S., & Spiegel, S. (2012). Structure of the first sphingosine 1-phosphate receptor. *Science Signaling*, *5*, pe23.

Patrussi, L., Capitani, N., Martini, V., Pizzi, M., Trimarco, V., Frezzato, F., et al. (2015). Enhanced chemokine receptor recycling and impaired S1P1 expression promote leukemic cell infiltration of lymph nodes in chronic lymphocytic leukemia. *Cancer Research*, *75*, 4153–4163.

Perla, A. S., Fratini, L., Cardoso, P. S., de Farias, C. B., da Cunha Jaeger, M., & Roesler, R. (2020). Fingolimod (FTY720) reduces viability and survival and increases histone H3 acetylation in medulloblastoma cells. *Pediatric Hematology and Oncology*, *37*, 170–175.

Rao, S., Cai, K. Q., Stadanlick, J. E., Greenberg-Kushnir, N., Solanki-Patel, N., Lee, S. Y., et al. (2016). Ribosomal protein Rpl22 controls the dissemination of T-cell lymphoma. *Cancer Research*, *76*, 3387–3396.

Rupp, T., Pelouin, O., Genest, L., Legrand, C., Froget, G., & Castagné, V. (2021). Therapeutic potential of Fingolimod in triple negative breast cancer preclinical models. *Translational Oncology*, *14*, 100926.

Schürer, S. C., Brown, S. J., Gonzalez-Cabrera, P. J., Schaeffer, M.-T., Chapman, J., Jo, E., et al. (2008). Ligand-binding pocket shape differences between sphingosine 1-phosphate (S1P) receptors S1P1 and S1P3 determine efficiency of chemical probe identification by ultrahigh-throughput screening. *ACS Chemical Biology*, *3*, 486–498.

Spiegel, S., & Milstien, S. (2011). The outs and the ins of sphingosine-1-phosphate in immunity. *Nature Reviews Immunology*, *11*, 403–415.

Stessin, A. M., Banu, M. A., Clausi, M. G., Berry, N., Boockvar, J. A., & Ryu, S. (2017). FTY720/fingolimod, an oral S1PR modulator, mitigates radiation induced cognitive deficits. *Neuroscience Letters*, *658*, 1–5.

Till, K. J., Pettitt, A. R., & Slupsky, J. R. (2015). Expression of functional sphingosine-1 phosphate receptor-1 Is reduced by B cell receptor signaling and increased by inhibition of PI3 kinase δ but not SYK or BTK in chronic lymphocytic leukemia cells. *The Journal of Immunology*, *194*, 2439–2446.

van Meer, G., Voelker, D. R., & Feigenson, G. W. (2008). Membrane lipids: Where they are and how they behave. *Nature Reviews Molecular Cell Biology*, *9*, 112–124.

Vrzalikova, K., Ibrahim, M., Vockerodt, M., Perry, T., Margielewska, S., Lupino, L., et al. (2018). S1PR1 drives a feedforward signalling loop to regulate BATF3 and the transcriptional programme of Hodgkin lymphoma cells. *Leukemia*, *32*, 214–223.

Weichand, B., Popp, R., Dziumbla, S., Mora, J., Strack, E., Elwakeel, E., et al. (2017). S1PR1 on tumor-associated macrophages promotes lymphangiogenesis and metastasis via NLRP3/IL-1β. *Journal of Experimental Medicine*, *214*, 2695–2713.

Wen, Z., Zhao, S., Liu, S., Liu, Y., Li, X., & Li, S. (2015). MicroRNA-148a inhibits migration and invasion of ovarian cancer cells via targeting sphingosine-1-phosphate receptor 1. *Molecular Medicine Reports*, *12*, 3775–3780.

Xiao, S., & Yang, J. (2019). Preclinical study of the antitumor effect of sphingosine-1-phosphate receptor 1 antibody (S1PR1-antibody) against human breast cancer cells. *Investigational New Drugs*, *37*, 57–64.

Xu, X.-Q., Huang, C.-M., Zhang, Y.-F., Chen, L., Cheng, H., & Wang, J.-M. (2016). S1PR1 mediates anti-apoptotic/pro-proliferative processes in human acute myeloid leukemia cells. *Molecular Medicine Reports*, *14*, 3369–3375.

Yoshida, Y., Nakada, M., Sugimoto, N., Harada, T., Hayashi, Y., Kita, D., et al. (2010). Sphingosine-1-phosphate receptor type 1 regulates glioma cell proliferation and correlates with patient survival. *International Journal of Cancer*, *126*, 2341–2352.

Young, N., Pearl, D. K., & Van Brocklyn, J. R. (2009). Sphingosine-1-phosphate regulates glioblastoma cell invasiveness through the urokinase plasminogen activator system and CCN1/Cyr61. *Molecular Cancer Research*, *7*, 23–32.

Young, N., & Van Brocklyn, J. R. (2007). Roles of sphingosine-1-phosphate (S1P) receptors in malignant behavior of glioma cells. Differential effects of S1P2 on cell migration and invasiveness. *Experimental Cell Research*, *313*, 1615–1627.

Zhang, L., & Wang, H. (2017). FTY720 inhibits the Nrf2/ARE pathway in human glioblastoma cell lines and sensitizes glioblastoma cells to temozolomide. *Pharmacological Reports*, *69*, 1186–1193.

Zhang, L., Wang, H., Zhu, J., Ding, K., & Xu, J. (2014). FTY720 reduces migration and invasion of human glioblastoma cell lines via inhibiting the PI3K/AKT/mTOR/p70S6K signaling pathway. *Tumor Biology*, *35*, 10707–10714.

Zhong, L., Xie, L., Yang, Z., Li, L., Song, S., Cao, D., et al. (2020). Prognostic value of S1PR1 and its correlation with immune infiltrates in breast and lung cancers. *BMC Cancer*, *20*, 766.

CHAPTER TEN

Phyto-drug conjugated nanomaterials enhance apoptotic activity in cancer

Karuppaiya Vimala and Soundarapandian Kannan*
Division of Cancer Nanomedicine, Department of Zoology, School of Life Science, Periyar University, Salem, Tamil Nadu, India
*Corresponding author: e-mail address: sk_protein@periyaruniversity.ac.in

Contents

Abstract

Cancer continues to be one of the leading causes of death worldwide and is a major obstacle to increased life expectancy. However, survival has not improved significantly with average cancer standard treatment strategies over the past few decades; survival

Advances in Protein Chemistry and Structural Biology, Volume 125
ISSN 1876-1623
https://doi.org/10.1016/bs.apcsb.2020.12.003

rates have remained low, with tumor metastasis, adverse drug reactions, and drug resistance. Therefore, substitute therapies are essential to treat this dreadful disease. Recently, research has shown that natural compounds in plants, such as phytochemicals, are extensively exploited for their anticarcinogenic potential. Phytochemicals may show their anticancer activity different cancer cell markers may alter molecular pathways, which promote in cellular events such as cell cycle arrest and apoptosis, regulate antioxidant status, cell proliferation, migration, invasion and toxicity. Although their outstanding anticancer activity, however, their pharmacological budding is hindered by their low aqueous solubility, poor bioavailability, and poor penetration into cells, hepatic disposition, narrow therapeutic index, and rapid uptake by normal tissues. In this situation, nanotechnology has developed novel inventions to increase the potential use of phytochemicals in anticancer therapy. Nanoparticles can improve the solubility and stability of phytochemicals, specific tumor cell/tissue targeting, enhanced cellular uptake, reduction of phytochemicals. Therapeutic doses of phytochemicals for a long time. Additional benefits include better blood stability, multifunctional design of nanocarriers and improvement in countermeasures. This review summarizes the advances in the use of nanoparticles for the treatment of cancer, as well as various nano-drug deliveries of phytochemicals against cancer. In particular, we are introducing several applications of nanoparticles in combination with phyto-drug for the treatment of cancer.

1. Introduction

Cancer is a severe metabolic syndrome and is considered one of the furthermost challenging public health complications but its comprises a number of phases including initiation, promotion and progress. According to Cancer Research UK, 17 million new cases were reported, of which 9.6 million died in 2018. If this continues, by 2040, there will be 27.5 million new cancer patients each year (Ashraf, 2020). A defining feature of cancer is that abnormal cells develop beyond their normal limits and rapidly form, and then invade adjacent parts of the body and spread to other organs (metastases), which are the prominent cause of cancer death (World Health Organization, 2010). Several therapeutic approaches have been used to treat tumors and the environments surrounding them. The following method has been used as conventional treatment against cancer such as surgical resection, radiotherapy, chemotherapy, and a combination of any of these treatments. Among them, conventional chemotherapeutics is still the main treatment for many cancers in advanced stages (Przystupski et al., 2019). However, chemotherapy has helped improve patients' cancer treatment. In most cases, cancer with a very progressive stage usually occurs and, in general, multitrack resistance takes place. An attempt has been made to target the surrounding environment of the tumor, since cancer cells depend primarily on oxygen and angiogenesis for survival and metastasis (Saggar, Yu, Tan, & Tannock, 2013). Radiation therapy is another method of eradicating cancer cells, because cancer cells are more likely to be damaged than normal cells (Prise, 2006). Gene therapy and immunotherapy have also been used to treat cancer (Amer, 2014; Topalian, Weiner, & Pardoll, 2011). However, there are challenges concomitant with treatment-such as systemic toxicity, low selectivity, and various adverse reactions (Nussbaumer, Bonnabry, Veuthey, & Sandrine, 2011). The use of compounds that target fast dividing cells usually results in cancer treatment, which can cause unwanted side effects on normal, fast dividing cells, such as hair follicles and epithelial cells in the digestive system. Additionally, one of the worst-case scenarios is that numerous cancer cells steadily develop resistance to conventional therapies. Therefore, the search for the best treatment response, new anticancer agents with better efficacy and fewer side effects can be developed and delivery strategies should be designed.

Human evolution uses the plant as a food, shelter, and widely used plant-based natural product as a medicine against various diseases. Modern medicines are mainly sourced from herb based on traditional knowledge and practices and good sources for developing new solutions for different diseases (Fridlender, Kapulnik, & Koltai, 2015). Approximately 25% of the major pharmaceutical compounds and their derivatives are derived from natural resources (Mohanty, Swamy, Sinniah, & Anuradha, 2017). Natural compounds with different molecular backgrounds provide a basis for the discovery of novel drugs. These natural remedies are used to treat many different health problems. Most commonly, these drugs are used as a remedy for irritable bowel syndrome, menstrual disorders, and anxiety, depression, hay fever and skin ailments. With more than 60% of currently used anticancer drugs derived from natural sources, there is growing evidence that phytochemical components in the diet have anticancer functions in a variety of cancers. Phytochemicals are an attractive source of bioactive molecules and new therapeutic agents that can be obtained from fruits, vegetables, grains and other plant parts, animals, marine organisms and microorganisms and proven to be suitable candidates for drugs with potential use as anti-cancer agents. Though, natural compounds are now being screened to treat many major diseases, including cancer, diabetes, cardiovascular, inflammatory and microbial diseases. Furthermore, several chemical phytochemicals isolated from medicinal plants have been shown to inhibit cell proliferation, induce apoptosis, cell cycle arrest, DNA repair, retarded metastasis, and angiogenesis. (Baena Ruiz & Salinas Hernández, 2016; DiMarco-Crook & Xiao, 2015;

Lall, Syed, Adhami, Khan, & Mukhtar, 2015; Wilken, Veena, Wang, & Srivatsan, 2011a, 2011b). Furthermore, natural products exhibit remarkable properties that are well tolerated in macromolecular specification and low toxicity and normal cells, such as abnormal chemical diversity, chemical and biological properties. Natural products are also tolerant of high doses of normal cells compared to chemotherapeutic drugs. The main advantages of natural drugs are lower toxicity and side effects, low-price, and good therapeutic potential. These make them favorable leads in the discovery of novel drugs. However, concerns associated with the biocompatibility, and toxicity of natural compounds presents a greater challenge of using them as medicine. Although incredible efforts in preclinical settings, the applicability of phytochemicals to humans has met with only limited success. Among the many reasons, including *in vivo* instability, poor bioavailability, and poor solubility, poor absorption in the body, issues with target-specific delivery, and tonic effectiveness and probable adverse effects of drugs, inefficacy, and low tissue distribution could probable be the main reason for the clinical failure. Therefore, using development of innovative drug delivery systems for targeting drugs to specific body parts could be an option that might solve these critical issues (Jahangirian, Lemraski, Webster, Rafee-Moghaddam, & Abdol-Lahi, 2017). Nanotechnology is shown to bridge the barrier of biological and physical sciences by applying nanostructures and nanophases at various fields of science (Liu, Tabakman, Welsher, & Dai, 2009). Nanomedicine can be defined as nanotechnology, or the use of materials between 1 and 100 nm, which implicates the utilization of nanodimensional materials including biosensors, microfluidics, drug delivery, and microarray tests to tissue engineering (Tran, DeGiovanni, Piel, & Rai, 2017), especially in nanomedicine and nano based drug delivery systems (Scheme 1), used for delivering natural agents have suggested to us that this technology may have considerable advantages over conventional therapies for cancer (Navya et al., 2019). One advantage of this nanotechnology is that drugs encapsulated in nanoparticles can be protected from destructive action of external media, prolong blood circulation and promote specific types of endocytosis, improve drug distribution, cellular uptake into tumor tissue, provide sustained release of a drug, and help in delivery of two or more drugs for combined therapy as compared with the non-encapsulated free drugs (Rawal & Patel, 2019; Tran et al., 2017). The properties of nanocarriers, including their nanoscale sizes, morphology, high surface-to-volume ratios, favorable drug release profiles, and targeting modifications, can allow them to better reach target tumor tissue and release drugs in a stable, controlled manner properties (Anirudhan & Nair, 2018), to increase

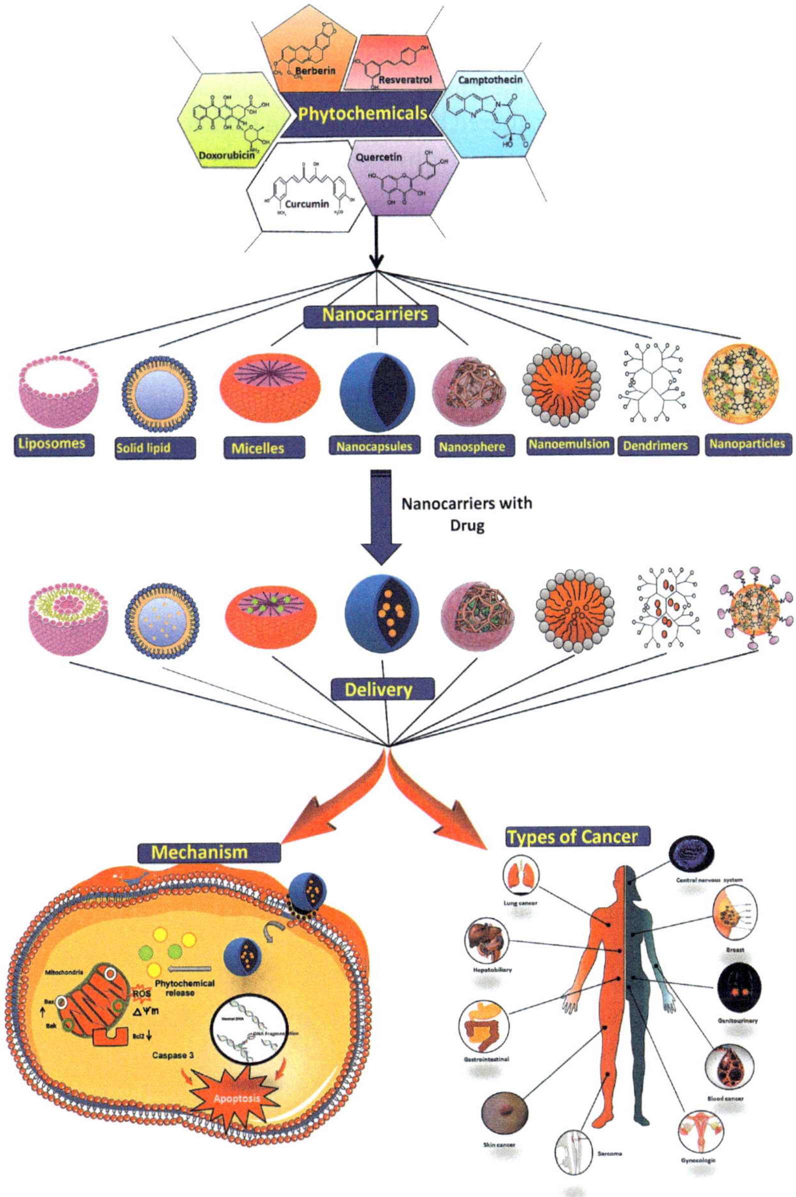

Scheme 1 Multi-faceted properties of phytodrugs used in prevention and treatment of cancer. The phytodrugs have been supplied to the patients in the form of cosmetics, food additives, supplements and pharmaceutics to enhance immunity, wound healing and cancer treatments at various levels to induce apoptosis and inhibit the cell proliferation.

the efficacy of treatment, reduce side effects, and overcome drug resistance. In this state, the development of nano-based drug delivery systems grasps promise for cancer therapy, because natural compounds fabricated at the nanometer scale exhibit drastically altered bioactivities and toxicity. In this review we discuss the recent updates of phytochemicals, phyto-drugs conjugated nanomaterials that their anticancer effects.

2. Nanoparticles (NPS) for cancer therapy: An insight

2.1 Inorganic nanomedicines

Due to its unique physicochemical properties, versatile development strategies, easy surface engineering and outstanding bioavailability, inorganic nanoparticles have been extensively studied as therapeutic agents for cancer treatment. Inorganic nanoparticles can be converted into variety chemical functional groups. This change allowed them antibodies, ligands and the drug should be combined curiosity, therefore, enables them to be broad applications like targeted drug delivery and most mainly diagnostic imaging (Mody, Siwale, Singh, & Mody, 2010). Imaging modalities such as MRI, CT scan, and ultrasound require a unique contrast agent, which ultimately led to the discovery of a different agent of different types of nanoparticles. Recent advances in inorganic nanoparticles, including magnetic nanoparticles, gold nanoparticles, carbon nanotubes, and multifunctional inorganic nanoparticle cancer treatments, are presented to show their *in vitro* and *in vivo* applications in cancer therapy (Scheme 1).

Research on green-integrated NPs has become a hot spot because of the increasing use of NPs in cancer treatment and management (Bhanumathi, Vimala, Shanthi, Thangaraj, & Kannan, 2017; Mary, Shanthi, Vimala, & Kannan, 2016; Nipun Babu & Kannan, 2012; Shanthi, Sreevani, Vimala, & Kannan, 2017; Sujitha & Kannan, 2013; Vimala, Sundarraj, Paulpandi, Venkatesan, & Kannan, 2014; Vivek et al., 2012, 2016). Because, various synthesis methods have been developed to prepares different nanostructures. However, this common method has raised concerns for environmental pollution and human health because of the high cytotoxicity of high-risk by products and manufactured nanomaterials. For this reason, the use of microorganisms, enzymes, fungi, fruits, plant extracts or agricultural waste has recently been used in the development of green chemistry and biological synthetic processes to produce metal or metal oxide nanoparticles. This method has many advantages with low cost, ease of use, large scale production, green and eco-friendly system without using any toxic chemicals. We summarize the many NPs currently used for anti-cancer therapies and discuss their uses in cancer management (Scheme 1).

2.2 Gold nanoparticles

Gold nanoparticles (AuNPs) are used in different domains, which play an important role in the biomedical field. Since of its good compatibility with human organisms, their low toxicity and high stability, their small dimensions and their ability to interact with a wide variety of substances, they are beneficial for controlled drug delivery and cancer treatment. Because of their unique and uniform surface plasmon resonance (SPR), gold nanoparticles and ability to combine with a large variety of therapeutic agents are considered to be an important part of medical applications and advantages over other metal particles in terms of bioavailability and cytotoxicity. Several attempts have been made to use the AuNPs as multiple functional gene carriers due to their excellent bioavailability, well-defined surface chemistry and simplicity. Molecular Imaging Using Fluorescence Resonance Energy Transfer (FRET). Recently, various biogenic AuNPs have been reported to have different therapeutic properties, such as earthworm (Kar, Murmu, Saha, Tandon, & Acharya, 2014), antibiotics (Rahaman Mollick et al., 2014), fungicides (Piruthiviraj, Margret, & Krishnamurthy, 2015) and anti-cancer (Geetha et al., 2013; Kajani, Bordbar, Esfahani, & Razmjou, 2016; Kamala Priya & Iyer, 2015; Zhang, Wang, & Webster, 2017). With regard to poor bioavailability and selectivity acute toxicity of gold-based anticancer complexes, (Zou et al., 2015) biogenic AuNPs without these defects, may show a promising chemotherapeutic potential for cancer treatment. In addition, several studies suggest intrinsic antitumor activity the property of AuNPs is that it selects cancer cells (Zhang et al., 2017). Recently, synthesized AuNPs using active compounds from plants have been reported to be non-toxic to normal human cells. Also, the collection process is simple and environmentally friendly (Govindaraju, Rengaraj, Arivazhagan, Huh, & Yun, 2018; Singh, Du, Singh, & Yi, 2018).

The most studied phyto-fabricated AuNPs are synthesized using Cur (Au-AuNPs). Cur is a good performance reducing agent forming functional AuNPs and maintaining consistency when combined with AuNPs (Singh et al., 2017). Cur-AuNPs direct an antioxidant property compared to free Cur and control cancer to some extent immunity and induces apoptosis in colon and breast cancer cell lines. Furthermore, AuNPs are synthesized using quercetin as a reducing agent, while Qc-AuNPs are involved in epithelial-mesenchymal transition, angiogenesis, and epidermal growth factor receptor (EGFR)/vascular endothelial growth factor receptor (VEGFR)-2-mediated kinase-3 kinase activity (PI3K)/protein kinase B (Akt)-induced apoptosis in breast cancer cells (MCF-7 and MDA-MB 231) (Balakrishnan et al., 2017). Similarly, AuNPs synthesized with EGCG also have anticancer efficacy induces (Chavva et al., 2019).

2.3 Magnetic nanoparticles

Recent advances in synthesis and surface modification technologies have led to the possibility of various new potential applications. Class of nanomaterials diagnostic and therapeutic techniques have the potential to revolutionize current medicine. In inorganic nanomedicines, the detection and treatment of cancer by magnetic nanoparticles (MNPs) has been the most widely investigated. Magnetic nanoparticles (MNPs) particularly iron oxide (also known as magnet) NPs and their multilayered resonant excellent magnetic properties, are an important class of biocompatible nanoparticles that are of great interest for drug delivery (Vivek et al., 2014) and their potential applications in diagnosis. Due to their ability to function at the cellular and molecular level of their biological systems and their unique physico-chemical properties, high saturation magnetization, MNPs are highly susceptible. Magnetic resonance imaging and the next generation of targeted drug delivery vehicles are being actively explored.

Castillo et al. combined PEG coated CPT mounted iron oxide super magnetic NPs using iron co-precipitation method Evaluated them under alkaline conditions and in H460 lungs Cancer cell line (Castillo, Mata, Casula, Sánchez-Alcázar, & Zaderenko, 2014). Super paramagnetic iron oxide NPs (SPIONs) are promising as delivery systems, especially because of their low toxicity and potential for use in cancer detection and treatment. Verma et al. generated magnetic core–shell NPs for aerosol delivery of quercetin (QUR) by nebulization and evaluated their cytotoxicity in lung cancer cell line A549 (Verma et al., 2013). Supermagnetic magnetic iron oxide nanoparticles (SPIONs) best differential agents for MRI (Zhou & Wei, 2016). Some iron oxide-based magnetic nanoparticles have been approved for use in clinical MRI, for example in pheromox (gastromarkin) which improves intestinal imaging (Tse et al., 2015). Yallapu et al. Developed curcumin (CUR) loaded magnetic nanoparticles (CUR-MNPs) for the treatment of pancreatic cancer. The developed CUR-MNPs showed quantitatively enhanced cellular uptake and cytotoxicity in Panc-1 pancreatic cancer cells compared with free CUR (Yallapu, Ebeling, Khan, et al., 2013). Magnetic nanoparticles showed enhanced survival time in mice with high serum bioavailability of CUR. Iron and silver magnetic nanoparticles loaded with artemisinin showed enhanced cellular uptake and cytotoxicity in HeLa cells. Due to the complete internalization and localization of the endosomes and lysosomes into acid packs, they release Fe^{2+} ions, which are converted to the toxic substances of artemisinin, which kill the cancer cells (Chen, Guo, Wang, et al., 2014).

2.4 Silver nanoparticles

Silver NPs are the most important nanomaterials in other metal NPs. It has been widely used in various biological applications, including anti-bacterial, anti-fungal, anti-viral, anti-inflammatory, angiogenic, and anticancer agents. Compared to other metal nanoparticles, Ag is the most accessible metal material for cancer treatment. In most administration of anti-cancer agents in AgNPs, most studies show loss of membrane integrity, antioxidant stress and apoptosis by showing cytotoxic effects against human cancer cells, resulting in cellular damage (Alshatwi, Athinarayanan, & Subbarayan, 2015; AshaRani, Low Kah Mun, Hande, & Valiyaveettil, 2009; Bhanumathi et al., 2017; Gopinath, Gogoi, Chattopadhyay, & Ghosh, 2008; Jeyaraj et al., 2011, 2014; Pimentel, Martínez, García, Gómez-García, & Palacios, 2013; Sinha, Kim, Nie, & Shin, 2006; Vasanth, Ilango, MohanKumar, Agrawal, & Dubey, 2014; Vivek et al., 2012; Ye & Song, 2015). The reactive properties of AgNPs have been demonstrated, and the composition of AgNPs affects the stability, interaction and toxicity. As for phytochemicals synthesized AgNPs, the interaction between AgNPs and serum protein is interesting except for anticancer activity. Both Kerr and EGCG-coated AkNPs can be incorporated into the serum protein as a relaxed complex, becoming a stable and viable biosensor. Each and every element in the complex has therapeutic potential (Abraham, Sharma, Bansal, & Shukla, 2018).

Other metals, such as copper and platinum, are also converted into raw materials to form phytochemical-inert metal NPs (Charbgoo, Ahmad, & Darroudi, 2017). However, most of these products lack inspiring findings regarding their respective cytotoxicity effect. Nanoparticle to metal NPs should be observed; some researchers state that metal NPs have very low nanotoxicity. Nevertheless, the use of metallic NPs should be cautious *in vivo* because toxicity mainly results from problem accumulation, so long-term toxicity should be taken seriously.

3. Lipid-based nanomedicines

Lipid-based nanomedicines such as liposomes, solid lipid nanoparticles (SLN), nano-structured lipid carriers (NLCs) and lipid micelles have shown some promising results for use in anticancer drug delivery. A particular drug accumulation in the tumor site and ensure lipid-based nanomedicines to offer tangible opportunities, enhance pharmacokinetic profile (short biological plasma half-life, absorption of oral stereotypes, etc.) and safety (such as toxic degradation compounds and decreased localized activity) of substrates.

3.1 Liposomes

Liposomes are defined as spherical and single lipid player vesicles, which are biocompatible and biodegradable microvesicles that are spontaneously obtained by dispersing phospholipids into aqueous medium. The biphasic property of liposomes is commonly used to create more lipophilic antitumor molecules in their lipid player(s), as well as hydrophilic chemotherapeutic agents in their internal aqueous compartments (Huwyler, Drewe, & Krähenbuhl, 2008; Samad, Sultana, & Aqil, 2007; Sapra & Allen, 2003). Phospholipid players are composed of natural or synthetic phospholipids that deal with the physicochemical properties, surface charge, composition, and size of liposomes. Modifying these properties will provide control over pharmacology and pharmacodynamics improves the therapeutic index of various agents (Muehlmann, Joanitti, Silva, Longo, & Azevedo, 2011; Yong, Gan, Sasmita, Mak, & Ling, 2018).

Liposomes have the potential to provide a wide variety of molecules to treat cancer (Malam, Loizidou, & Seifalian, 2009). Mullauer, van Bloois, Daalhuisen, et al. (2011) Betulinic acid (BA) linked liposomes have been used to reduce the progression of lung and colon cancer in female atomic naked Foxn1 mice. Also, the Betulinic acid (BA) behavior, body weight and histological examination of rats treated with loaded liposomes showed no serious systemic toxicity. Curcumin's liposomal complex with active-cyclodextrin has been used to induce the cytotoxicity of *in vitro* A549 lung cancer cells at low concentrations (Wu et al., 2016) with high water solubility compared to free active agent. The use of cyclodextrin as a liposomal carrier may, thus, offer the ability to avoid solubility problems of hydrophobic natural products such as curcumin (Rahman, Cao, Steadman, Wei, & Parekh, 2012). de Pace, Liu, Sun, et al. (2013) concocted epigallocatechin-3-gallate-loaded nanoliposomes for enhanced efficacy against MCF-7 breast cancer cells. The developed EGCG-nanoliposomes showed significantly improved EGCG stability and enhanced stable release of EGCG. EGCG-nanoliposomes showed significantly stronger antitumor activity against MGF-7 breast cancer cells compared to free EGCG. Similarly, Dadgar, Alavi, Esfahani, et al. (2013) evaluated artemisinin loaded PEGylated nanoliposomes and their cytotoxic effect on MCF-7 breast cancer cell line. Their results showed that PEGylated nonoliposomal formulation improved the level of cytotoxicity rather than the standard drug. Furthermore, the final formulation showed significantly improved stability and prolonged release of artemisinin. Numerous research groups have developed berberine liposomes using different lipid mixtures and evaluated them on selective cancer cell lines. Results indicated enhanced cytotoxicity compared with free berberine (Lin et al., 2014; Sailor, Seth, Parmar, Chauhan, & Javia, 2015; Wen et al., 2011). Vesinblastine (VNP) conjugated cationic liposomes modified by peanut agglutinin (PNA) significantly enhanced VNP in Lewis lung cancer cells (LLD cells) compared to VINP-liposomes and cellular growth is also enhanced. PNA-VINP-cationic liposomes exhibited significantly enhanced induction of apoptosis by activating proapoptotic pathways and apoptotic enzymes, and showed stronger antitumor efficacy against LLT cancer cells compared to VINP-liposomes (Li, He, Zhou, et al., 2015). Liposomes enhance the bioavailability, cellular uptake and stability of anticancer drugs, including cisplatin and curcumin, in the treatment of cervical cancer (Saengkrit, Saesoo, Srinuanchai, Phunpee, & Ruktanonchai, 2014). Transferrin-targeted liposomes containing paclitaxel were found to be very effective in cervical cancer cells (Saengkrit et al., 2014). Liposomes targeting folic acid and transferrin have shown enhanced efficacy of doxorubicin delivery in the human cervical cancer cell line (HeLa) (Sriraman, Salzano, Sarisozen, & Torchilin, 2016). The encapsulated liposomes of pectin coated puffalin also exhibited excellent stability, mucoadhesive properties due to pectin coating, and continuous release of puffalin (Li, Zhang, Hao, et al., 2014; Li, Zhao, Duan, et al., 2014). Developed pectin coated puffalin-linked liposomes exhibited significantly higher cytotoxicity against SW480 colon cancer cells by cell cycle arrest in the G0/G1 phase.

3.2 Solid lipid nanoparticles

Solid lipid nanoparticles (SLNs) are widely used as nanocarrier because they have evolved into an efficient, non-toxic, and versatile pulp carrier system for many drugs and this avoids some of the disadvantages of liposomes and polymeric nanoparticles. SLNs are made of solid lipids (i.e., fatty acids; mono, di, and triglycerides; phospholipids; etc.), which are stable at body temperature and room temperature, exhibit high compatibility and biodegradability, and have low risk of acute and chronic Most of the toxicity Solid lipids are stabilized using bioactive compatible surfactants. The production of SLNPs is relatively inexpensive and suitable for large scale production. Furthermore, because SLNs have a solid lipid core instead of aquatic core, they provide better protection against the chemical degradation of anticancer agents than nanoparticles, and facilitate the continuous release of anticancer agents due to the zero-order degradation kinetics of the solid lipid matrix (Martins, Sarmento, Ferreira, et al., 2007; Mehnert & Mader, 2001) Storage and drug leakage problems are fewer in SLNPs than in liposomes (Ekambaram, Sathali, & Priyanka, 2012; Hu, Liu, Zhao, et al., 2013). Cucurbitacin B-mounted solid lipid nanoparticles (Cu B-SLN) have been developed for the treatment of hepatic cancer. Developed cucurbitacin B-loaded solid lipid nanoparticles revealed significantly enhanced cellular uptake and cytotoxicity of cucurbitacin B in human hepatic cancer cells compared to free Cu B solution. In another study, Zhu, Qin, Wang, et al. (2009) fused Podophyllotoxin (PPT) 293D conjugated solid lipid nanoparticles (PPT-SLN) to enhance the *in vitro* anticancer activity of PPT against human embryonic kidney cells and HeLa human cervical cancer cells. PPT-SLNs exhibited significantly enhanced cytotoxicity toward 293T and HeLa cells compared to free PPT and lower for normal cells. Curcumin-enriched SLNs have better anticancer activity than free curcumin in the use of *in vitro* cell culture models and animal studies (Rompicharla et al., 2017; Sun et al., 2013). Similar studies have been conducted with curcumin-enriched NLCs (Beloqui et al., 2016; Chen, Pan, Jiang, et al., 2016; Chen, Pan, Jiang, Li, & Jin, 2016; Godugu, Doddapaneni, Safe, & Singh, 2016; Jiang, Geng, Liu, Li, & Cao, 2016; Wang, Chen, et al., 2017; Wang, Zhang, et al., 2017; Yang, Liu, & Liu, 2017). SLN and NLCs have been used to synthesize, protect, and deliver many hydrophobic phytochemicals, including beta-carotene (Cornacchia & Roos, 2011; Gomes, Simplicio, Souto, Cardoso, & Pinho, 2013; Hentschel, Gramdorf, Muller, & Kurz, 2008; Nik, Langmaid, & Wright, 2012; Oliveira, Michelon, Furtado, Sinigaglia-Coimbra, & Cunha, 2016; Pan, Tikekar, & Nitin, 2016; Qian, Decker, Xiao, & McClements, 2013; Salminen, Gommel, Leuenberger, & Weiss, 2016), ω-3 fatty acids (Awad, Helgason, Weiss, Decker, & McClements, 2009;

Salminen, Aulbach, Leuenberger, Tedeschi, & Weiss, 2014; Salminen, Helgason, Kristinsson, Kristbergsson, & Weiss, 2013), lycopene (Okonogi & Riangjanapatee, 2014; Riangjanapatee, Muller, Keck, & Okonogi, 2013), oil-soluble vitamins (Kong, Xia, & Liu, 2011; Liu, Wu, & Fang, 2010; Mohammadi, Pezeshki, Abbasi, Ghanbarzadeh, & Hamishehkar, 2017; Sabzichi et al., 2017; Xia & Kong, 2011), and resveratrol (Carlotti, Sapino, Ugazio, Gallarate, & Morel, 2012; Jose et al., 2014; Loureiro et al., 2017; Wang, Zhang, et al., 2017). Furthermore, Wang, Li, Wang, et al. (2014) developed berberine conjugated solid lipid nanoparticles for enhanced anticancer activity against MCF-7, HepG2 and A549 cancer cells. Berberin-loaded SLN exhibited significantly higher cellular uptake of berberine in MCF-7 cancer cells due to persistent release of berberine from SLN. Further, developed SLN exhibited significantly higher cytotoxicity against all cancer cells. Recently, emodin solid lipid nanoparticles (E-SLN) have been incorporated to improve the efficacy of emodin against breast cancer. E-SLN exhibited significantly higher cytotoxicity against human breast cancer MCF-7 cells and human hepatoma HepG2 cells compared to free emodin solution, while E-SLN exhibited a higher expression of human mammary epithelial MCF-10A showed no toxicity on cells (Chen, Wang, Zhang, et al., 2015).

4. Polymer-based nanomedicines

Polymer-based nanoparticles, polymeric nanoparticles, polymeric micelles, polymer drug conjugates, and dendrimers to improve the therapeutic efficacy of anticancer drugs have been extensively investigated. The comprehensive aim of chemically converting polymers into novel nanocarriers leads to a versatile drug delivery system for anticancer therapeutics (Prabhu, Patravale, & Joshi, 2015).

4.1 Polymeric nanoparticles

Polymeric nanoparticles (NPs) are solid colloidal systems in which the anticancer drug is dissolved, incorporated, or trapped in a volume polymer matrix. Polymeric NPs are often chosen as carrier systems of pharmaceuticals because of its stability in storage. Poly (lactide-co-glycolide) (PLGA), polylactide (PLA), polycaprolactone (PCL), polyglycolide, poly (D, L-lactide) and chitosan Polymers have been developed for the passive and active delivery of therapeutic moieties (Prabhu et al., 2015). Chen, Pan, Jiang,

Li, and Jin (2016), Chen, Pan, Jiang, et al. (2016) developed folic acid conjugated vincristine (VCR) loaded PLGA-mPEG nanoparticles (FA-VCR-PLGA-mPEG-NPs) to improve anticancer efficacy of VCR against folic acid receptor overexpressed MCF-7 cells. The PEG-PLGA-Rg3 formation of ginsenocytes has been shown by Geng et al. Metallometrics were used to reduce protein expression levels of proteinase-9 (MMP-9), HIF-1, and vascular endothelial growth factor (VEGF) (Geng, Fan, Gao, Yu, & Hua, 2016). Yang, Liu, Yan, et al. (2016) developed biotin receptor targeted camptothecin loaded biotin-F127-PLA block co-polymeric nanoparticles for treatment of hepatocellular carcinoma. Biotin CPT. Targeted CPT-NPs showed significantly higher *in vitro* antitumor efficacy against H22 hepatocellular carcinoma cells compared to non-targeted CPT-NPs. Further, developed biotin receptor targeted camptothecin loaded biotin-F127-PLA block co-polymeric nanoparticles for treatment of hepatocellular carcinoma. Biotin receptor targeted CPT-NPs exhibited an initial burst release followed by sustained release of CPT. Targeted CPT-NPs showed significantly higher *in-vitro* antitumor efficacy against H22 hepatocellular carcinoma cells compared to non-targeted CPT-NPs. In another study, glycyrrhizic acid 10-hydroxygamptothecin-loaded bovine serum albumin targeted nanoparticles (GL-BSA-HCBDNPs) were observed in SMMC 7721 hepatocellular carcinoma cells, it showed significantly higher cellular growth. Furthermore, GL-BSA-HCPD-NPs indicate higher cytotoxicity and free drug against SMMC 7721 cells compared to uncoated NPs. Furthermore, utilizing tamoxifen (Tam), biodegradable antibody acting agent formed polymeric nanoparticles (NPs) was created to accomplish targeted on conveyance just as sustained release of the medication against breast cancer cells (Vivek et al., 2014).

5. Phytochemical characteristics

In general, there are many different bioactive phytochemicals in foods, supplements and medications due to health benefits including carotenoids, flavonoids, essential fatty acids, peptides and glucosinolates (Gupta, 2016). These phytochemicals differ in their molecular weights, functional groups, charges, and polarities, leading to differences in their solubility, distribution, physical states, and chemical stability. The properties of specific phytochemicals are dealt with elsewhere in this special issue and are therefore not considered here again. However, knowledge of the specific molecular and physicochemical properties of phytochemicals is important in designing appropriate phytochemical oral delivery systems (McClements, 2017).

5.1 Limitations of phytochemical delivery using nanocarriers

As discussed above, nanotechnology has proven to be very helpful in the healthcare industry, particularly in the treatment of cancer. However, many studies in the past have listed the limitations that nanotechnology-based therapies face. For example, the most discussed range of drug delivery based on nanoparticles is safety and toxicity issues. NPs are well known for their ability to deliver various drugs or compounds to target tissues in the human body. This is due to their very small size and other unique properties. However, their unique properties are also a major cause of many toxic effects.

5.2 Physiochemical characterization of NPs

The main physicochemical properties of NPs are molecular weight, shape, size, composition, surface properties, charge, and solubility and aggregation behavior. Characterization of NPs has always been an important issue because variation between physico-chemical properties can affect the production of NPs in pharmaceutical products. This is particularly important since it can also affect the products of NPs after administration. In addition to the limitations mentioned above; environmental concerns have also been voiced development of nanomedicine.

5.3 Phytochemicals as anticancer agents

Many natural agents are used for cancer prevention and treatment (Sundarraj et al., 2012), mainly because of their low toxicity profiles and multimodal activities. Phytochemicals are being investigated as food components for anti-cancer activities. Moreover, the combination of phytochemicals with standard chemo-therapeutic drugs is very promising in providing synergistic effect. Based on their biochemical appearance, phytochemicals such as quercetin, piperine, curcumin, lycopene, quinacrine (QC), epigallocatechin gallate (EGCG), apigenin, resveratrol (RES), 6-gingerol (6G) were used. Their chemical structures are shown in Scheme 2.

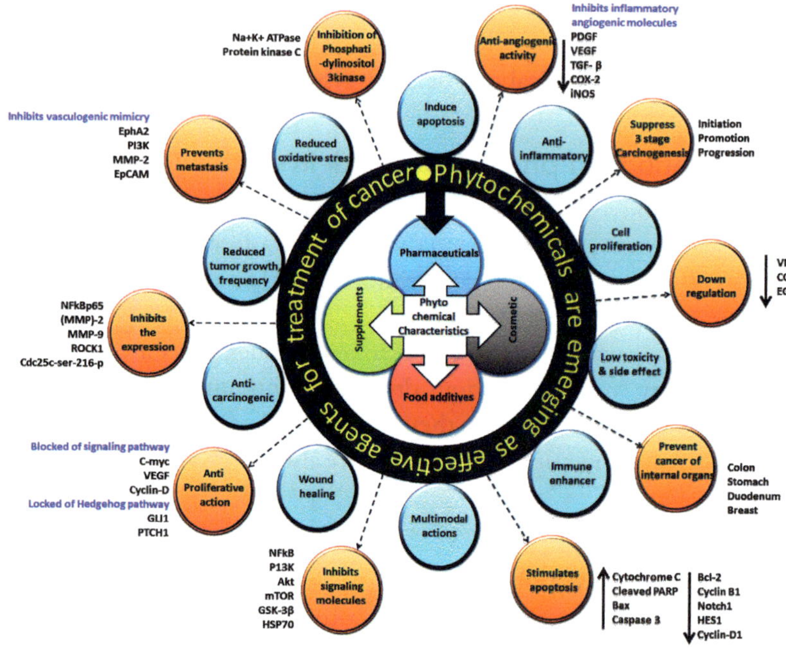

Scheme 2 Nanoformulation of phytodrugs in the form of nano-capsules, nano-emulsions, micelles, liposome to deliver the active compounds in a slow and steady manner to target various types of cancers. The delivered phytodrugs could effectively induce apoptosis by changing the mitochondrial membrane potential to release cytochrome-c to functionalize the BCl2 family members.

6. Phytochemicals conjugated with NPs as nanomedicine

6.1 Quercetin

Quercetin is a widely distributed natural flavonoid and an important component of various foods derived from fruits and vegetables such as onions, broccoli and apples (Boots, Haenen, & Bast, 2008). Quercetin's anti-cancer properties have been extensively investigated in a variety of cells, including ovarian, breast, and stomach cancers. Minaei et al. investigated its potential use in the preparation of doxorubicin-induced apoptosis by producing mixed NPs by mixing QC and lecithin. The data from this study demonstrated that the combination of nano-QC and doxorubicin increased the toxic effects of doxorubicin in human MCF-7 breast cancer cells. Moreover, this formulation provided improved drug loading, durability and continuous release of both agents (Minaei, Sabzichi, Ramezani, Hamishehkar, & Samadi, 2016).

6.2 Piperine

Piperine, also known as (E, E)-1-piperoylpiperidine and (E, E)-1-[5-(1, 3-benzodioxol-5-y1)-1-oxo-2, 4-pentdienyl] piperidine suppresses tumor growth and metastasis *in vitro* as well as *in vivo* (Lai et al., 2012).

6.3 6-Gingerol (6G)

6-Gingerol (1-(40-hydroxy-30-methoxyphenyl)-5-hydroxy-3-dexanone), one of the most important natural compounds isolated from ginger rhizomes, has attracted many research interests in its wide range of biological activities such as anti-inflammatory, antitumor and many others. Manatunga et al. (2017) reported a new pH-sensitive sodium alginate, hydroxyapatite flavored-coated iron oxide nanoparticle composite (IONP/HAp-NaAlg) loaded with two anticancer drugs, 6G or Cur. The manufactured nanocarrier system has shown high coupling efficiency and durable pH-controlled release capability, ensuring the release of 6G or Cur in a targeted and controlled manner for treating cancer. In another important study, the toxic effects of incorporating 6G on PEGylated nanosystems were tested in breast cancer cell line T47D (Behroozeh et al., 2018).

6.4 (-)-Epigallocatechin-3-gallate (EGCG)

EGCG nanoparticles have shown efficient retardation of tumor growth in mice (Ray, Kumar, & Gupta, 2013) and promoted DNA damage in lymphocytes (Alotaibi, Bhatnagar, Najafzadeh, Gupta, & Anderson, 2013). EGCG is a rich phytochemical of green tea that affects tumor proliferation, growth and metastasis. Research evidence from *in vitro* and *in vivo* studies demonstrates the potential of this particular natural compound in preventing or preventing many types of cancer. EGCG retained its biological function by incorporating PLA-PEG NPs 10-fold dose benefit *in vitro* and *in vivo* in human prostate cancer cells compared to 22Rγ1 prostate carcinoma cell line *in vitro*. Recently, a study showed that self-assembling of PEG-EGCG with sunitinib (SU) leads to the formation of stable micellar nanocomplexes (SU-MNC), which have higher anticancer efficacy than conventional SU formulations on human renal cell carcinoma-xenograft mice. (Yongvongsoontorn et al., 2019).

6.5 Curcumin

Curcumin (1,7-bis (4-hydroxy-3-methoxyphenyl)-1,6-heptadine-3,5-dione) is a bioactive compound extracted from the root of turmeric. It has been used in medical treatment for centuries because of its antioxidant, anti-inflammatory, analgesic, antiseptic and antimalarial properties (Aggarwal, Sundaram, Malani, & Ichikawa, 2007; Joe, Vijaykumar, & Lokesh, 2004; Wilken et al., 2011a; Wilken et al., 2011b). Recent research shows that curcumin suppresses all three stages of cancer, namely the reduction of nuclear factor kappa B (NF-κB) and subsequently various NF-κB-related pro-inflammatory pathways (Goel & Aggarwal, 2010).

Nano-CUR modulated cancer-related miRNAs, transcription factors, and proteins in cervical cancer cells (Zaman et al., 2016). Curcumin suppresses the c-Jun/AP-1 signal transduction pathway (Han, Keum, Seo, & Surh, 2002; Kang et al., 2004). Higher doses of curcumin reduced VEGF, COX-2, and EGFR in cask-matched mice (Yoysungnoen-Chintana, Bhattarakosol, & Patumraj, 2014). Curcumin also prevents skin cancer and prevents cancer of the internal organs such as the colon, stomach, duodenum and breast of rodents after oral administration (Kawamori et al., 1999). Furthermore, researchers have developed an effective delivery system of Cur by loading it with D, L-PLGA NPs, and coating it with chitosan and PEG to achieve the optimal therapeutic effect. *In vitro*, cellular studies reveal a reduction in the migration and invasion potential of metastatic pancreatic cancer, which means that the novel formulation exhibits excellent cytotoxicity and apoptosis-inducing ability (Arya, Das, & Sahoo, 2018). Some recently developed phytochemical loaded nanomedicines for cancer treatment are summarized in Table 1.

Table 1 Phytochemicals conjugated with different types NPs for treatment of cancer.

Nanomedicine	Phytochemicals	Type of cancer	References
Liposomes	Liposomes Berberine (BER)	Liver	Lin, Kuo, Hsu, et al. (2013)
	Oleanolic acid (OA)	Cervical	Tang, Gao, Zhao, et al. (2013)
	Thymoquinone (TQ)	Breast	Odeh, Ismail, Abu-Dahab, et al. (2012)
	Ursolic acid (UA)	Non-Specific	Caldeira de Araújo Lopes, Vinícius Melo Novais, Salviano Teixeira, et al. (2013)

Table 1 Phytochemicals conjugated with different types NPs for treatment of cancer.—cont'd

Nanomedicine	Phytochemicals	Type of cancer	References
SLNs	Camptothecin (CPT)	Glioma	Martins, Sarmento, Nunes, et al. (2013)
	Cucurbitacin B	Liver	Wang, Yang, Wang, et al. (2010), Wang, Zhao, Hu, et al. (2010)
NLCs	Curcumin (CUR)	Brain	Chen, Pan, Jiang, Li, and Jin (2016), Chen, Pan, Jiang, et al. (2016)
	Quercetin	Breast	Sun, Nie, Pan, et al. (2014)
	Vincristine (VCR)	Breast	Gao, Zhang, Xu, et al. (2017)
Lipid micelles	Gambogic acid (GA)	Non-specific	Cai, Qiu, Xiang, et al. (2014)
Polymeric nanoparticles (PNPs)	Bufalin (Bu)	Breast	Tian, Yin, Zhang, et al. (2014)
	Epigallocatechin-3-gallate (EGCG)	Melanoma	Siddiqui, Bharali, Nihal, et al. (2014)
	Ellagic acid (EA)	Breast	Shirode, Bharali, Nallanthighal, et al. (2015)
	Emodin (E)	Liver	Liu, Gao, Xu, et al. (2016)
	Ferulic acid (FA)	Lung	Merlin, Prasad, Shibil, et al. (2012)
	Genistein (GEN)	Cervical	Zhang, Liu, Zeng, et al. (2015)
	Morin (M)	Lung	Abbad, Wang, Waddad, et al. (2015)
	Resveratrol (RSV)	Lung	Karthikeyan, Prasad, Ganamani, et al. (2013)
	Silbinin (SIL)	Prostate	Pooja, Babu Bikkina, Kulhari, et al. (2014)
	Naringenin (NAR)	Cervical	Krishnakumar, Sulfikkarali, Prasad, et al. (2011)

Continued

Table 1 Phytochemicals conjugated with different types NPs for treatment of cancer.—cont'd

Nanomedicine	Phytochemicals	Type of cancer	References
Polymeric micelles	Combretastatin A4	Human umbilical vein endothelial cells (HUVECs)	Wang, Yang, et al. (2010), Wang, Zhao, et al. (2010)
	Curcumin (CUR)	Colon	Gou, Men, Shi, et al. (2011)
	Gambogic acid (GA)	MDR-Breast	Saxena and Hussain (2012)
	Luteolin (Lu)	Non-specific	Qiu, Gao, Wang, et al. (2013)
	Quercetin (QT)	MDR-Breast	Zhao, Shi, Zou, et al. (2011)
Polymer drug conjugates	Curcumin (CUR)	Non-specific	Fang, Zhang, Xie, et al. (2016)
Dendrimers	3,4-difluorobenzylidene curcumin	Pancreatic	Kesharwani, Xie, Banerjee, et al. (2015)
Magnetic nanoparticles (MNPs)	Curcumin (CUR)	Breast	Yallapu, Othman, Curtis, et al. (2012)
	Curcumin (CUR)	Breast	Mancarella, Greco, Baldassarre, et al. (2015)
Gold nanoparticles (GNPs)	Apigenin (Ap)	Skin	Rajendran, Dhandapani, Anantanarayanan, et al. (2015)
	Epigallocatechin-3-gallate (EGCG)	Bladder	Hsieh, Lu, Chen, et al. (2012)
Carbon nanotubes (CNTs)	Betulinic acid	Lung	Tan, Karthivashan, Arulselvan, et al. (2014)
	Curcumin (CUR)	Prostate	Li, Zhang, et al. (2014), Li, Zhao, et al. (2014)

7. Phytochemical-based composite nanomedicines

Collective chemotherapy has been developed to combat some of the limitations of single chemotherapy drugs, such as the development of drug resistance, the main side effects, and the limited regimen of clinical use. Fundamental principles of concurrent administration include chemotherapeutic drug and phytochemical; (i) Use anticancer agents with non-toxic agents so that each anticancer agent can be administered to the maximum extent possible; (ii) Anticancer drug resistance by combining anticancer agents with different mechanisms of action and minimal crossover. Preparing nanomedicines for co-encapsulation or co-delivery opens new possibilities to overcome drug resistance, reduce the dose of each drug, and achieve a synergistic or additive effect for cancer treatments. Furthermore, additional benefits include improved patient compliance due to decreased number of drug administrations and drug dosage, resulting in reduced toxicity to healthy tissues (Parhi, Mohanty, & Sahoo, 2012).

Recently, PLGA-casein nanoparticles coated with paclitaxel (PTX) and epigallocatechin gallate (EGCG) have been fabricated to overcome the effects of PTX. Synthesized nanoparticles exhibited a synergistic *in vitro* cytotoxic effect on PTX-resistant MDA-MB-231 breast cancer cells (Narayanan, Mony, Vijaykumar, et al., 2015). In another study, Soni, Kaur, and Tikoo (2015) fabricated PLGA nanoparticles loaded with thymoquinone (TQ) and paclitaxel (PTX) for the combined inhibition of MCF-7 breast cancer cells.

8. Conclusion and future perspectives

A number of natural phytochemicals are potential anticancer drugs. It is widely accepted that nanotechnology may be the future direction for cancer treatment. It has been demonstrated that nanoparticle phyto-drug delivery is one of the promising techniques for the treatment of borderline cancer. Therefore, combining phytochemicals with nanotechnology is a promising approach. However, the challenges of nanotechnology are still not fully resolved. Conversely, the translation of nanomedicine from phytochemical or synthetic chemical research benchmark has been progressing slower than expected due to safety concerns, mainly for medical systems, and most importantly, for the cost-effectiveness of the nanomedicines after the cost of production and its cost. If these issues can be solved, phytochemical-based nanomedicines will become an option for cancer treatment in the future.

References

Abbad, S., Wang, C., Waddad, A. Y., et al. (2015). Preparation, *in vitro* and *in vivo* evaluation of polymeric nanoparticles based on hyaluronic acid-poly(butyl cyanoacrylate) and D-lphatocopheryl polyethylene glycol 1000 succinate for tumor-targeted delivery of morin hydrate. *International Journal of Nanomedicine, 10*, 305–320.

Abraham, A. N., Sharma, T. K., Bansal, V., & Shukla, R. (2018). Phytochemicals as dynamic surface ligands to control nanoparticle-protein interactions. *ACS Omega, 3*, 2220–2229.

Aggarwal, B. B., Sundaram, C., Malani, N., & Ichikawa, H. (2007). Curcumin: The Indian solid gold. *Advances in Experimental Medicine and Biology, 595*, 1–75.

Alotaibi, A., Bhatnagar, P., Najafzadeh, M., Gupta, K. C., & Anderson, D. (2013). Tea phenols in bulk and nanoparticle form modify DNA damage in human lymphocytes from colon cancer patients and healthy individuals treated *in vitro* with platinum-based chemotherapeutic drugs. *Nanomedicine, 8*(3), 389–401.

Alshatwi, A., Athinarayanan, J., & Subbarayan, P. V. (2015). Green synthesis of platinum nanoparticles that induce cell death and G2/M-phase cell cycle arrest in human cervical cancer cells. *Journal of Materials Science. Materials in Medicine, 26*(1), 5330–5341.

Amer, M. H. (2014). Gene therapy for cancer: Present status and future perspective. *Molecular and Cellular Therapies, 2*(1), 27–88.

Anirudhan, T. S., & Nair, A. S. (2018). Temperature and ultrasound sensitive gate-keepers for the controlled release of chemotherapeutic drugs from mesoporous silica nanoparticles. *Journal of Materials Chemistry B, 6*, 428–439.

Arya, G., Das, M., & Sahoo, S. K. (2018). Evaluation of curcumin loaded chitosan/PEG blended PLGA nanoparticles for effective treatment of pancreatic cancer. *Biomedicine & Pharmacotherapy, 102*, 555–566.

AshaRani, P. V., Low Kah Mun, G., Hande, M. P., & Valiyaveettil, S. (2009). Cytotoxicity and genotoxicity of silver NPs in human cells. *ACS Nano, 3*(2), 279–290.

Ashraf, M. A. (2020). Phytochemicals as potential anticancer drugs: Time to ponder nature's bounty. *BioMed Research International, 2020*, 1–7.

Awad, T. S., Helgason, T., Weiss, J., Decker, E. A., & McClements, D. J. (2009). Effect of omega-3 fatty acids on crystallization, polymorphic transformation and stability of tripalmitin solid lipid nanoparticle suspensions. *Crystal Growth & Design, 9*(8), 3405–3411.

Baena Ruiz, R., & Salinas Hernández, P. (2016). Cancer chemoprevention by dietary phytochemicals: Epidemiological evidence. *Maturitas, 94*, 13–19.

Balakrishnan, S., Mukherjee, S., Das, S., Bhat, F. A., Raja Singh, P., Patra, C. R., et al. (2017). Gold nanoparticles–conjugated quercetin induces apoptosis via inhibition of EGFR/PI3K/Akt–mediated pathway in breast cancer cell lines (MCF-7 and MDA-MB-231). *Cell Biochemistry and Function, 35*, 217–231.

Behroozeh, A., Mazloumi Tabrizi, M., Kazemi, S. M., Choupani, E., Kabiri, N., Ilbeigi, D., et al. (2018). Evaluation the anti-cancer effect of PEGylated Nano-niosomal Gingerol, on breast cancer cell lines (T47D), *in-vitro*. *Asian Pacific Journal of Cancer Prevention, 19*, 645–648.

Beloqui, A., Memvanga, P. B., Coco, R., Reimondez-Troitino, S., Alhouayek, M., Muccioli, G. G., et al. (2016). A comparative study of curcumin-loaded lipid-based nanocarriers in the treatment of inflammatory bowel disease. *Colloids and Surfaces. B, Biointerfaces, 143*, 327–335.

Bhanumathi, R., Vimala, K., Shanthi, K., Thangaraj, R., & Kannan, S. (2017). Bioformulation of silver nanoparticles as berberine carrier cum anticancer agent against breast cancer. *New Journal of Chemistry, 41*, 14466.

Boots, A. W., Haenen, G. R., & Bast, A. (2008). Health effects of quercetin: From antioxidant to nutraceutical. *European Journal of Pharmacology, 585*(2–3), 325–337.

Cai, L., Qiu, N., Xiang, M., et al. (2014). Improving aqueous solubility and antitumor effects by nanosized gambogic acid-mPEG$_{2000}$ micelles. *International Journal of Nanomedicine, 9*, 243–255.

Caldeira de Araújo Lopes, S., Vinícius Melo Novais, M., Salviano Teixeira, C., et al. (2013). Preparation, physicochemical characterization, and cell viability evaluation of long circulating and pH-sensitive liposomes containing ursolic acid. *BioMed Research International, 2013*, 1–7.

Carlotti, M. E., Sapino, S., Ugazio, E., Gallarate, M., & Morel, S. (2012). Resveratrol in solid lipid nanoparticles. *Journal of Dispersion Science and Technology, 33*(4), 465–471.

Castillo, P. M., Mata, M., Casula, M. F., Sánchez-Alcázar, J. A., & Zaderenko, A. P. (2014). PEGylated versus non-PEGylated magnetic nanoparticles as camptothecin delivery system. *Beilstein Journal of Nanotechnology, 5*, 1312–1319.

Charbgoo, F., Ahmad, M. B., & Darroudi, M. (2017). Cerium oxide nanoparticles: Green synthesis and biological applications. *International Journal of Nanomedicine, 12*, 1401–1413.

Chavva, S. R., Deshmukh, S. K., Kanchanapally, R., Tyagi, N., Coym, J. W., Singh, A. P., et al. (2019). Epigallocatechin gallate-gold nanoparticles exhibit superior antitumor activity compared to conventional gold nanoparticles: Potential synergistic interactions. *Nanomaterials (Basel), 9*, 396–430.

Chen, J., Guo, Z., Wang, H. B., et al. (2014). Multifunctional mesoporous nanoparticles as pH responsive Fe_2+ reservoirs and artemisinin vehicles for synergistic inhibition of tumor growth. *Biomaterials, 35*(24), 6498–6507.

Chen, Y. H., Pan, L. Z., Jiang, M., Li, D., & Jin, L. J. (2016). Nanostructured lipid carriers enhance the bioavailability and brain cancer inhibitory efficacy of curcumin both *in vitro* and *in vivo*. *Drug Delivery, 23*(4), 1383–1392.

Chen, Y., Pan, L., Jiang, M., et al. (2016). Nanostructured lipid carriers enhance the bioavailability and brain cancer inhibitory efficacy of curcumin both *in vitro* and *in vivo*. *Drug Delivery, 23*(4), 1383–1392.

Chen, R., Wang, S., Zhang, J., et al. (2015). Aloe-emodin loaded solid lipid nanoparticles: Formulation design and *in vitro* anti-cancer study. *Drug Delivery, 22*(5), 666–674.

Cornacchia, L., & Roos, Y. H. (2011). State of dispersed lipid carrier and interface composition as determinants of beta-carotene stability in oil-in-water emulsions. *Journal of Food Science, 76*(8), C1211–C1218.

Dadgar, N., Alavi, S. E., Esfahani, M. K., et al. (2013). Study of toxicity effect of pegylated nanoliposomal artemisinin on breast cancer cell line. *Indian Journal of Clinical Biochemistry, 28*(4), 410–412.

de Pace, R. C., Liu, X., Sun, M., et al. (2013). Anticancer activities of (-)-epigallocatechin-3-gallate encapsulated nanoliposomes in MCF7 breast cancer cells. *Journal of Liposome Research, 23*(3), 187–196.

DiMarco-Crook, C., & Xiao, H. (2015). Diet-based strategies for cancer chemoprevention: The role of combination regimens using dietary bioactive components. *Annual Review of Food Science and Technology, 6*, 505–526.

Ekambaram, P., Sathali, A. A. H., & Priyanka, K. (2012). Solid lipid nanoparticles: A review. *Scientific Reviews and Chemical Communications, 2*, 80–102.

Fang, X. B., Zhang, J. M., Xie, X., et al. (2016). pH-sensitive micelles based on acid-labile pluronic F68-curcumin conjugates for improved tumor intracellular drug delivery. *International Journal of Pharmaceutics, 502*(1–2), 28–37.

Fridlender, M., Kapulnik, Y., & Koltai, H. (2015). Plant derived substances with anti-cancer activity: From folklore to practice. *Frontiers in Plant Science, 6*, 799–838.

Gao, X., Zhang, J., Xu, Q., et al. (2017). Hyaluronic acid-coated cationic nanostructured lipid carriers for oral vincristine sulfate delivery. *Drug Development and Industrial Pharmacy, 43*(4), 661–667.

Geetha, R., Ashokkumar, T., Tamilselvan, S., Govindaraju, K., Sadiq, M., & Singaravelu, G. (2013). Green synthesis of gold nanoparticles and their anticancer activity. *Cancer Nanotechnology*, *4*, 91–98.

Geng, L., Fan, J., Gao, Q. L., Yu, J., & Hua, B. J. (2016). Preliminary study for the roles and mechanisms of 20(R)- ginsenoside Rg3 and PEG-PLGA-Rg3 NPs in the Lewis lung cancer mice. *Beijing Da Xue Xue Bao. Yi Xue Ban*, *48*(3), 496–501.

Godugu, C., Doddapaneni, R., Safe, S. H., & Singh, M. (2016). Novel diindolylmethane derivatives based NLC formulations to improve the oral bioavailability and anticancer effects in triple negative breast cancer. *European Journal of Pharmaceutics and Biopharmaceutics*, *108*, 168–179.

Goel, A., & Aggarwal, B. B. (2010). Curcumin, the golden spice from Indian saffron, is a chemosensitizer and radiosensitizer for tumors and chemoprotector and radioprotector for normal organs. *Nutrition and Cancer*, *62*(7), 919–930.

Gomes, G. V. L., Simplicio, I. A. S., Souto, E. B., Cardoso, L. P., & Pinho, S. C. (2013). Development of a lipid particle for beta-carotene encapsulation using a blend of tristearin and sunflower oil: Choice of lipid matrix and evaluation of shelf life of dispersions. *Food Technology and Biotechnology*, *51*(3), 383–391.

Gopinath, P., Gogoi, S. K., Chattopadhyay, A., & Ghosh, S. S. (2008). Implications of silver nanoparticle induced cell apoptosis for *in vitro* gene therapy. *Nanotechnology*, *19*(7), 75104–75114.

Gou, M., Men, K., Shi, H., et al. (2011). Curcumin-loaded biodegradable polymeric micelles for colon cancer therapy *in vitro* and *in vivo*. *Nanoscale*, *3*(4), 1558–1567.

Govindaraju, S., Rengaraj, A., Arivazhagan, R., Huh, Y. S., & Yun, K. (2018). Curcumin-conjugated gold clusters for bioimaging and anticancer applications. *Bioconjugate Chemistry*, *29*, 363–370.

Gupta, R. C. (2016). *Nutraceuticals: Efficacy, safety and toxicity*. London, UK: Academic Press.

Han, S. S., Keum, Y. S., Seo, H. J., & Surh, Y. J. (2002). Curcumin suppresses activation of NF-κB and AP-1 induced by phorbol ester in cultured human promyelocytic leukemia cells. *BMB Reports*, *35*(3), 337–342.

Hentschel, A., Gramdorf, S., Muller, R. H., & Kurz, T. (2008). Beta-carotene-loaded nanostructured Lipid carriers. *Journal of Food Science*, *73*(2), N1–N6.

Hsieh, D. S., Lu, H. C., Chen, C. C., et al. (2012). The preparation and characterization of gold conjugated polyphenol nanoparticles as a novel delivery system. *International Journal of Nanomedicine*, 7, 1623–1633.

Hu, H., Liu, D., Zhao, X., et al. (2013). Preparation, characterization, cellular uptake and evaluation *in vivo* of solid lipid nanoparticles loaded with cucurbitacin B. *Drug Development and Industrial Pharmacy*, *39*(5), 770–779.

Huwyler, J., Drewe, J., & Krähenbuhl, S. (2008). Tumor targeting using liposomal antineoplastic drugs. *International Journal of Nanomedicine*, *3*(1), 21–29.

Jahangirian, H., Lemraski, E. G., Webster, T. J., Rafee-Moghaddam, R., & Abdol-Lahi, Y. (2017). A review of drug delivery systems based on nanotechnology and green chemistry: Green nanomedicine. *International Journal of Nanomedicine*, *12*, 2957–2978.

Jeyaraj, M., Arun, R., Sathishkumar, G., MubarakAli, D., Rajesh, M., Sivanandhan, G., et al. (2014). An evidence on G2/M arrest, DNA damage and caspase mediated apoptotic effect of biosynthesized gold nanoparticles on human cervical carcinoma cells (HeLa). *Materials Research Bulletin*, *52*, 15–24.

Jeyaraj, M., Sathishkumar, G., Sivanandhan, G., MubarakAli, D., Rajesh, M., Arun, R., et al. (2011). Biogenic silver NPs for cancer treatment: An experimental report. *Colloids and Surfaces B: Biointerfaces*, *106*, 86–92.

Jiang, H., Geng, D. M., Liu, H. Q., Li, Z. R., & Cao, J. (2016). Co-delivery of etoposide and curcumin by lipid nanoparticulate drug delivery system for the treatment of gastric tumors. *Drug Delivery*, *23*(9), 3665–3673.

Joe, B., Vijaykumar, M., & Lokesh, B. R. (2004). Biological properties of curcumin-cellular and molecular mechanisms of action. *Critical Reviews in Food Science and Nutrition*, *44*, 97–111.

Jose, S., Anju, S. S., Cinu, T. A., Aleykutty, N. A., Thomas, S., & Souto, E. B. (2014). *In vivo* pharmacokinetics and biodistribution of resveratrol-loaded solid lipid nanoparticles for brain delivery. *International Journal of Pharmaceutics*, *474*(1–2), 6–13.

Kajani, A. A., Bordbar, A., Esfahani, S. H. Z., & Razmjou, A. (2016). Gold nanoparticles as potent anticancer agent: Green synthesis, characterization, and *in vitro* study. *RSC Advances*, 1–12.

Kamala Priya, M. R., & Iyer, P. R. (2015). Anticancer studies of the synthesized gold nanoparticles against MCF 7 breast cancer cell lines. *Applied Nanoscience*, *5*, 443–448.

Kang, G., Kong, P. J., Yuh, Y. J., Lim, S. Y., Yim, S. V., Chun, W., et al. (2004). Curcumin suppresses lipopolysaccharide-induced cyclooxygenase-2 expression by inhibiting activator protein 1 and nuclear factor κB bindings in BV2 microglial cells. *Journal of Pharmacological Sciences*, *94*(3), 325–328.

Kar, P. K., Murmu, S., Saha, S., Tandon, V., & Acharya, K. (2014). Anthelmintic efficacy of gold nanoparticles derived from a phytopathogenic fungus, Nigrospora oryzae. *PLoS One*, *9*, e8469–e8478.

Karthikeyan, S., Prasad, N. R., Ganamani, A., et al. (2013). Anticancer activity of resveratrol-loaded gelatin nanoparticles on NCI-H460 non-small cell lung cancer cells. *Biomedicine & Preventive Nutrition*, *3*, 64–73.

Kawamori, T., Lubet, R., Steele, V. E., Kelloff, G. J., Kaskey, R. B., Rao, C. V., et al. (1999). Chemopreventive effect of curcumin, a naturally occurring anti-inflammatory agent, during the promotion/progression stages of colon cancer. *Cancer Research*, *59*(3), 597–601.

Kesharwani, P., Xie, L., Banerjee, S., et al. (2015). Hyaluronic acid-conjugated polyamidoamine dendrimers for targeted delivery of 3,4-difluorobenzylidene curcumin to CD44 overexpressing pancreatic cancer cells. *Colloids and Surfaces. B, Biointerfaces*, *136*, 413–423.

Kong, R., Xia, Q., & Liu, G. Y. (2011). Preparation and characterization of vitamin A palmitate-loaded nanostructured lipid carriers as delivery systems for food products. *Advanced Materials Research*, *236*(238), 1818–1823.

Krishnakumar, N., Sulfikkarali, N., Prasad, N. R., et al. (2011). Enhanced anticancer activity of naringenin-loaded nanoparticles in human cervical (HeLa) cancer cells. *Biomedicine & Preventive Nutrition*, *1*, 223–231.

Lai, L. H., Fu, Q. H., Liu, Y., Jiang, K., Guo, Q. M., Chen, Q. Y., et al. (2012). Piperine suppresses tumor growth and metastasis *in vitro* and *in vivo* in a 4T1 murine breast cancer model. *Acta Pharmacologica Sinica*, *33*(4), 523–530.

Lall, R. K., Syed, D. N., Adhami, V. M., Khan, M. I., & Mukhtar, H. (2015). Dietary polyphenols in prevention and treatment of prostate cancer. *International Journal of Molecular Sciences*, *16*, 3350–3376.

Li, X. T., He, M. L., Zhou, Z. Y., et al. (2015). The antitumor activity of PNA modified vinblastine cationic liposomes on Lewis lung tumor cells: *In vitro* and *in vivo* evaluation. *International Journal of Pharmaceutics*, *487*(1–2), 223–233.

Li, H., Zhang, N., Hao, Y., et al. (2014). Formulation of curcumin delivery with functionalized single-walled carbon nanotubes: Characteristics and anticancer effects *in vitro*. *Drug Delivery*, *21*(5), 379–387.

Li, Y., Zhao, H., Duan, L. R., et al. (2014). Preparation, characterization and evaluation of bufalin liposomes coated with citrus pectin. *Colloids and Surfaces A: Physicochemical and Engineering Aspects*, *444*, 54–62.

Lin, Y. C., Kuo, J. Y., Hsu, C. C., et al. (2013). Optimizing manufacture of liposomal berberine with evaluation of its antihepatoma effects in a murine xenograft model. *International Journal of Pharmaceutics*, *441*(1–2), 381–388.

Lin, H., Li, L., Thanh, X. N., Ahmed, M. E. A., Mario, G., & Guang, Y. (2014). Chitosan-coated nano-liposomes for the oral delivery of berberine hydrochloride. *Journal of Materials Chemistry B, 41*, 7149–7159.

Liu, H., Gao, M., Xu, H., et al. (2016). A promising emodin-loaded poly (lactic-co-glycolic acid)-d- α-tocopheryl polyethylene glycol 1000 succinate nanoparticles for liver cancer therapy. *Pharmaceutical Research, 33*(1), 217–236.

Liu, Z., Tabakman, S., Welsher, K., & Dai, H. (2009). Carbon nanotubes in biology and medicine: *In vitro* and *in vivo* detection, imaging and drug delivery. *Nano Research, 2*, 85–120.

Liu, C. H., Wu, C. T., & Fang, J. Y. (2010). Characterization and formulation optimization of solid lipid nanoparticles in vitamin K1 delivery. *Drug Development and Industrial Pharmacy, 36*(7), 751–761.

Loureiro, J. A., Andrade, S., Duarte, A., Neves, A. R., Queiroz, J. F., Nunes, C., et al. (2017). Resveratrol and grape extract-loaded solid lipid nanoparticles for the treatment of alzheimer's disease. *Molecules, 22*(2), 277–293.

Malam, Y., Loizidou, M., & Seifalian, A. M. (2009). Liposomes and nanoparticles: Nanosized vehicles for drug delivery in cancer. *Trends in Pharmacological Sciences, 30*(11), 592–599.

Manatunga, D. C., de Silva, R. M., de Silva, K. M. N., de Silva, N., Bhandari, S., Yap, Y. K., et al. (2017). pH responsive controlled release of anti-cancer hydrophobic drugs from sodium alginate and hydroxyapatite bi-coated iron oxide nanoparticles. *European Journal of Pharmaceutics and Biopharmaceutics, 117*, 29–38.

Mancarella, S., Greco, V., Baldassarre, F., et al. (2015). Polymer coated magnetic nanoparticles for curcumin delivery to cancer cells. *Macromolecular Bioscience, 15*(10), 1365–1374.

Martins, S., Sarmento, B., Ferreira, D. C., et al. (2007). Lipid-based colloidal carriers for peptide and protein delivery—Liposomes versus lipid nanoparticles. *International Journal of Nanomedicine, 2*(4), 595–607.

Martins, S. M., Sarmento, B., Nunes, C., et al. (2013). Brain targeting effect of camptothecin-loaded solid lipid nanoparticles in rat after intravenous administration. *European Journal of Pharmaceutics and Biopharmaceutics, 85*(3), 488–502.

Mary, T. A., Shanthi, K., Vimala, K., & Kannan, S. (2016). PEG functionalized selenium nanoparticles as a carrier of crocin to achieve anticancer synergism. *RSC Advances, 6*, 22936–22949.

McClements, D. J. (2017). Delivery by design (DbD): A standardized approach to the development of efficacious nanoparticle- and microparticle-based delivery systems. *Comprehensive Reviews in Food Science and Food Safety, 17*, 200–219.

Mehnert, W., & Mader, K. (2001). Solid lipid nanoparticles: Production, characterization and applications. *Advanced Drug Delivery Reviews, 47*(2–3), 165–196.

Merlin, J. P. J., Prasad, N. R., Shibil, S. M. A., et al. (2012). Ferulic acid loaded Poly-D, L-lactide-coglycolide nanoparticles: Systematic study of particle size, drug encapsulation efficiency and anticancer effect in non-small cell lung carcinoma cell line *in vitro*. *Biomedicine & Preventive Nutrition, 2*, 69–76.

Minaei, A., Sabzichi, M., Ramezani, F., Hamishehkar, H., & Samadi, N. (2016). Co-delivery with nano-quercetin enhances doxorubicin-mediated cytotoxicity against MCF-7 cells. *Molecular Biology Reports, 43*, 99–105.

Mody, V., Siwale, R., Singh, A., & Mody, H. (2010). Introduction to metallic NPs. *Journal of Pharmacy & Bioallied Sciences, 2*(4), 282–291.

Mohammadi, M., Pezeshki, A., Abbasi, M. M., Ghanbarzadeh, B., & Hamishehkar, H. (2017). Vitamin D-3-loaded nanostructured lipid carriers as a potential approach for fortifying food beverages; *in vitro* and *in vivo* evaluation. *Advanced Pharmaceutical Bulletin, 7*(1), 61–71.

Mohanty, S. K., Swamy, M. K., Sinniah, U. R., & Anuradha, M. (2017). Leptadenia reticulata (Retz.) Wight & Arn. (Jivanti): Botanical, agronomical, phyto-chemical, pharmacological, and biotechnological aspects. *Molecules*, *22*(6), 1019–1071.

Muehlmann, L. A., Joanitti, G. A., Silva, J. R., Longo, J. P. F., & Azevedo, R. B. (2011). Liposomal photosensitizers: Potential platforms for anticancer photodynamic therapy. *Brazilian Journal of Medical and Biological Research*, *44*(8), 729–737.

Mullauer, F. B., van Bloois, L., Daalhuisen, J. B., et al. (2011). Betulinic acid delivered in liposomes reduces growth of human lung and colon cancers in mice without causing systemic toxicity. *Anti-Cancer Drugs*, *22*(3), 223–233.

Narayanan, S., Mony, U., Vijaykumar, D. K., et al. (2015). Sequential release of epigallocatechin gallate and paclitaxel from PLGA-casein core/shell nanoparticles sensitizes drug-resistant breast cancer cells. *Nanomedicine*, *11*(6), 1399–1406.

Navya, P. N., Kaphle, A., Srinivas, S. P., Bhargava, S. K., Rotello, V. M., & Daima, H. K. (2019). Current trends and challenges in cancer management and therapy using designer nanomaterials. *Nano Convergence*, *6*, 23–53.

Nik, A. M., Langmaid, S., & Wright, A. J. (2012). Digestibility and beta-carotene release from lipid nanodispersions depend on dispersed phase crystallinity and interfacial properties. *Food & Function*, *3*(3), 234–245.

Nipun Babu, V., & Kannan, S. (2012). Enhanced delivery of baicalein using cinnamaldehyde cross-linked chitosan nanoparticle inducing apoptosis. *International Journal of Biological Macromolecules*, *51*, 1103–1108.

Nussbaumer, S., Bonnabry, P., Veuthey, J. L., & Sandrine, F. (2011). Analysis of anticancer drugs: A review. *Talanta*, *85*, 2265–2289.

Odeh, F., Ismail, S. I., Abu-Dahab, R., et al. (2012). Thymoquinone in liposomes: A study of loading efficiency and biological activity towards breast cancer. *Drug Delivery*, *19*(8), 371–377.

Okonogi, S., & Riangjanapatee, P. (2014). Potential technique for tiny crystalline detection in lycopene-loaded SLN and NLC development. *Drug Development and Industrial Pharmacy*, *40*(10), 1378–1385.

Oliveira, D. R. B., Michelon, M., Furtado, G. D., Sinigaglia-Coimbra, R., & Cunha, R. L. (2016). Betacarotene-loaded nanostructured lipid carriers produced by solvent displacement method. *Food Research International*, *90*, 139–146.

Pan, Y. J., Tikekar, R. V., & Nitin, N. (2016). Distribution of a model bioactive within solid lipid nanoparticles and nanostructured lipid carriers influences its loading efficiency and oxidative stability. *International Journal of Pharmaceutics*, *511*(1), 322–330.

Parhi, P., Mohanty, C., & Sahoo, S. K. (2012). Nanotechnology-based combinational drug delivery: An emerging approach for cancer therapy. *Drug Discovery Today*, *17*(17–18), 1044–1052.

Pimentel, R. C., Martínez, E. S. M., García, A. M., Gómez-García, C., & Palacios, Q. G. A. (2013). Silver nanoparticles nanocarriers, synthesis and toxic effect on cervical cancer cell lines. *BioNanoScience*, *3*(2), 198–207.

Piruthiviraj, P., Margret, A., & Krishnamurthy, P. P. (2015). Gold nanoparticles synthesized by *Brassica oleracea* (Broccoli) acting as antimicrobial agents against human pathogenic bacteria and fungi. *Applied Nanoscience*, *6*, 467–473.

Pooja, D., Babu Bikkina, D. J., Kulhari, H., et al. (2014). Fabrication, characterization and bio evaluation of silibinin loaded chitosan nanoparticles. *International Journal of Biological Macromolecules*, *69*, 267–273.

Prabhu, R. H., Patravale, V. B., & Joshi, M. D. (2015). Polymeric nanoparticles for targeted treatment in oncology: Current insights. *International Journal of Nanomedicine*, *10*, 1001–1018.

Prise, K. M. (2006). New advances in radiation biology. *Occupational Medicine*, *56*(3), 156–161.

Przystupski, D., Niemczura, M. J., Górska, A., Supplitt, S., Kotowski, K., Wawryka, P., et al. (2019). In search of panace-review of recent studies concerning nature-derived anticancer agents. *Nutrients*, *11*(6), 1426–1464.

Qian, C., Decker, E. A., Xiao, H., & McClements, D. J. (2013). Impact of lipid nanoparticle physical state on particle aggregation and beta-carotene degradation: Potential limitations of solid lipid nanoparticles. *Food Research International*, *52*(1), 342–349.

Qiu, J. F., Gao, X., Wang, B. L., et al. (2013). Preparation and characterization of monomethoxy poly(ethylene glycol)-poly(ε-caprolactone) micelles for the solubilization and *in vivo* delivery of luteolin. *International Journal of Nanomedicine*, *8*, 3061–3069.

Rahaman Mollick, M. M., Bhowmick, B., Mondal, D., Maity, D., Rana, D., Dash, S. K., et al. (2014). Anticancer (*in vitro*) and antimicrobial effect of gold nanoparticles synthesized using *Abelmoschus esculentus* (L.) pulp extract via a green route. *RSC Advances*, *4*, 37838–37848.

Rahman, S., Cao, S., Steadman, K. J., Wei, M., & Parekh, H. S. (2012). Native and β-cyclodextrin-enclosed curcumin: Entrapment within liposomes and their *in vitro* cytotoxicity in lung and colon cancer. *Drug Delivery*, *19*(7), 346–353.

Rajendran, I., Dhandapani, H., Anantanarayanan, R., et al. (2015). Apigenin mediated gold nanoparticle synthesis and their anti-cancer effect on human epidermoid carcinoma (A431) cells. *RSC Advances*, *5*, 51055–51066.

Rawal, S., & Patel, M. M. (2019). Threatening cancer with nanoparticle aided combination oncotherapy. *Journal of Controlled Release*, *301*, 76–109.

Ray, L., Kumar, P., & Gupta, K. C. (2013). The activity against Ehrlich's ascites tumors of doxorubicin contained in self assembled, cell receptor targeted nanoparticle with simultaneous oral delivery of the green tea polyphenol epigallocatechin-3-gallate. *Biomaterials*, *34*(12), 3064–3076.

Riangjanapatee, P., Muller, R. H., Keck, C. M., & Okonogi, S. (2013). Development of lycopene loaded nanostructured lipid carriers: Effect of rice oil and cholesterol. *Pharmazie*, *68*(9), 723–731.

Rompicharla, S. V. K., Bhatt, H., Shah, A., Komanduri, N., Vijayasarathy, D., Ghosh, B., et al. (2017). Formulation optimization, characterization, and evaluation of *in vitro* cytotoxic potential of curcumin loaded solid lipid nanoparticles for improved anticancer activity. *Chemistry and Physics of Lipids*, *208*, 10–18.

Sabzichi, M., Mohammadian, J., Mohammadi, M., Jahanfar, F., Pour, A. A. M., Hamishehkar, H., et al. (2017). Vitamin D-loaded nanostructured lipid carrier (nlc): A new strategy for enhancing efficacy of doxorubicin in breast cancer treatment. *Nutrition and Cancer*, *69*(6), 840–848.

Saengkrit, N., Saesoo, S., Srinuanchai, W., Phunpee, S., & Ruktanonchai, U. R. (2014). Influence of curcumin-loaded cationic liposome on anticancer activity for cervical cancer therapy. *Colloids and Surfaces. B, Biointerfaces*, *114*, 349–356.

Saggar, J. K., Yu, M., Tan, Q., & Tannock, I. F. (2013). The tumor microenvironment and strategies to improve drug distribution. *Frontiers in Oncology*, *3*, 154–161.

Sailor, G., Seth, A. K., Parmar, G., Chauhan, S., & Javia, A. (2015). Formulation and *in vitro* evaluation of berberine containing liposome optimized by 32 full factorial designs. *Journal of Applied Pharmaceutical Science*, *5*, 23–28.

Salminen, H., Aulbach, S., Leuenberger, B. H., Tedeschi, C., & Weiss, J. (2014). Influence of surfactant composition on physical and oxidative stability of Quillaja saponin-stabilized lipid particles with encapsulated omega-3 fish oil. *Colloids and Surfaces B: Biointerfaces*, *122*, 46–55.

Salminen, H., Gommel, C., Leuenberger, B. H., & Weiss, J. (2016). Influence of encapsulated functional lipids on crystal structure and chemical stability in solid lipid nanoparticles: Towards bioactive-based design of delivery systems. *Food Chemistry*, *190*, 928–937.

Salminen, H., Helgason, T., Kristinsson, B., Kristbergsson, K., & Weiss, J. (2013). Formation of solid shell nanoparticles with liquid omega-3 fatty acid core. *Food Chemistry*, *141*(3), 2934–2943.

Samad, A., Sultana, Y., & Aqil, M. (2007). Liposomal drug delivery systems: An update review. *Current Drug Delivery*, *4*(4), 297–305.

Sapra, P., & Allen, T. M. (2003). Ligand-targeted liposomal anticancer drugs. *Progress in Lipid Research*, *42*(5), 439–462.

Saxena, V., & Hussain, M. D. (2012). Poloxamer 407/TPGS mixed micelles for delivery of gambogic acid to breast and multidrug-resistant cancer. *International Journal of Nanomedicine*, 7, 713–721.

Shanthi, K., Sreevani, V., Vimala, K., & Kannan, K. (2017). Cytotoxic effect of palladium nanoparticles synthesized from *Syzygium aromaticum* aqueous extracts and induction of apoptosis in cervical carcinoma. *Proceedings of the National Academy of Sciences, India Section B: Biological Sciences*, *87*(4), 1101–1112.

Shirode, A. B., Bharali, D. J., Nallanthighal, S., et al. (2015). Nanoencapsulation of pomegranate bioactive compounds for breast cancer chemoprevention. *International Journal of Nanomedicine*, *10*, 475–484.

Siddiqui, I. A., Bharali, D. J., Nihal, M., et al. (2014). Excellent anti-proliferative and pro-apoptotic effects of (-)-epigallocatechin-3-gallate encapsulated in chitosan nanoparticles on human melanoma cell growth both *in vitro* and *in vivo*. *Nanomedicine*, *10*(8), 1619–1626.

Singh, L. M., Chakraborty, B., Pal, R., Nath, A., Pal, S., Shahidur Rahman, D., et al. (2017). Comparative study on the antioxidant and immunomodulatory properties of curcumin conjugated gold nanospheres and free curcumin. *Journal of Applied Pharmaceutical Science*, 7, 56–63.

Singh, H., Du, J., Singh, P., & Yi, T. H. (2018). Ecofriendly synthesis of silver and gold nanoparticles by Euphrasia officinalis leaf extract and its biomedical applications. *Artificial Cells, Nanomedicine, and Biotechnology*, *46*, 1163–1170.

Sinha, R., Kim, G. J., Nie, S., & Shin, D. M. (2006). Nanotechnology in cancer therapeutics: Bioconjugated nanoparticles for drug delivery. *Molecular Cancer Therapeutics*, *5*(8), 1909–1917.

Soni, P., Kaur, J., & Tikoo, K. (2015). Dual drug-loaded paclitaxel–thymoquinone nanoparticles for effective breast cancer therapy. *Journal of Nanoparticle Research*, *17*, 1–12.

Sriraman, S. K., Salzano, G., Sarisozen, C., & Torchilin, V. (2016). Anti-cancer activity of doxorubicin-loaded liposomes co-modified with transferrin and folic acid. *European Journal of Pharmaceutics and Biopharmaceutics*, *105*, 40–49.

Sujitha, M. V., & Kannan, S. (2013). Green synthesis of gold nanoparticles using Citrus fruits (Citrus limon, Citrus reticulata and Citrus sinensis) aqueous extract and its characterization. *Spectrochim Acta Part A*, *102*, 19–23.

Sun, J. B., Bi, C., Chan, H. M., Sun, S. P., Zhang, Q. W., & Zheng, Y. (2013). Curcumin-loaded solid lipid nanoparticles have prolonged *in vitro* antitumour activity, cellular uptake and improved *in vivo* bioavailability. *Colloids and Surfaces B: Biointerfaces*, *111*, 367–375.

Sun, M., Nie, S., Pan, X., et al. (2014). Quercetin-nanostructured lipid carriers: Characteristics and anti-breast cancer activities *in vitro*. *Colloids and Surfaces. B, Biointerfaces*, *113*, 15–24.

Sundarraj, S., Thangam, R., Sreevani, V., Kaveri, K., Gunasekaran, P., Achiraman, S., et al. (2012). γ-Sitosterol from *Acacia nilotica* L. induces G2/M cell cycle arrest and apoptosis through c-Myc suppression in MCF-7 and A549 cells. *Journal of Ethnopharmacology*, *141*, 803–809.

Tan, J. M., Karthivashan, G., Arulselvan, P., et al. (2014). Sustained release and cytotoxicity evaluation of carbon nanotube-mediated drug delivery system for betulinic acid. *Journal of Nanomaterials*, *2014*, 862148.

Tang, S., Gao, D., Zhao, T., et al. (2013). An evaluation of the anti-tumor efficacy of oleanolic acid loaded PEGylated liposomes. *Nanotechnology*, *24*(23), 235102.

Tian, X., Yin, H., Zhang, S., et al. (2014). Bufalin loaded biotinylated chitosan nanoparticles: An efficient drug delivery system for targeted chemotherapy against breast carcinoma. *European Journal of Pharmaceutics and Biopharmaceutics*, *87*(3), 445–453.

Topalian, S. L., Weiner, G. J., & Pardoll, D. M. (2011). Cancer immunotherapy comes of age. *Journal of Clinical Oncology*, *29*(36), 4828–4836.

Tran, S., DeGiovanni, P., Piel, B., & Rai, P. (2017). Cancer nanomedicine: A review of recent success in drug delivery. *Clinical and Translational Medicine*, *6*, 44–67.

Tse, B. W. C., Cowin, G. J., Soekmadji, C., Jovanovic, L., Vasireddy, R. S., Ling, M. T., et al. (2015). PSMA-targeting iron oxide magnetic nanoparticles enhance MRI of preclinical prostate cancer. *Nanomedicine*, *10*(3), 375–386.

Vasanth, K., Ilango, K., MohanKumar, R., Agrawal, A., & Dubey, G. P. (2014). Anticancer activity of *Moringa oleifera* mediated silver nanoparticles on human cervical carcinoma cells by apoptosis induction. *Colloids and Surfaces. B, Biointerfaces*, *117*, 354–359.

Verma, N. K., Crosbie-Staunton, K., Satti, A., Gallagher, S., Ryan, K. B., Doody, T., et al. (2013). Magnetic core-shell nanoparticles for drug delivery by nebulisation. *Journal of Nanbiotechnology*, *11*, 1–12.

Vimala, K., Sundarraj, S., Paulpandi, M., Venkatesan, S., & Kannan, S. (2014). Green synthesized doxorubicin loaded zinc oxide nanoparticles regulates the Bax and Bcl-2 expression in breast and colon carcinoma. *Process Biochemistry*, *49*, 160–172.

Vivek, R., Thangam, R., Muthuchelian, K., Gunasekaran, P., Kaveri, K., & Kannan, S. (2012). Green biosynthesis of silver nanoparticles from *Annona squamosa* leaf extract and its in vitro cytotoxic effect on MCF-7 cells. *Process Biochemistry*, *47*, 2405–2410.

Vivek, R., Thangam, R., NipunBabu, V., Rejeeth, C., Sivasubramanian, S., Palani Gunasekaran, P., et al. (2014). Multifunctional HER2-antibody conjugated polymeric nanocarrier- based drug delivery system for multi-drug-resistant breast cancer therapy. *ACS Applied Materials & Interfaces*, *6*, 6469–6480.

Vivek, R., Thangam, R., Rajesh Kumar, S., Rejeeth, C., Sivasubramanian, S., Vincent, S., et al. (2016). HER2 targeted breast cancer therapy with switchable "off/on" multifunctional "smart" magnetic polymer core–shell nanocomposites. *ACS Applied Materials & Interfaces*, *8*, 2262–2279.

Wang, F. L., Chen, J., Dai, W. T., He, Z. M., Zhai, D. D., & Chen, W. D. (2017). Pharmacokinetic studies and anticancer activity of curcumin-loaded nanostructured lipid carriers. *Acta Pharmaceutica*, *67*(3), 357–371.

Wang, L., Li, H., Wang, S., et al. (2014). Enhancing the antitumor activity of berberine hydrochloride by solid lipid nanoparticle encapsulation. *AAPS PharmSciTech*, *15*(4), 834–844.

Wang, Y., Yang, T., Wang, X., et al. (2010). Targeted polymeric micelle system for delivery of combretastatin A4 to tumor vasculature *in vitro*. *Pharmaceutical Research*, *27*(9), 1861–1868.

Wang, W. R., Zhang, L. Y., Chen, T. T., Guo, W., Bao, X. X., Wang, D. D., et al. (2017). Anticancer effects of resveratrol-loaded solid lipid nanoparticles on human breast cancer cells. *Molecules*, *22*(11), 1814–1825.

Wang, W., Zhao, X., Hu, H., et al. (2010). Galactosylated solid lipid nanoparticles with cucurbitacin B improves the liver targetability. *Drug Delivery*, *17*(3), 114–122.

Wen, W., Qian, H., Xu, T., Hanquiang, L., Junqi, T., Ying, Y., et al. (2011). Berberine induces cell death in human hepatoma cells *in vitro* by downregulating CD147. *Cancer Science*, *102*, 1287–1292.

Wilken, R., Veena, M. S., Wang, M. B., & Srivatsan, E. S. (2011a). Curcumin: A review of anticancer properties and therapeutic activity in head and neck squamous cell carcinoma. *Molecular Cancer*, *10*, 12–78.

Wilken, R., Veena, M. S., Wang, M. B., & Srivatsan, E. S. (2011b). Curcumin: A review of anti-cancer properties and therapeutic activity in head and neck squamous cell carcinoma. *Molecular Cancer, 2011*(10), 12–31.

World Health Organization. (2010). *NMH fact sheet January 2010*. Geneva: WHO.

Wu, G. Q., Chai, K. Q., Zhu, X. M., Jiang, H., Wang, X., Xue, Q., et al. (2016). Anti-cancer effects of curcumin on lung cancer through the inhibition of EZH2 and NOTCH1. *Oncotarget*, 7(18), 26535–26550.

Xia, Q., & Kong, R. (2011). Freeze-drying and characterization of vitamin A palmitate-loaded nanostructured lipid carriers (NLC). *Materials Science Forum*, *694*, 365–369.

Yallapu, M. M., Ebeling, M. C., Khan, S., et al. (2013). Novel curcumin-loaded magnetic nanoparticles for pancreatic cancer treatment. *Molecular Cancer Therapeutics*, *12*(8), 1471–1480.

Yallapu, M. M., Othman, S. F., Curtis, E. T., et al. (2012). Curcumin-loaded magnetic nanoparticles for breast cancer therapeutics and imaging applications. *International Journal of Nanomedicine*, 7, 1761–1779.

Yang, T. S., Liu, T. T., & Liu, H. I. (2017). Effects of aroma compounds and lipid composition on release of functional substances encapsulated in nanostructured lipid carriers lipolyzed by lipase. *Food Hydrocolloids*, *62*, 280–287.

Yang, A., Liu, Z., Yan, B., et al. (2016). Preparation of camptothecin-loaded targeting nanoparticles and their antitumor effects on hepatocellular carcinoma cell line H22. *Drug Delivery*, *23*(5), 1699–1706.

Ye, L., & Song, Q. (2015). Promising potency of retinoic acid-poly (ethylene glycol)-thiol gold nanoparticle conjugates for cervical cancer treatment. *International Journal of Clinical and Experimental Medicine*, *8*(7), 10501–10507.

Yong, P. Y. W., Gan, P. H., Sasmita, A. O., Mak, S. T., & Ling, A. P. K. (2018). Nanoparticles as carriers of phytochemicals: Recent applications against lung cancer. *International Journal of Research in Biomedicine and Biotechnology*, 7(1), 1–11.

Yongvongsoontorn, N., Chung, J. E., Gao, S. J., Bae, K. H., Yamashita, A., Tan, M. H., et al. (2019). Carrier-enhanced anticancer efficacy of sunitinib-loaded green tea-based micellar nanocomplex beyond tumor-targeted delivery. *ACS Nano*, *13*, 7591–7602.

Yoysungnoen-Chintana, P., Bhattarakosol, P., & Patumraj, S. (2014). Antitumor and anti-angiogenic activities of curcumin in cervical cancer xenografts in nude mice. *BioMed Research International*, *2014*, 1–12.

Zaman, M. S., Chauhan, N., Yallapu, M. M., Gara, R. K., Maher, D. M., Kumari, S., et al. (2016). Curcumin nanoformulation for cervical cancer treatment. *Scientific Reports*, *6*, 20051–20065.

Zhang, H., Liu, G., Zeng, X., et al. (2015). Fabrication of genistein-loaded biodegradable TPGS-b- PCL nanoparticles for improved therapeutic effects in cervical cancer cells. *International Journal of Nanomedicine*, *10*, 2461–2473.

Zhang, J., Wang, M., & Webster, T. J. (2017). Growth process and anticancer properties of gold nanorods. *Journal of Biomedical Materials Research. Part A*, *105*, 2616–2621.

Zhao, L., Shi, Y., Zou, S., et al. (2011). Formulation and *in vitro* evaluation of quercetin loaded polymeric micelles composed of pluronic P123 and D-a-tocopheryl polyethylene glycol succinate. *Journal of Biomedical Nanotechnology*, 7(3), 358–365.

Zhou, Q., & Wei, Y. (2016). For better or worse, iron overload by superparamagnetic iron oxide nanoparticles as a MRI contrast agent for chronic liver diseases. *Chemical Research in Toxicology*, *30*(1), 73–80.

Zhu, R. R., Qin, L. L., Wang, M., et al. (2009). Preparation, characterization, and anti-tumor property of podophyllotoxin-loaded solid lipid nanoparticles. *Nanotechnology*, *20*(5), 055702–055709.

Zou, T., Ching, A., Lum, T., Lok, C. N., Zhang, J. J., & Che, C. M. (2015). Chemical biology of anticancer gold (iii) and gold (i) complexes. *Chemical Society Reviews*, *44*, 8786–8801.

9780323853156